Data-Driven Energy Management and Tariff Optimization in Power Systems

IEEE Press
445 Hoes Lane
Piscataway, NJ 08854

Data-Driven Energy Management and Tariff Optimization in Power Systems

Shaping the Future of Electricity Distribution through Analytics

Edited by

Hamidreza Arasteh
Niroo Research Institute, Tehran, Iran

and

Center for Research on Microgrids (CROM), Huanjiang Laboratory
Zhuji, Shaoxing, Zhejiang, China

Pierluigi Siano
University of Salerno, Fisciano SA, Italy

Niki Moslemi
Niroo Research Institute, Tehran, Iran

Josep M. Guerrero
Center for Research on Microgrids (CROM), Huanjiang Laboratory
Zhuji, Shaoxing, Zhejiang, China

Zhejiang University, Hangzhou, Zhejiang, China

and

Department of Electrical Engineering, University of Valladolid, Valladolid, Spain

IEEE Press
WILEY

Published by John Wiley & Sons, Inc., Hoboken, New Jersey.

Library of Congress Cataloging-in-Publication Data Applied for:

Hardback ISBN: 9781394290277

Cover Design: Wiley
Cover Image: © Quang Ngoc Nguyen/Getty Images

Set in 9.5/12.5pt STIXTwoText by Straive, Chennai, India
Printed and bound by CPI Group (UK) Ltd, Croydon, CR0 4YY

C9781394290277_261025

To the global community of scholars and innovators committed to advancing a sustainable energy future;
To you, dear reader, whose curiosity and engagement make this work meaningful;
To the professional colleagues whose valuable contributions enhanced this book;
And to our publishers, IEEE Press and John Wiley & Sons, for their continued support.

Contents

About the Editors

Hamidreza Arasteh was born in Zanjan, Iran, in 1988, and is an Electrical Engineer, specializing in various domains of power systems. His academic journey commenced with a BSc from Tabriz University, Tabriz, Iran, in 2010, followed by an MSc from Tarbiat Modares University (TMU), Tehran, Iran, in 2012, and a PhD from Shahid Beheshti University (SBU), Tehran, Iran, in 2017, all in electrical engineering. He has been recognized for his academic prowess, earning distinctions such as the top graduate of both his BSc and MSc programs. He has also been acknowledged for his scholarly contributions, notably achieving placement among the top 1% of reviewers in Engineering on Publons' global reviewer database from 2017 to 2018. Since 2017, he has been a part of the "Power Systems Operation and Planning Research Department" at the "Niroo Research Institute" (NRI). Throughout his tenure, he has undertaken various roles within the institute, ranging from serving as a Research Member and Project Manager/Supervisor to leading as the Head of the Microgrid research track. In addition to his academic contributions, he undertook an industrial sabbatical at the Office of Economy, Investment, and Regulation of the Water and Electricity Market within the Ministry of Energy of Iran, demonstrating his commitment to bridging the gap between academia and industry. Currently, he also serves as a Researcher at the Center for Research on Microgrids (CROM), Huanjiang Laboratory, Zhuji, Shaoxing, Zhejiang, China.

His research interests span a wide spectrum of power systems, including renewable integration, demand response, smart grid, microgrids, data analysis, energy management, electricity markets, and optimization. He has received numerous awards throughout his career, recognizing his exceptional achievements and dedication to advancing the field of electrical engineering.

Pierluigi Siano received the MSc degree in electronic engineering and the PhD degree in information and electrical engineering from the University of Salerno, Salerno, Italy, in 2001 and 2006, respectively. Since 2021, he has been a Distinguished Visiting Professor in the Department of Electrical and Electronic Engineering Science, University of Johannesburg. He is currently a Professor and the Scientific Director of the Smart Grids and Smart Cities Laboratory, Department of Management and Innovation Systems, University of Salerno. His research interests include demand response, energy management, the integration of distributed energy resources in smart grids, electricity markets, and the planning and management of power systems. In these research fields, he has coauthored more than 700 articles, including more than 410 in international journals that have received more than 17,900 citations on Scopus, with an H-index of 64. From 2019 to 2022, he was awarded as a Highly Cited Researcher in Engineering by the Web of Science Group. He has been the Chair of the IES TC on Smart Grids. He is an Editor of the *Power and Energy Society Section of IEEE Access*, *IEEE Transactions on Power Systems*, *IEEE Transactions on Industrial Informatics*, *IEEE Transactions on Industrial Electronics*, and *IEEE Systems Journal*.

Niki Moslemi, born in Babol, Iran, in 1975, holds a BSc degree from the University of Tehran, Tehran, Iran, attained in 1993, and an MSc degree from the Isfahan University of Technology, Isfahan, Iran, earned in 1999, both in electrical engineering. Since 1998, she has been a researcher at the "Niroo Research Institute," where she also served as the head of the "Power Systems Operation and Planning Research Department." Her diverse research interests span several domains within electrical engineering, encompassing resiliency, load forecasting, power quality, data analysis, expansion planning of power systems, and power system protection.

Josep M. Guerrero received the BSc degree in telecommunications engineering, the MSc degree in electronics engineering, and the PhD degree from the Technical University of Catalonia, Barcelona, Spain.

Since 2011, he has been a Full Professor with AAU Energy, Aalborg University, Denmark, where he is responsible for the Microgrid Research Program. In 2019, he became a Villum Investigator by the Villum Fonden, which supports the Center for Research on Microgrids (CROM) at Aalborg University, with Professor Guerrero as the Founder and Director of the same center (www.crom.energy.aau.dk). In 2020, he initiated neuroscience studies and research. As a result, in 2022, he received the MSc degree in Psychobiology and Cognitive Neuroscience from the Institute of Neuroscience (INc) at the Autonomous University of Barcelona. From 2023 to 2024, he was with the Technical University of Catalonia as an ICREA Research Professor. In 2025, he became a Distinguished Senior Researcher at the Department of Electrical Engineering, University of Valladolid, Spain.

His research interests are oriented to different microgrid frameworks like energy microgrids, hydrogen and biomass, water micronets, biological systems, seaport microgrids and electrical ships, airport microgrids and more electrical aircraft, space microgrids, and smart medical systems. In these fields, he has researched distributed and cyber-physical energy systems, cybersecurity for microgrids and smart grids, neuroscience-inspired artificial intelligence for energy systems, machine learning and applications using signal processing, bioinformatics, bioinspired computing, and quantum computing and quantum communication for complex energy networks.

Professor Guerrero is an Associate Editor for several IEEE Transactions. He has published more than 1,000 journal papers in the fields of microgrids and renewable energy systems, which have been cited more than 120,000 times. During 10 consecutive years, from 2014 to 2023, he was awarded by Clarivate Analytics as a Highly Cited Researcher. From 2020 to 2024, he was listed in the world's top 2% of scientists by Stanford/Elsevier. In 2021, he received the IEEE Bimal Bose Award for Industrial Electronics Applications in Energy Systems, for his pioneering contributions to renewable-energy-based microgrids. In 2022, he received the IEEE PES Douglas M. Staszesky Distribution Automation Award, for contributions to making the hierarchical control of microgrid systems a practical reality. In 2023, he was the recipient of IEEE Modeling and Control Technical Achievement Award for contributions to modeling and control of power-electronics-based microgrids. In 2024, he was ranked among the top 100 highly cited sustainability researchers. Since 2023, he has been with Zhejiang University, Hangzhou, Zhejiang, China, and also is the Director of the Center for Research on Microgrids (CROM), Huanjiang Laboratory, Zhuji, Shaoxing, China.

List of Contributors

Mohammad Taghi Ameli
Department of Electrical Engineering
Shahid Beheshti University
Tehran, Iran

and

Electrical Networks Research Institute
Shahid Beheshti University
Tehran, Iran

Mohammad Mehdi Amiri
Department of Electrical Engineering
Shahid Beheshti University
Tehran, Iran

and

Electrical Networks Research Institute
Shahid Beheshti University
Tehran, Iran

Hamidreza Arasteh
Power Systems Operation and Planning
Research Department
Niroo Research Institute
Tehran, Iran

and

Center for Research on Microgrids (CROM)
Huanjiang Laboratory
Zhuji, Shaoxing, Zhejiang, China

Sasan Azad
Department of Electrical Engineering
Shahid Beheshti University
Tehran, Iran

and

Electrical Networks Research Institute
Shahid Beheshti University
Tehran, Iran

Maryam Azimi
Faculty of Electrical and Computer
Engineering
Tarbiat Modares University
Tehran, Iran

Salah Bahramara
Department of Electrical Engineering
Sa. C., Islamic Azad University
Sanandaj, Iran

Hamed Delkhosh
Faculty of Electrical and Computer
Engineering
Tarbiat Modares University
Tehran, Iran

Afshin Ebrahimi
Department of Electrical Engineering
Sahand University of Technology
Tabriz, Iran

Mahdi Ghaedi
Faculty of Electrical and Computer Engineering
Tarbiat Modares University
Tehran, Iran

Khabat Ghamari
Iran Meteorological Organization
Tehran, Iran

Josep M. Guerrero
Center for Research on Microgrids (CROM)
Huanjiang Laboratory
Zhuji, Shaoxing, Zhejiang, China

and

Zhejiang University
Hangzhou, Zhejiang, China

Farkhondeh Jabari
Power Systems Operation and Planning Research Department
Niroo Research Institute (NRI)
Tehran, Iran

Hamid Jabari
Department of Mechanical Engineering
Sahand University of Technology
Tabriz, Iran

Seyed Mohsen Hashemi
Power System Operation and Planning Research Group
Niroo Research Institute (NRI)
Tehran, Iran

Alireza Mansoori
Faculty of Electrical and Computer Engineering
Tarbiat Modares University
Tehran, Iran

Abbas Marini
Power System Operation and Planning Research Group
Niroo Research Institute (NRI)
Tehran, Iran

Majid Miri Larimi
Faculty of Electrical and Computer Engineering
Tarbiat Modares University
Tehran, Iran

Niki Moslemi
Power Systems Operation and Planning Research Department
Niroo Research Institute
Tehran, Iran

Sobhan Naderian
James Watt School of Engineering
University of Glasgow
Glasgow, Scotland

Javad Nezafat Namini
Power Systems Operation and Planning Research Department
Niroo Research Institute (NRI)
Tehran, Iran

Ali Paeizi
Department of Electrical Engineering
Shahid Beheshti University
Tehran, Iran

and

Electrical Networks Research Institute
Shahid Beheshti University
Tehran, Iran

Mohsen Parsa Moghaddam
Faculty of Electrical and Computer Engineering
Tarbiat Modares University
Tehran, Iran

Nazanin Pourmoradi
Department of Electrical Engineering
Shahid Beheshti University
Tehran, Iran

and

Electrical Networks Research Institute
Shahid Beheshti University
Tehran, Iran

Seyed Pouya Ramezanzadeh
Faculty of Electrical and Computer Engineering
Tarbiat Modares University
Tehran, Iran

Habibollah Raoufi
Power Systems Operation and Planning Research Department
Niroo Research Institute (NRI)
Tehran, Iran

Asrin Seyedzahedi
Department of Science
Campus of Bijar
University of Kurdistan
Bijar, Iran

Miadreza Shafie-khah
Research and Innovation Division
Nowocert
Dublin, Ireland

Ardalan Shafiei-Ghazani
Department of Mechanical Engineering
Sahand University of Technology
Tabriz, Iran

Mohammad Reza Sheibani
Power Systems Operation and Planning Research Department
Niroo Research Institute (NRI)
Tehran, Iran

Pierluigi Siano
Department of Management and Innovation Systems
University of Salerno
Fisciano, Italy

Reza Zamani
Faculty of Electrical and Computer Engineering
Tarbiat Modares University
Tehran, Iran

Preface

The Energy Transformation Imperative

The global energy sector is undergoing a significant transformation. The combination of climate urgency, renewable energy adoption, and digital innovation is fundamentally reshaping global power systems. Traditional grid management approaches, designed for predictable, centralized generation, now face unprecedented challenges such as variable renewables, electrified transportation, and actively engaged energy consumers. Equally revolutionary solutions are required for these paradigm shifts.

At the heart of this energy revolution lies data – the critical resource enabling smarter, more adaptive power systems. Modern grids generate immense volumes of operational data through a wide variety of smart devices. When leveraged through advanced analytics and artificial intelligence, this data becomes a powerful tool for optimizing energy management, designing equitable tariffs, and building resilient infrastructure.

This book addresses the urgent need for practical, data-driven solutions in power systems. We present a comprehensive framework that translates theoretical data science concepts into actionable strategies for grid operators, utility managers, and energy researchers. Our method equips professionals to handle the challenging shift toward sustainable, efficient, and consumer-centric energy systems by fusing solid theoretical underpinnings with practical implementation insights.

The Urgent Need for Data-Driven Energy Solutions

This book was inspired by a number of important insights regarding the changing energy landscape. First, the rapid decarbonization of power systems has introduced new layers of complexity. Renewable energy sources like wind and solar, while environmentally beneficial, introduce variability and uncertainty that challenge conventional grid management approaches. Second, the digitalization of energy infrastructure through smart meters, IoT devices, and advanced sensors has created an explosion of data – yet many utilities struggle to extract actionable insights from this wealth of information. Third, evolving consumer expectations and the rise of prosumers, demand more sophisticated, equitable tariff structures that can accommodate diverse usage patterns while maintaining grid stability.

Our work on this book began with a realization that although there are numerous studies dedicated to either power systems engineering or data science, there are few that successfully connect both of these disciplines in a technically accurate and practically useful way. This gap in the literature was particularly clear to us in our own research and industry work, where we always encountered professionals eager to implement data-driven solutions but lacking comprehensive resources to guide their efforts.

Comprehensive Coverage for a Complex Challenge

This book is structured to provide readers with a complete journey through the data-driven transformation of power systems. We begin with fundamental concepts in Chapter 1, establishing the basic building blocks of power system analytics. Chapter 2 addresses predictive modeling and machine learning applications. The heart of the book (Chapters 3–5) addresses two critical applications: demand-side management and tariff optimization. These sections reflect our conviction that the true power of data analytics lies not just in understanding energy systems, but in actively shaping their operation. We explore innovative demand response strategies enabled by granular consumption data, and present cutting-edge approaches to tariff design considering different aspects like economic efficiency and equity. Renewable energy integration forms the focus of Chapters 6–8, where we examine how machine learning can address the variability challenge through improved forecasting, optimal scheduling, and enhanced grid stability measures. Power system resiliency, as well as data challenge and solutions, are addressed in Chapter 9. Chapter 10 establishes the foundation with advanced non-intrusive load monitoring techniques, demonstrating how deep learning approaches can enhance the accuracy in appliance-level energy disaggregation. Chapter 11 then provides a comprehensive framework for cyber-physical security, systematically analyzing preventive measures and mitigation strategies against increasingly sophisticated grid threats. Finally, Chapter 12 addresses AI-driven under-voltage load shedding solutions, illustrating how adaptive algorithms can maintain grid stability during critical events.

Meeting the Needs of Diverse Stakeholders

The transformative potential of data-driven energy management extends across the entire electricity value chain, and accordingly, this book is designed to serve multiple audiences:

- For power system engineers and utility professionals, we provide the tools to utilize operating data to drive better decision-making.
- Energy economists and policymakers will find valuable guidance in our treatment of tariff design and energy management. These sections demonstrate how data analytics can inform more efficient and equitable market structures.
- Researchers and academics will benefit from our comprehensive coverage of emerging techniques, complete with references to cutting-edge literature.
- Students in electrical engineering, energy systems, and data science programs will find the structured progression from fundamentals to advanced topics ideal for building expertise.

Pedagogical Approach and Supplementary Resources

Recognizing that different readers will approach this material with varying backgrounds and objectives, we've designed the book to support multiple learning paths:

- **Systematic Study:** Readers new to the field can progress sequentially through the chapters to build comprehensive understanding.
- **Topic-Focused Exploration:** Recognizing the diverse needs of professionals, the book's design enables direct access to specialized content. Readers can efficiently locate and engage with topic-specific chapters. This structure optimizes the learning experience for time-constrained practitioners, allowing them to immediately apply relevant methodologies while maintaining conceptual coherence.

Acknowledgments and Collaborative Spirit

This book is the culmination of years of study, collaboration among industries, and sharing of information. We are especially grateful to our many colleagues who have contributed their time and skill through chapter contributions, case studies, and helpful critique.

We are deeply grateful for the invaluable insights and expertise shared by our collaborators across industry, academia, and research institutions worldwide. These partnerships have kept our work grounded in real-world challenges and practical constraints.

A Call to Action for the Energy Transition

As we conclude this preface, we wish to emphasize that the energy transition is not merely a technical challenge. The tools and techniques presented in this book are more than academic exercises; they represent powerful levers for creating a more sustainable, resilient, and equitable energy future.

To engineers and technologists, we offer these methods to enhance grid operations and integrate clean energy. To policymakers and regulators, we provide evidence-based approaches to design better markets and tariffs. To researchers and students, we present questions and future directions that demand innovative solutions.

The path forward will require collaboration across disciplines and borders. It will demand both technical and non-technical excellence.

We hope this book will serve as both a practical guide and a source of inspiration for all those working to shape the future of electricity.

8 July 2025

Hamidreza Arasteh
Power Systems Operation and Planning Research Department, Niroo Research Institute, Tehran, Iran
Center for Research on Microgrids (CROM), Huanjiang Laboratory, Zhuji, Shaoxing, Zhejiang, China

Pierluigi Siano
Department of Management & Innovation Systems, University of Salerno, Via Giovanni Paolo II, 132, Fisciano, Italy

Niki Moslemi
Power Systems Operation and Planning Research Department, Niroo Research Institute, Tehran, Iran

Josep M. Guerrero
Center for Research on Microgrids (CROM), Huanjiang Laboratory, Zhuji, Shaoxing, Zhejiang, China
Zhejiang University, Hangzhou, Zhejiang, China
Department of Electrical Engineering, University of Valladolid, Valladolid, Spain

1

Fundamentals of Power System Data and Analytics

Seyed Pouya Ramezanzadeh, Mohsen Parsa Moghaddam, and Reza Zamani

Faculty of Electrical and Computer Engineering, Tarbiat Modares University, Tehran, Iran

1.1 Introduction

Imagine a world where power systems can predict outages before they happen, optimize energy distribution in real time, and seamlessly integrate renewable energy sources. This image is no longer a distant dream but a rapidly approaching reality, thanks to the transformative power of data analytics. The convergence of digital technologies and power systems heralds a new era of data-based operation in the energy sector [1]. As traditional power grids evolve into interconnected, data-rich ecosystems, data analytics becomes increasingly critical.

Data analytics enables grid operators and market participants to glean actionable insights from vast volumes of data, empowering proactive grid management, demand response optimization, and risk mitigation strategies [2]. By leveraging advanced data analytics techniques, power systems can enhance reliability, optimize energy efficiency, and support the integration of renewable energy sources, paving the way for a sustainable energy future.

As adopting renewable energy sources grows, it introduces more significant variability into the electrical grid, complicating the balance between supply and demand [3]. Innovative solutions must be developed to manage this challenge and enhance system flexibility on both the supply and demand sides. One of the most promising strategies involves the integration of smart grids, supported by big data analytics, to strengthen the resilience and management of power systems [4]. Smart grids allow for more adaptive energy distribution by enabling two-way communication and power flow between consumers and utilities.

A vital element of this transformation is the widespread deployment of smart meters, which have been installed extensively in recent years. These devices monitor near-real-time energy usage at customer sites and relay this information back to utility companies [5]. This system supports a dynamic interaction between consumers and the grid, facilitating better energy management.

The shift toward intelligent grids has been partly driven by the deregulation of power systems, which has moved away from traditional, vertically integrated utility models [6, 7]. By decentralizing energy production and distribution, smart grids can more effectively respond to fluctuating energy demands, incorporating renewable energy into the system more efficiently.

Big data plays a crucial role in the operation of smart grids, particularly in processing the vast amounts of information generated by these systems [8]. Big data analytics applications include real-time monitoring and management of energy consumption, automatic billing systems, predictive forecasting for load and energy generation, and identifying faults or anomalies in the grid [9].

Data-Driven Energy Management and Tariff Optimization in Power Systems: Shaping the Future of Electricity Distribution through Analytics, First Edition.
Edited by Hamidreza Arasteh, Pierluigi Siano, Niki Moslemi, and Josep M. Guerrero.

This data-driven approach improves the grid's capacity to respond to unpredictability and increases the overall efficiency of the energy system.

The characteristics of the data collected by intelligent grids align with the core attributes of big data: volume, velocity, and variety. Additionally, it introduces other critical "V" characteristics, such as veracity (accuracy of the data), visibility (transparency), and value (the actionable insights drawn from the data). However, these benefits come with significant challenges, particularly in the areas of security and privacy [10], which must be carefully managed to fully realize the potential of smart grid technologies [11].

This chapter explores the digital transformation of power systems, the structure of data-driven power systems, and the pivotal role of data-driven decision-making. It delves into state-of-the-art data analytics techniques explicitly tailored for power systems and presents real-world examples and case studies. Finally, it addresses this rapidly evolving field's challenges and future directions.

To understand the full potential of data analytics in power systems, it is essential to first explore the current landscape and the transformative changes underway in the power industry. The following section provides a comprehensive background on these topics.

1.2 Background

Digital trends have gradually changed traditional power systems over the last decade. Therefore, it is necessary to understand these digital trends and their impact on power systems. Figure 1.1 shows current trends in power systems and the expected results they will cause when the digital transformation enables the power industry to respond to digital trends.

The following sections discuss concepts, opportunities, and challenges of present and future power systems; transformation in the power industry; and drivers and barriers to such transformation.

1.2.1 Concept, Opportunities, and Challenges of Present and Future Power Systems

Traditionally composed of generation, transmission, and distribution networks, legacy power systems are designed to deliver electricity from power plants to end consumers reliably and

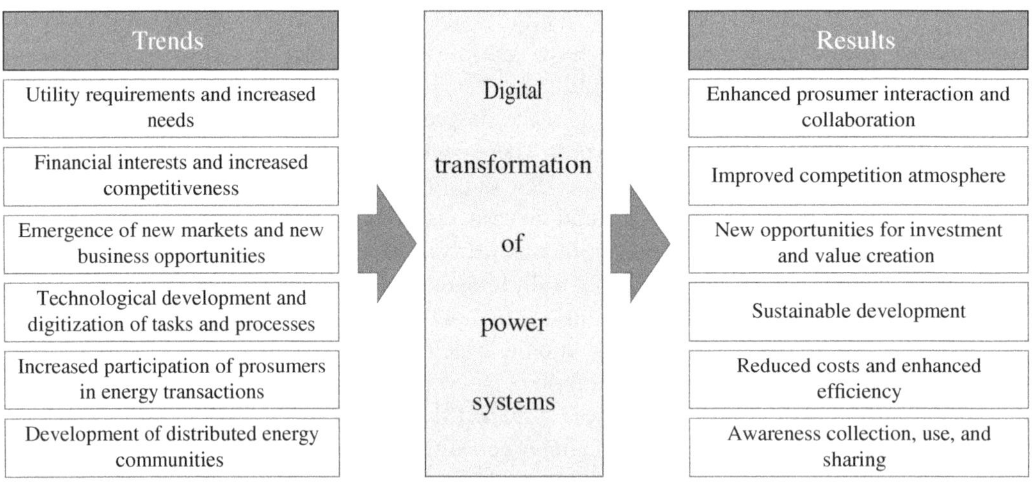

Figure 1.1 Digital trends and their results in the power industry.

efficiently [12]. Modernizing power systems opens new opportunities for an increased role of distributed sources, improved participation of stakeholders, and the integration of renewable energy sources, ultimately contributing to a more sustainable energy future [13]. Advanced data analytics can optimize power flow, reduce operational costs, and improve demand forecasting, making the grid more adaptable and resilient [14]. Meanwhile, these groundbreaking digital technologies expose power systems to new challenges.

However, this transformation comes with challenges such as maintaining grid stability with intermittent renewable sources, ensuring cybersecurity, and managing the vast amounts of data generated by smart devices [15, 16]. Balancing the benefits of modernization with these challenges is critical for the future of power systems.

Figure 1.2 shows critical opportunities and challenges of future power systems. It is worth mentioning that some unknown opportunities and challenges will emerge in the operational phase of digital transformation.

1.2.2 Transformation in the Power Industry

The power industry is undergoing a significant transformation, driven by a convergence of technological advancements, environmental imperatives, regulatory changes, and evolving consumer expectations [17, 18]. This transformation is not merely a shift in operational practices but a profound rethinking of how energy is generated, distributed, and consumed and how different parts of the energy sector work together [19]. The philosophy behind this transformation centers on sustainability, resilience, and decentralization.

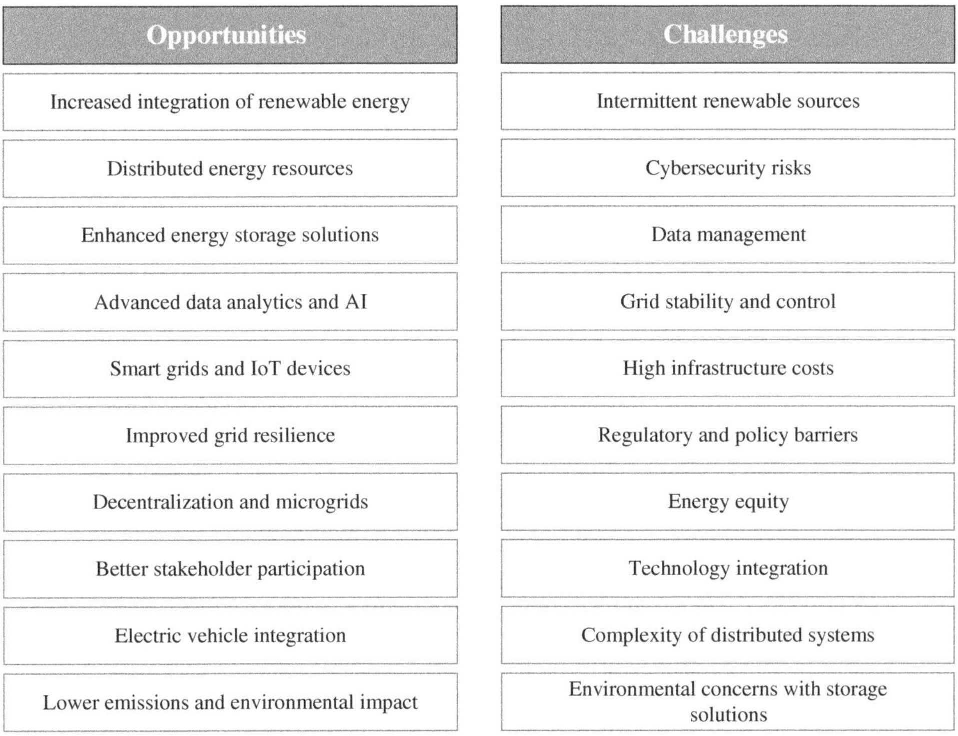

Figure 1.2 Opportunities and challenges of future power systems.

Digitalization revolutionizes the power industry by enabling real-time monitoring, predictive maintenance, and optimized energy management through advanced data analytics. Integrating digital communication technologies with the electrical grid has led to smart grids. Smart grids allow for more efficient and reliable electricity delivery, enhanced by the capability to dynamically respond to changes in demand and supply [2]. Digitalization of power systems relies on the process by which data is gathered, prepared, and used for decision-making.

Decentralization involves generating power closer to where it is used, reducing transmission losses, and enhancing grid resilience [20]. Distributed energy resources like solar panels and battery storage play a crucial role in this shift, allowing consumers to become prosumers who both produce and consume electricity. Historically, power systems were designed with a centralized generation model, where large power plants produced electricity, which was then transmitted over long distances to consumers. This model was efficient for many decades but is increasingly seen as inadequate in addressing the challenges of modern energy demands and environmental concerns. The new philosophy emphasizes a more distributed and dynamic approach, integrating renewable energy sources, enhancing grid flexibility, and prioritizing energy efficiency.

One of the primary drivers of transformation is the integration of renewable energy sources such as wind, solar, and hydropower. These sources are inherently variable and decentralized, requiring a shift from the traditional central control to a more flexible and adaptive grid management system [21]. For instance, the rise of solar power has led to millions of households and businesses installing photovoltaic panels, turning them from mere consumers into prosumers who both produce and consume electricity and data [22]. Data is the core of the transformation by which consumers can manage their participation behavior in energy markets, and utilities can establish new profit opportunities [23].

Advanced data analytics and smart grid technologies are crucial in managing the complexities introduced by renewable energy. Smart grids use sensors, advanced metering infrastructure, and data analytics to monitor and manage real-time energy flow. For example, predictive analytics can forecast energy demand and supply patterns, allowing grid operators to balance loads more effectively and prevent outages [24]. This level of real-time control and optimization was unimaginable in traditional grid systems and represents a paradigm shift in energy management.

Energy storage systems like batteries are also pivotal in the power industry's transformation. They address the intermittency of renewable energy by storing the excess power generated during peak production times and releasing it when demand is high or production is low [25]. Large-scale battery installations, like those seen in South Australia and California [26, 27], demonstrate how energy storage can enhance grid stability and reliability, making renewable energy a more viable and dependable option.

Another aspect of the transformation is the increasing role of digital technologies and automation. The Internet of Things (IoT) enables the connection and coordination of various power system components, from generation to distribution to end-user devices [28]. Smart thermostats, connected appliances, and automated demand response systems empower consumers to manage their energy use more efficiently [8], contributing to overall system efficiency and reducing peak demand.

The regulatory environment is also evolving to support this transformation [29]. Policies and incentives are being designed to encourage the adoption of renewable energy, energy storage, and energy efficiency measures. For instance, feed-in tariffs, renewable portfolio standards, and tax credits for solar installations are some of the mechanisms that have spurred the growth of clean energy [11]. Regulatory frameworks are also being updated to accommodate and promote distributed generation and to ensure grid reliability in the face of increasing renewable integration.

The philosophy of resilience is another cornerstone of the power industry's transformation. Climate change and extreme weather events pose significant risks to the traditional power grid. The new approach emphasizes building a more resilient infrastructure that can withstand and quickly recover from disruptions [14]. Microgrids are a prime example of this philosophy in action. These localized grids can operate independently of the central grid [30], providing critical power during emergencies and enhancing the overall robustness of the energy system.

Economic factors also play a significant role in driving transformation. The cost of renewable energy technologies has dropped dramatically over the past decade, making them competitive with traditional fossil fuel sources [10]. The declining costs of solar panels, wind turbines, and batteries are making renewable energy an environmentally sound choice and economically viable one. This economic feasibility accelerates the adoption of renewables, further driving the power industry's transformation. Figure 1.3 indicates a transformation in the power industry, from the legacy system to the emerging power industry.

Consumer behavior and expectations are evolving as well. There is a growing demand for cleaner, more sustainable energy options. Consumers are increasingly aware of their environmental impact and are seeking ways to reduce their carbon footprint. This shift in consumer preferences pushes utilities and energy providers to innovate and offer more sustainable solutions [31].

In conclusion, the power industry's transformation is a multifaceted and dynamic process driven by the need for sustainability, resilience, and efficiency. This transformation's philosophy emphasizes decentralization, renewable energy integration, advanced data analytics, and digital technologies. Examples such as the widespread adoption of solar power, smart grids, energy storage, regulatory support, and evolving consumer behavior illustrate the profound changes. This transformation represents not just an upgrade of the existing system but a fundamental reimagining of how we generate, distribute, and consume energy, paving the way for a more sustainable and resilient future.

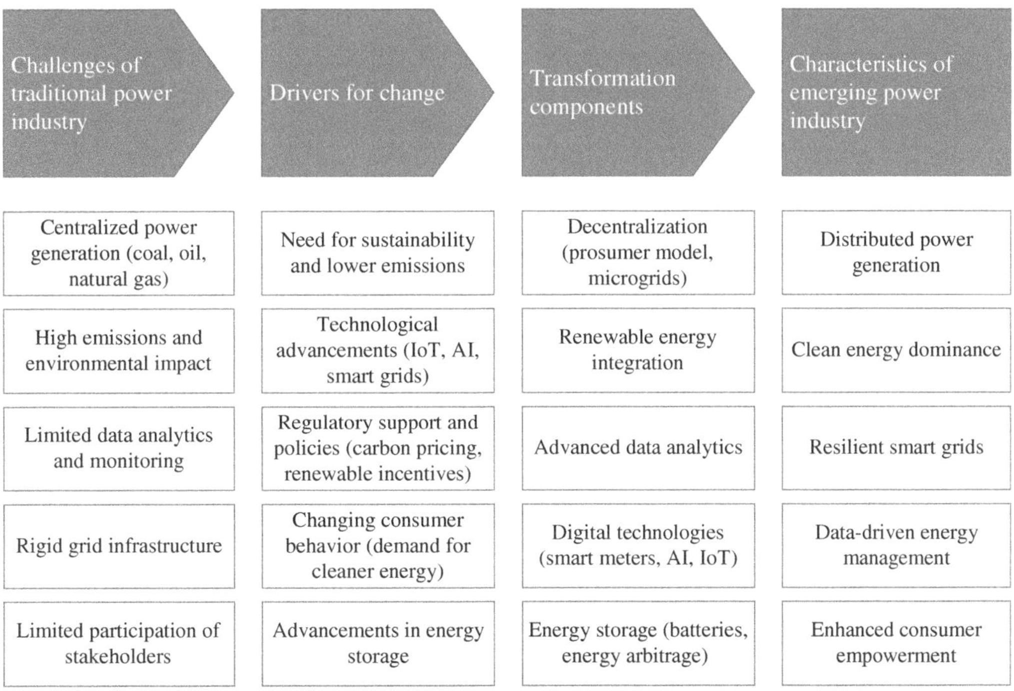

Figure 1.3 Transformation in the power industry from the legacy system to the emerging power industry.

1.2.3 Drivers and Barriers

Advances in sensor technology, IoT, and machine learning are pivotal in driving the digital transformation of power systems. These technologies enable the collection and analysis of vast amounts of data, providing insights that can improve grid operations and maintenance.

Supportive policies and regulatory frameworks are essential to encourage investment in modern power systems and ensure a stable transition. Governments and regulatory bodies play a crucial role in setting standards, providing incentives for renewable energy adoption, and ensuring the cybersecurity of digitalized power systems.

The evolving electricity market, with increasing competition and the growing demand for renewable energy, is a significant driver of transformation. Market dynamics encourage innovation and investment in new technologies, driving the continuous improvement of power systems.

The digital transformation of power systems is driven by advancements in sensor technology, IoT, machine learning, and big data analytics, enabling better grid monitoring, efficiency, and renewable energy integration. Supportive regulatory frameworks and decarbonization goals further accelerate this shift, promoting the development of smart grids. However, barriers such as high initial investment costs, legacy infrastructure, cybersecurity risks, regulatory uncertainty, and skill gaps hinder the pace of transformation. Addressing these challenges ensures a stable, secure, and efficient digital power system transition.

Regulatory frameworks and policies are crucial in addressing the barriers to data analytics in power systems. Well-defined regulations can standardize data management practices, ensuring data quality and interoperability across different systems, thus overcoming issues related to legacy infrastructure. Cybersecurity policies that establish stringent guidelines for data protection and enforce robust security measures can mitigate the risks of cyberattacks. Policies promoting workforce development in data science and analytics through education, training programs, and incentives can help bridge the skill gap. Clear, supportive regulations can also ease concerns about data privacy, creating a balanced environment for innovation while protecting consumer rights. Furthermore, government incentives and financial support can reduce the costs of transitioning to advanced digital infrastructures, making it easier for utilities to invest in data analytics technologies. Drivers and barriers to the power system's digital transformation are demonstrated in Figure 1.4.

1.3 Data-rich Power Systems

Data-rich power systems represent a fundamental shift from conventional power systems, leveraging vast amounts of data generated by modern technology to enhance the efficiency, reliability, and sustainability of electricity generation, distribution, and consumption. The difference between these two paradigms is stark, with data-rich systems being characterized by real-time data collection, advanced analytics, and proactive management [32], whereas conventional systems rely on static, less dynamic approaches.

In conventional power systems, data collection was limited and typically involved periodic manual readings from meters and centralized control systems. The information flow was slow, often lagging behind real-time events, making it difficult to respond quickly to emerging issues or optimize operations effectively [33]. Decision-making was largely reactive, based on historical data and scheduled maintenance, leading to inefficiencies and increased risks of outages.

In contrast, data-rich power systems utilize many sensors, smart meters, phasor measurement units (PMUs), and IoT devices that continuously monitor and report various parameters such as

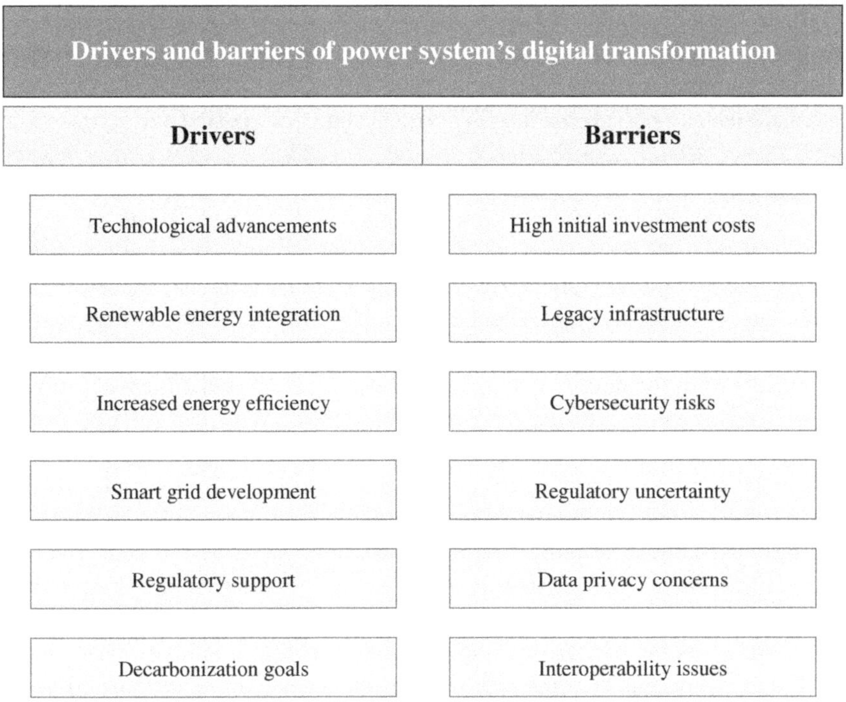

Figure 1.4 Drivers and barriers to power system's digital transformation.

voltage, current, frequency, temperature, and equipment health. This real-time data flow enables a more dynamic and responsive grid management system. For example, smart meters provide granular data on electricity consumption patterns at the household level, allowing utilities to offer more accurate billing, detect power theft, and design targeted demand response programs to reduce peak loads [33].

The integration of advanced data analytics is a cornerstone of data-rich power systems. These analytics tools process and analyze the grid's vast datasets, extracting valuable insights that drive informed decision-making. Predictive analytics, for instance, can forecast electricity demand with high accuracy, helping utilities to balance supply and demand more effectively. In conventional systems, demand forecasting was often based on historical consumption patterns and could not account for sudden changes in usage behavior or weather conditions.

Machine learning and artificial intelligence (AI) further enhance the capabilities of data-rich power systems. These technologies can identify patterns and anomalies in the data that would be impossible to detect manually [34]. For example, AI algorithms can predict equipment failures by analyzing data from sensors on transformers, generators, and other critical infrastructure [35]. This allows for predictive maintenance, where potential issues are addressed before they lead to costly and disruptive failures. In conventional systems, maintenance schedules were typically based on fixed intervals or after-the-fact repairs, leading to unnecessary maintenance or increased risk of unexpected failures.

Data-rich power systems also enable more effective integration of renewable energy sources. The intermittent nature of solar and wind power presents significant challenges for conventional grids, which were designed for steady, predictable power flows from large, centralized plants [36]. With real-time data and advanced forecasting models, data-rich systems can better predict renewable generation patterns and adjust grid operations accordingly. For instance, if a sudden drop in wind power generation is detected, the system can quickly ramp up other generation resources or deploy stored energy to maintain balance [37].

Energy storage systems, such as batteries, are another critical component of data-rich power systems. They help manage the variability of renewable energy by storing excess power generated during periods of high production and releasing it during high demand or low production periods. Advanced data analytics optimize these batteries' charging and discharging cycles, ensuring that they are used most effectively to stabilize the grid [38]. Conventional power systems could not integrate and manage such dynamic resources efficiently. Also, data visibility of battery status can help prosumers decide when energy transactions can benefit them [39]. With access to energy market data and real-time prices, prosumers can optimize their consumption behavior. Demand response programs benefit immensely from the data-rich approach. These programs encourage consumers to reduce or shift their electricity usage during peak demand periods. With real-time data from smart meters and connected devices, utilities can process the data and send targeted signals to consumers, offering incentives to reduce usage when the grid is under stress [40]. For example, a utility might provide a discount for customers who agree to lower their air conditioning use during a heat wave. In conventional systems, demand response programs were often broad and less effective due to the lack of precise, real-time data. Another significant advantage of data-rich power systems is enhanced grid security and resilience. Continuous monitoring and data analysis can detect cyber threats, physical attacks, and other security breaches in real time, enabling a rapid response to mitigate these threats [41]. In conventional systems, security measures were often reactive and based on periodic assessments rather than continuous monitoring.

Regulatory and policy frameworks are also evolving to support data-rich power systems. Governments and regulatory bodies recognize the value of data and analytics in improving grid operations and are implementing policies that encourage deploying smart grid technologies [42]. For example, regulations may mandate the installation of smart meters in new residential and commercial buildings or provide incentives for utilities to invest in data analytics infrastructure. The consumer experience is significantly enhanced in data-rich power systems. Real-time data empowers consumers to make more informed decisions about their energy use [43]. For instance, home energy management systems can provide consumers with insights into their consumption patterns, helping them to identify energy-saving opportunities and reduce their bills. In conventional systems, consumers have little visibility into their real-time energy usage and have limited means to control it.

In conclusion, data-rich power systems represent a transformative evolution from conventional power systems, driven by real-time data collection, advanced analytics, and proactive management. This paradigm shift enables greater efficiency, reliability, and sustainability in electricity generation, distribution, and consumption. Examples such as smart meters, predictive maintenance, renewable energy integration, energy storage optimization, demand response programs, enhanced grid security, and improved consumer experiences illustrate the profound impact of data-rich power systems. This transformation addresses the limitations of conventional systems and paves the way for a more resilient and adaptable energy future. To summarize this section, Figure 1.5 indicates the data analytics process needed to transform power grids successfully.

1.3.1 Data Sources and Types

Every aspect of a fully digitalized power system is correlated with the source and type of the relevant data. There are several sources for data in future power systems, including [44–46]:

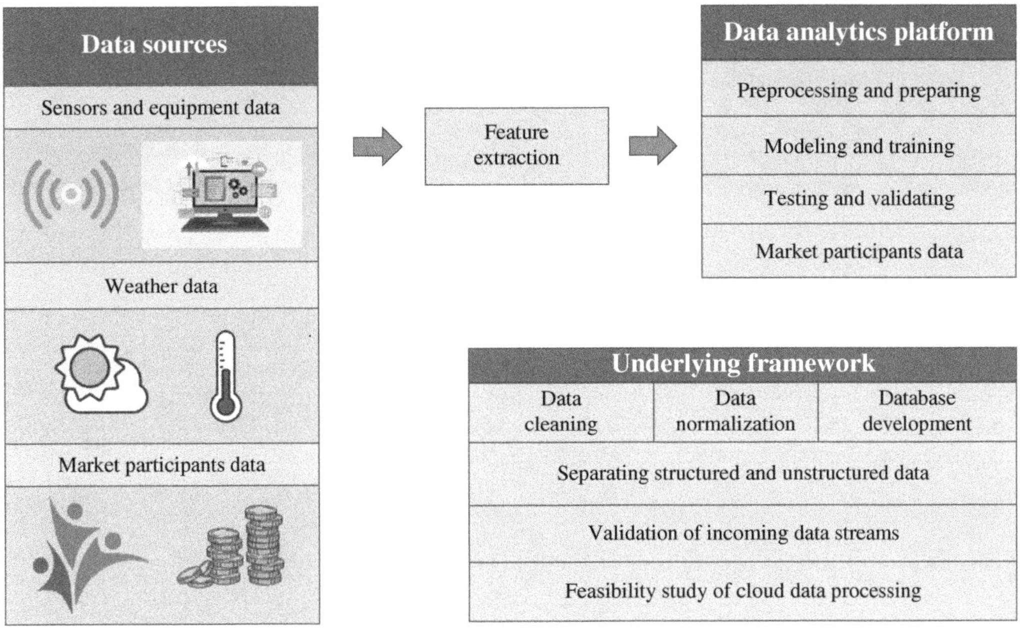

Figure 1.5 Big data analytics process.

- Smart meters are advanced devices that record energy consumption in real time and communicate this information to the consumer and the utility provider. They provide detailed data on electricity usage patterns, enabling more efficient energy management and billing. Smart meters help utilities optimize their operations and reduce peak demand by offering insights into how and when electricity is used.
- Supervisory control and data acquisition (SCADA) systems are used to monitor and control industrial processes, including power systems. They collect data from various sensors and instruments across the grid, providing real-time insights into system performance and facilitating rapid response to issues. SCADA systems are essential for ensuring the reliability and efficiency of power systems.
- PMUs, also known as synchrophasors, measure the electrical waves on an electricity grid to monitor the grid's health. They provide high-resolution voltage, current, and frequency data, which is crucial for maintaining grid stability and detecting anomalies. The data from PMUs allow grid operators to make informed decisions to enhance the reliability and security of the power system.
- IoT devices, including sensors and actuators, are increasingly being deployed in power systems to monitor and control various aspects of the grid. They enable real-time data collection and remote management, enhancing power systems' overall efficiency and reliability. IoT devices contribute to developing intelligent grids by providing granular data and enabling advanced analytics.

Every data-based optimization or machine learning algorithm method needs a specific data format. For instance, for detecting anomalies in the consumption pattern of consumers through analyzing their demand data from Advance Metering Infrastructure (AMI), the type and format of every piece of data should be the same.

Figure 1.6 shows different sources and types of data. The data can be gathered from web services, load monitoring, equipment parameters, regional economic centers, smart meters, SCADA, PMUs, and IoT devices. The type of data can be structured, unstructured, or semi-structured.

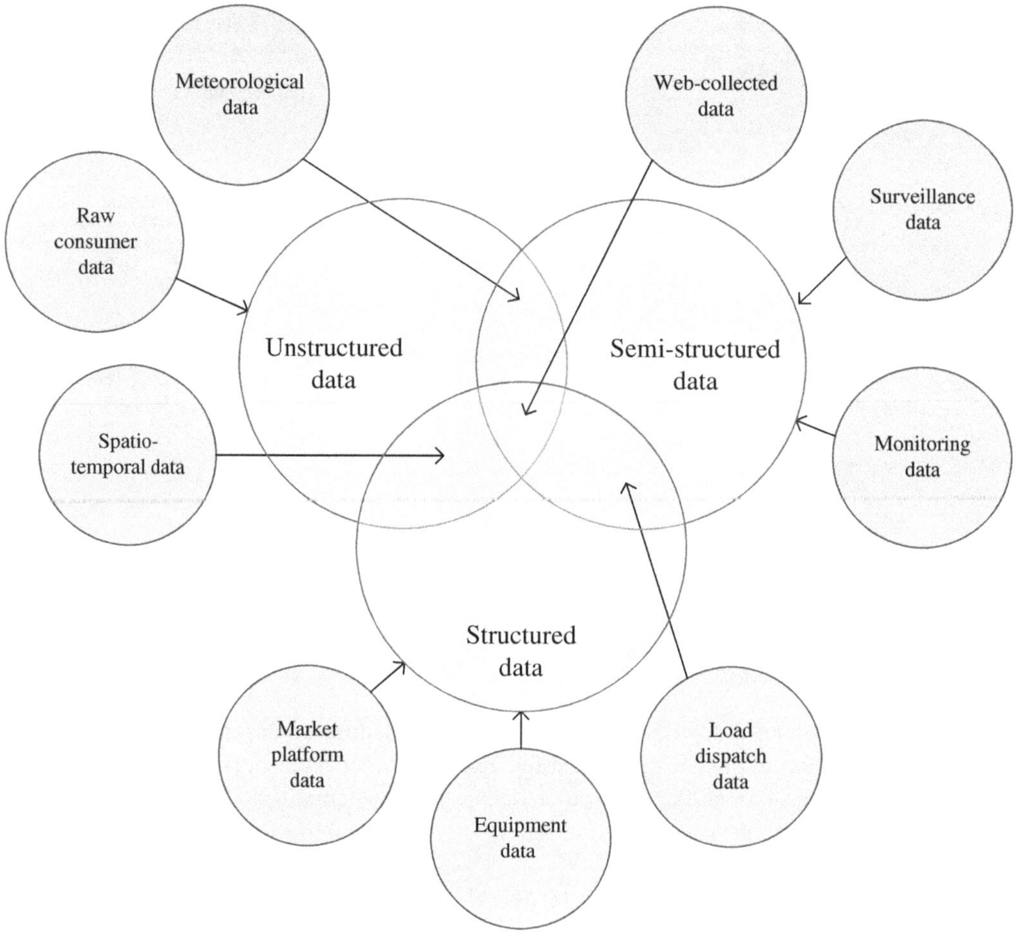

Figure 1.6 Data sources and types.

1.3.2 Data Structure

The data collected from power systems can be broadly categorized into operational, event, and historical data [45]. Operational data includes real-time measurements such as voltage, current, and frequency. Event data captures specific occurrences like outages and faults. Historical data consists of archived records used for trend analysis and forecasting.

Data from power systems is typically stored in standardized formats such as the Common Information Model and IEC 61850 [3]. These standards ensure interoperability between different systems and facilitate data integration from various sources. Standardized data formats enable seamless communication and analysis across other parts of the power system. The data processing structure needed for big data services is indicated in Figure 1.7.

Effective data management involves collecting, storing, and retrieving data to ensure its accuracy, security, and accessibility. Power systems use databases and data warehouses to store vast amounts of data, which can be accessed for analysis and decision-making. Data integrity and security are paramount for the reliable operation of power systems.

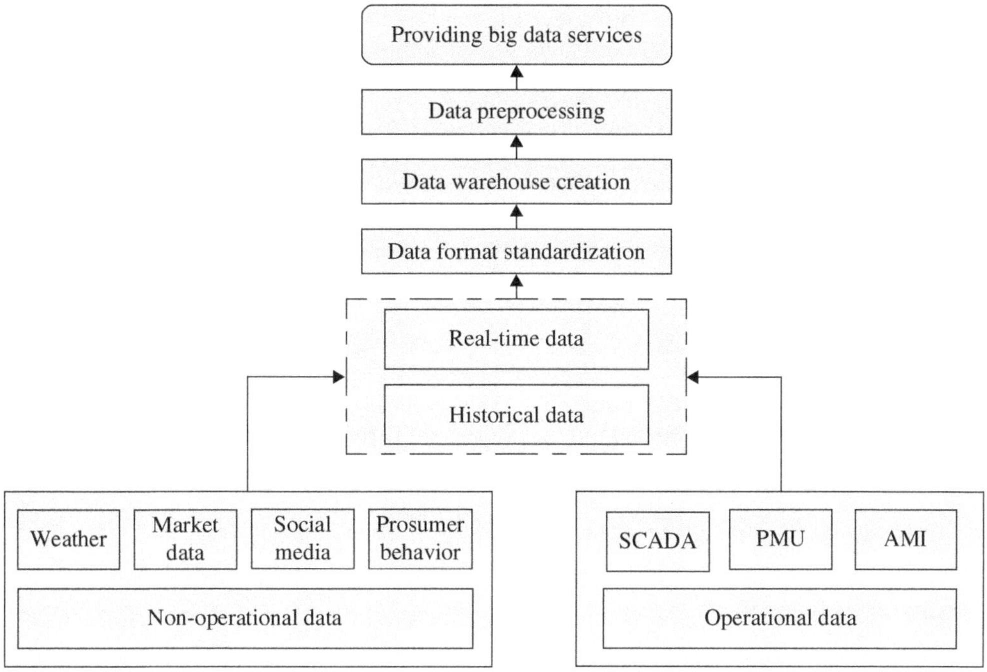

Figure 1.7 Data processing structure needed for big data services.

1.4 Data Analytics in Power Systems

Data analytics in power systems has revolutionized how electricity can be traded among different parties by enabling the implementation of distributed and local energy communities [20]. The primary difference between novel approaches and conventional methods lies in utilizing vast amounts of real-time data and advanced analytical techniques to optimize operations, enhance reliability, and integrate renewable energy sources more effectively.

Conventional approaches in power systems were largely reactive and relied on historical data and periodic manual readings. These methods had limited capability to predict or quickly respond to real-time events, leading to inefficiencies and increased risks of outages. Decision-making was based on fixed schedules for maintenance and operations, often resulting in either over-maintenance or unexpected failures.

Novel approaches leverage real-time data from smart meters, sensors, and IoT devices, enabling continuous monitoring and proactive management of the grid. Advanced analytics, including machine learning and AI, process this data to provide predictive insights and automate decision-making. For example, predictive maintenance uses real-time data to foresee equipment failures and schedule maintenance before issues arise, significantly reducing downtime and costs compared to reactive maintenance in conventional systems. Furthermore, integrating renewable energy sources is much more efficient with data analytics. Conventional systems struggled with the intermittency of renewables due to a lack of real-time data and forecasting capabilities. Novel approaches use advanced forecasting models to predict renewable generation patterns and adjust grid operations dynamically, ensuring a stable supply. Demand response programs also benefit from data analytics. Traditional methods were broad and less effective, lacking precise, real-time data to guide consumer behavior. Modern data-driven approaches can send targeted signals to consumers based on real-time grid conditions, optimizing energy usage and reducing peak demand more effectively.

In summary, novel approaches in data analytics for power systems offer a proactive, real-time, and data-driven framework that significantly enhances efficiency, reliability, and sustainability compared to the reactive and static nature of conventional methods. This transformation not only addresses the limitations of the traditional systems but also enables a more resilient and adaptable power grid.

1.4.1 What Is Data Analytics?

Data analytics involves systematic computational analysis of information or statistics. In power systems, it refers to using data-driven techniques to extract valuable insights, identify patterns, and make informed decisions to improve system performance and reliability. As power systems become more complex and data-rich, data analytics will play a crucial role in ensuring the efficient operation of the grid and stakeholders' participation in multiple energy markets.

1.4.2 Analytics Techniques

Data analytics contains various techniques, such as:

- **Descriptive Analytics**: Descriptive analytics focuses on summarizing historical data to understand what has happened in the past. In power systems, it is used to analyze historical load data, monitor equipment performance, and identify operational trends. By providing insights into past performance, descriptive analytics helps grid operators make informed decisions and improve operational efficiency.
- **Predictive Analytics**: Predictive analytics uses historical data and statistical algorithms to forecast future events. It is applied in load forecasting, predictive maintenance of equipment, and anticipating system failures in power systems. Accurate load forecasting enables utilities to balance supply and demand effectively, while predictive maintenance reduces downtime and extends the lifespan of critical infrastructure.
- **Prescriptive Analytics**: Prescriptive analytics goes beyond prediction to recommend actions based on data insights. It is used in power systems to optimize energy dispatch, manage demand response programs, and improve decision-making processes. Prescriptive analytics helps utilities enhance grid stability and efficiency by suggesting the best course of action.
- **Machine Learning**: Machine learning algorithms can automatically identify patterns and anomalies in large datasets. In power systems, machine learning is used for fault detection, predictive maintenance, and optimizing grid operations. For example, machine learning models can analyze sensor data to detect early signs of equipment failure, allowing for timely intervention.
- **Big Data Analytics**: Big data analytics involves processing and analyzing vast amounts of data generated by power systems. It helps in real-time monitoring, enhancing situational awareness, and improving the overall efficiency of grid operations. By leveraging big data analytics, utilities can gain deeper insights into grid performance, detect issues more quickly, and respond more effectively to changing conditions.

Figure 1.8 depicts different techniques that can be incorporated to enhance the operation indices of power grids.

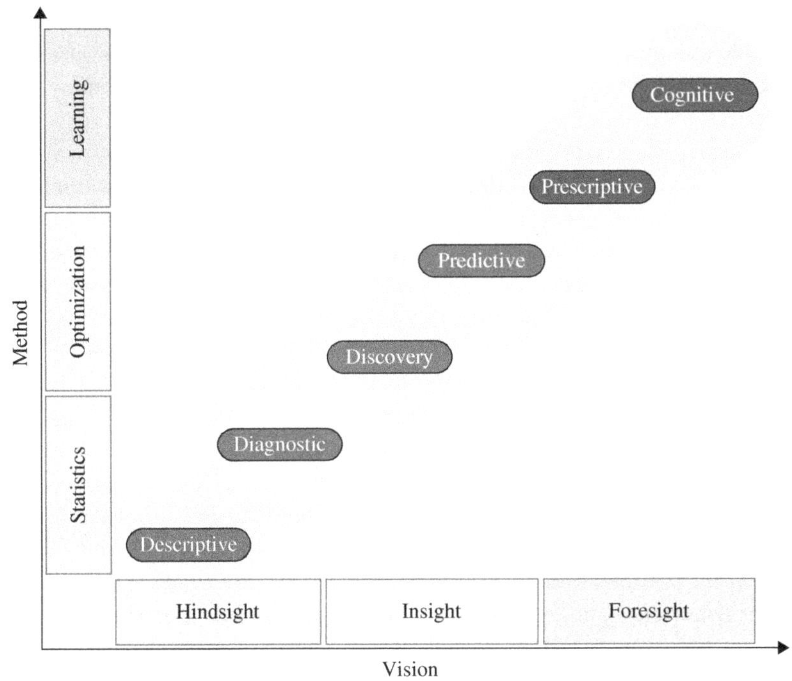

Figure 1.8 Different techniques to enhance operation indices of power grids.

1.5 Data Analytics-Based Decision-Making in Future Power Systems

Data analytics-based decision-making is revolutionizing the future of power systems, fundamentally altering how electricity is traded as a product. This approach contrasts sharply with conventional methods, offering significant efficiency, reliability, and sustainability improvements. Understanding the characteristics and implications of data analytics-based decision-making illuminates its transformative potential and highlights the advantages of traditional decision-making processes.

At the heart of data analytics-based decision-making is the utilization of real-time data collection. Modern power systems are equipped with smart meters, sensors, and IoT devices that continuously monitor and report on various parameters such as voltage, current, frequency, and equipment status. This constant stream of data enables instant monitoring of grid conditions, consumer usage patterns, and the health of critical infrastructure, providing a comprehensive and up-to-date picture of the power system's state.

Advanced predictive analytics are crucial in this new approach [47]. Machine learning and AI can forecast demand, supply, and potential failures. By analyzing historical and real-time data, these tools can anticipate issues before they arise. For instance, predictive models can forecast peak demand periods, allowing utilities to adjust generation and distribution proactively. Similarly, predictive maintenance uses data to identify equipment nearing scheduled maintenance before disruptions occur [48]. This shift from planned to predictive maintenance reduces downtime and extends the lifespan of equipment.

Enhanced grid flexibility is another critical characteristic of data analytics-based decision-making. Integrating renewable energy sources, such as solar and wind power, introduces variability that conventional systems struggle to manage. Data-rich power systems can predict renewable energy output and adjust operations dynamically to maintain grid stability [49]. Demand response programs, which adjust consumer usage patterns in real time based on grid conditions, are also more effective with data analytics. For example, utilities can send targeted signals to consumers to reduce usage during peak times, optimizing energy consumption and reducing strain on the grid.

Optimized resource allocation is facilitated by data analytics. By analyzing data with advanced AI algorithms, utilities can optimize power generation and distribution, balancing loads more effectively to minimize energy wastage [50]. Real-time data allows for quick adjustments to changing conditions, ensuring that resources are used efficiently and effectively.

Improved decision support is perhaps one of the most significant benefits of data analytics-based decision-making. Decision-makers are provided with detailed insights and actionable information, enhancing situational, spatial, and temporal awareness and supporting more informed and timely decisions [51]. For example, during an extreme weather event, real-time data can help grid operators manage load and distribution to prevent outages and ensure a reliable power supply.

The implications of data analytics-based decision-making in power systems are profound – first, increased efficiency results from optimized operations and reduced energy losses. Operational costs are lowered, and power generation and distribution efficiency are improved [48]. Enhanced reliability is achieved through proactive maintenance and real-time monitoring, decreasing the likelihood of outages and enhancing grid stability.

Sustainable energy integration is significantly bolstered by data analytics. The ability to predict and manage the output of renewable energy sources makes them more viable and reliable. Energy storage systems that are managed through advanced data analytics can store excess renewable energy and release it during high-demand or low-production periods, ensuring a consistent power supply.

Consumer empowerment is another important implication. Real-time data gives consumers insights into their energy usage, promoting energy-saving behaviors and enabling more personalized energy management solutions. Dynamic pricing models, supported by real-time data, allow consumers to reduce their electricity bills by shifting their usage to off-peak times [52].

Regulatory and policy support is also enhanced through data analytics. Detailed reporting and data transparency support regulatory compliance and influence policymaking by providing evidence-based insights into grid operations and energy usage patterns [5]. For instance, governments can design more effective energy policies based on real-time data, promoting renewable energy adoption and improving grid reliability.

Comparing data analytics-based decision-making to traditional methods highlights several key differences. Traditional decision-making in power systems is often reactive, relying on historical data and fixed schedules [47]. In contrast, data analytics-based decision-making is proactive, using real-time data to anticipate and address issues before they escalate. Conventional methods depend heavily on manual processes and human intervention, while data analytics automates many decision-making processes, increasing efficiency and reducing the need for manual oversight.

Traditional approaches are typically static, with fixed rules and protocols that may not adapt well to changing conditions. Data analytics, however, enables dynamic and flexible decision-making, adapting to real-time changes in grid conditions and consumer behavior. Furthermore, conventional decision-making often relies on limited data points, leading to less informed decisions. Data analytics leverages vast amounts of data from multiple sources, providing a comprehensive and accurate basis for decision-making. Unlike traditional systems' broad strategies, this allows for personalized approaches, tailoring solutions to specific consumer patterns and needs.

In conclusion, data analytics-based decision-making in future power systems represents a significant advancement over traditional methods. It is characterized by real-time data collection, advanced predictive analytics, proactive maintenance, enhanced grid flexibility, optimized resource allocation, and improved decision support. These features lead to increased efficiency, enhanced reliability, sustainable energy integration, consumer empowerment, and better regulatory and policy support. This approach marks a fundamental shift from conventional methods' reactive, manual, and static nature, paving the way for a more resilient, adaptable, and efficient power grid.

1.5.1 Decision Framework

A data-driven decision-making framework integrates data analytics into power systems' operational and strategic processes. It allows utilities to make informed decisions based on real-time and historical data, improving monitoring and controllability of a multilayer digitalized power grid. The framework includes data collection and processing, analytics, visualization, and action. Data is collected from various sources, processed for quality and consistency, analyzed to extract insights, visualized for better understanding, and used to inform decisions.

Implementing a data-driven decision-making framework requires investment in technology, training for personnel, and the development of clear protocols and guidelines. It also involves collaboration between different departments within the utility to ensure data is used effectively.

1.5.1.1 Uncertainty Issues

Power systems face various uncertainties, including demand fluctuations, renewable energy variability, and unforeseen equipment failures. These uncertainties can impact the reliability and stability of the grid. Techniques such as probabilistic forecasting, scenario analysis, and robust optimization are used to handle uncertainties. These methods help assess risks and make decisions that account for potential variations in system behavior.

Utilities use probabilistic load forecasting to predict demand under different scenarios, allowing for better planning and resource allocation. Similarly, robust optimization techniques are applied to ensure grid stability despite the variability of renewable energy sources.

1.5.1.2 Behavioral Analytics

Behavioral analytics involves analyzing data on consumer behavior to understand and predict energy usage patterns [53]. It helps utilities design more effective demand response programs and energy efficiency initiatives. Utilities can use behavioral analytics to segment customers based on their usage patterns and target them with personalized energy-saving recommendations [54]. It can also be used to predict the impact of pricing changes on consumption behavior.

Behavioral analytics can improve customer satisfaction, reduce peak demand, and increase energy efficiency [55]. Utilities can optimize their operations and enhance grid reliability by understanding and influencing consumer behavior.

1.5.1.3 Policy Mechanisms

Policy mechanisms are critical in creating an enabling environment for data-driven decision-making. They provide guidelines, standards, and incentives for utilities to adopt advanced data analytics technologies. Examples include regulations mandating the use of smart meters, incentives for investing in data analytics infrastructure, and data interoperability and cybersecurity standards. These policies help ensure data analytics is effectively integrated into power system operations.

Supportive policies can accelerate the adoption of data analytics, leading to more flexible and robust power systems. These policies also encourage innovation and investment in new technologies, driving continuous improvement in grid performance.

1.5.2 Computational Aspects

Handling the vast amounts of data that power systems generate requires significant computational resources. Challenges include data storage, processing speed, and integrating heterogeneous data sources [56]. Solutions include cloud computing, high-performance computing (HPC) systems, and advanced data management techniques. These technologies enable efficient processing and analysis of large datasets, providing timely insights for decision-making.

Emerging trends such as edge computing, which involves processing data closer to its source [57], and the use of AI for automated data analysis are expected to further enhance the capabilities of data analytics in power systems [58].

1.6 Conclusion

This chapter explores the critical role of data analytics in modern power systems. At first, a discussion about data-rich power systems was presented, highlighting the various data sources and types, such as smart meters, SCADA systems, PMUs, and IoT devices. We also delved into the data structure, covering different types of data, formats, and data management practices.

Next, data analytics in power systems have been studied, defining its importance and the different techniques used, including descriptive, predictive, and prescriptive analytics, as well as machine learning and big data analytics. Then, data analytics-based decision-making in future power systems is discussed, outlining a data-driven decision framework, techniques for handling uncertainty, the role of behavioral analytics, the importance of policy mechanisms, and the computational aspects involved.

1.7 Future Trends and Challenges

Data analytics is essential for enhancing power systems' efficiency, reliability, observability, and controllability. Providing real-time insights and predictive capabilities allows for more efficient operation and proactive issue management. As renewable energy sources become more prevalent, data analytics becomes crucial for managing variability and ensuring grid stability. Additionally, data analytics drives innovation in power systems, encouraging the development of new technologies and solutions to address emerging challenges. Furthermore, behavioral analytics helps utilities understand and influence consumer behavior, improving customer engagement and satisfaction.

To fully harness the potential of data analytics in power systems, utilities and stakeholders must invest in technology and infrastructure. This includes deploying advanced metering infrastructure, upgrading data management systems, and adopting cloud and edge computing solutions. Training and capacity building are also vital, as developing the skills and expertise of personnel in data analytics is crucial for successful implementation. Collaboration between utilities, technology providers, policymakers, and research institutions is essential for driving progress and facilitating the sharing of knowledge, resources, and best practices. Finally, continued research and development efforts are needed to advance data analytics techniques and address emerging challenges in power systems. Focus areas should include improving predictive models, enhancing data security, and developing new applications for data analytics.

References

1 Voropai, N. (2020). Electric power system transformations: a review of main prospects and challenges. *Energies* 13 (21): 5639.

2 Syed, D., Zainab, A., Refaat, S.S., and Abu-Rub, H. (2020). Smart grid big data analytics: survey of technologies, techniques, and applications. *IEEE Access* 9: 59564–59585.

3 Marot, A., Kelly, A., Naglic, M. et al. (2022). Perspectives on future power system control centers for energy transition. *Journal of Modern Power Systems and Clean Energy* 10 (2): 328–344.

4 Metchkarski, N., Stanev, R., and Tzolov, A. (2019). Information and communication technologies potential for future power systems control. *2019 11th Electrical Engineering Faculty Conference (BulEF),* Varna, Bulgaria (11–14 September 2019). IEEE.

5 Burda, Y.D., Volkova, I.O., Gavrikova, E.V., and Kosygina, A.V. (2019). Digitalization and ways for the development of the electric energy industry with the participation of consumers: new challenges for shaping the investment climate. *Journal of Siberian Federal University, Humanities & Social Sciences* 12 (4): 545–564.

6 Dong, Z.Y. and Zhang, Y. (2021). Interdisciplinary vision of the digitalized future energy systems. *IEEE Open Access Journal of Power and Energy* 8: 557–569.

7 Vrba, P., Marík, V., Siano, P., and Leitão, P. (2014). A review of agent and service-oriented concepts applied to intelligent energy systems. *IEEE Transactions on Industrial Informatics* 10 (3): 1890–1903.

8 Pau, M., Mirz, M., Dinkelbach, J. et al. (2022). A service oriented architecture for the digitalization and automation of distribution grids. *IEEE Access* 10: 37050–37063.

9 Baidya, S., Potdar, V., Ray, P.P., and Nandi, C. (2021). Reviewing the opportunities, challenges, and future directions for the digitalization of energy. *Energy Research & Social Science* 81: 102243.

10 Siuta-Tokarska, B., Kruk, S., Krzemiński, P. et al. (2022). Digitalisation of enterprises in the energy sector: drivers – business models – prospective directions of changes. *Energies* 15 (23): 8962.

11 Küfeoglu, S., Liu, G., Anaya, K.L., and Pollitt, M.G. (2019). Digitalisation and new business models in energy sector. Report number: 1956. University of Cambridge.

12 McDonald, J.D. (2024). A holistic approach to becoming a data-driven utility. In: *Big Data Application in Power Systems* (ed. R. Arghandeh and Y. Zhou), 3–35. Elsevier.

13 Dekeyrel, S. and Fessler, M. (2024). Digitalisation: An enabler for the clean energy transition. *Journal of Energy & Natural Resources Law* 42 (2): 185–209.

14 Gao, W. and Gorinevsky, D. (2020). Probabilistic modeling for optimization of resource mix with variable generation and storage. *IEEE Transactions on Power Systems* 35 (5): 4036–4045.

15 Judson, E. (2024). Governing the GB digital energy revolution to support net zero by 2050, Doctoral Theses, University of Exeter.

16 Nakicenovic, N. (2022). Perspectives on the pervasive energy-systems transformations. *Oxford Open Energy* 1: oiab005.

17 Löbbe, S. and Hackbarth, A. (2017). The transformation of the German electricity sector and the emergence of new business models in distributed energy systems. In: *Innovation and Disruption at the Grid's Edge* (ed. F. Sioshansi), 287–318. Elsevier.

18 Małkowska, A., Urbaniec, M., and Kosała, M. (2021). The impact of digital transformation on European countries: insights from a comparative analysis. *Equilibrium. Quarterly Journal of Economics and Economic Policy* 16 (2): 325–355.

19 Gopstein, A., Nguyen, C., O'Fallon, C. et al. (2021). NIST framework and roadmap for smart grid interoperability standards, release 4.0. Department of Commerce, National Institute of Standards and Technology.

20 Moghaddam, M.P., Zamani, R., Alhelou, H.H., and Siano, P. (2022). *Decentralized Frameworks for Future Power Systems: Operation, Planning and Control Perspectives*. Academic Press.

21 Benedettini, S., Brugnetta, G., Fumiatti, F. et al. (2019). Assessment and roadmap for the digital transformation of the energy sector towards an innovative internal market, in Assessment and roadmap for the digital transformation of the energy sector towards an innovative internal energy market. Publications Office of the European Union.

22 Maroufkhani, P., Desouza, K.C., Perrons, R.K., and Iranmanesh, M. (2022). Digital transformation in the resource and energy sectors: a systematic review. *Resources Policy* 76: 102622.

23 Khalid, M. (2024). Energy 4.0: AI-enabled digital transformation for sustainable power networks. *Computers & Industrial Engineering* 110253.

24 Zhao, Y., Xia, S., Zhang, J. et al. (2021). Effect of the digital transformation of power system on renewable energy utilization in China. *IEEE Access* 9: 96201–96209.

25 Çolak, M. and Irmak, E. (2023). A state-of-the-art review on electric power systems and digital transformation. *Electric Power Components & Systems* 51 (11): 1089–1112.

26 Shang, Y., Raza, S.A., Huo, Z. et al. (2023). Does enterprise digital transformation contribute to the carbon emission reduction? Micro-level evidence from China. *International Review of Economics and Finance* 86: 1–13.

27 Cao, L., Hu, P., Li, X. et al. (2023). Digital technologies for net-zero energy transition: a preliminary study. *Carbon Neutrality* 2 (1): 7.

28 Doczi, S. (2019). China power system transformation: assessing the benefit of optimised operations and advanced flexibility options. Florence School of Regulation, European University Institute.

29 Han, Y., Lam, J.C.K., and Li, V.O.K. (2021). A Bayesian LSTM model to evaluate the effects of air pollution control regulations in Beijing, China. *Environmental Science & Policy* 115: 26–34.

30 Gao, W., Tayal, D., and Gorinevsky, D. (2019). Probabilistic planning of minigrid with renewables and storage in western Australia. *2019 IEEE Power & Energy Society General Meeting (PESGM)*, Atlanta, GA (4–8 August 2019). IEEE.

31 Sioshansi, F.P., Zamani, R., and Moghaddam, M.P. (2022). Energy transformation and decentralization in future power systems. In: *Decentralized Frameworks for Future Power Systems* (ed. M.P. Moghaddam, R. Zamani, H.H. Alhelou, and P. Siano), 1–18. Elsevier.

32 El Bassam, N., Schlichting, M., and Pagani, D. (2021). *Distributed Renewable Energies for Off-Grid Communities: Empowering a Sustainable, Competitive, and Secure Twenty-First Century*. Elsevier.

33 Nazari, Z. and Musilek, P. (2023). Impact of digital transformation on the energy sector: a review. *Algorithms* 16 (4): 211.

34 Wang, X., Yao, Z., and Papaefthymiou, M. (2023). A real-time electrical load forecasting and unsupervised anomaly detection framework. *Applied Energy* 330: 120279.

35 Chen, Z., Dong, Z.Y., and Xu, Y. (2024). Anomaly-detection-based learning for real-time data processing in non-intrusive load monitoring. *Energy Conversion and Economics.*
36 Zhou, W., O'Neill, E., Moncaster, A. et al. (2020). Forecasting urban residential stock turnover dynamics using system dynamics and Bayesian model averaging. *Applied Energy* 275: 115388.
37 Heggarty, T., Bourmaud, J.-Y., Girard, R., and Kariniotakis, G. (2020). Quantifying power system flexibility provision. *Applied Energy* 279: 115852.
38 Rana, M.M., Uddin, M., Sarkar, M.R. et al. (2023). Applications of energy storage systems in power grids with and without renewable energy integration – A comprehensive review. *Journal of Energy Storage* 68: 107811.
39 Sayed, E.T., Olabi, A.G., Alami, A.H. et al. (2023). Renewable energy and energy storage systems. *Energies* 16 (3): 1415.
40 Boopathy, P., Liyanage, M., Deepa, N. et al. (2024). Deep learning for intelligent demand response and smart grids: a comprehensive survey. *Computer Science Review* 51: 100617.
41 Alimi, O.A., Ouahada, K., and Abu-Mahfouz, A.M. (2020). A review of machine learning approaches to power system security and stability. *IEEE Access* 8: 113512–113531.
42 Heymann, F., Milojevic, T., Covatariu, A., and Verma, P. (2023). Digitalization in decarbonizing electricity systems – Phenomena, regional aspects, stakeholders, use cases, challenges and policy options. *Energy* 262: 125521.
43 Fang, D. and Wang, P. (2023). Optimal real-time pricing and electricity package by retail electric providers based on social learning. *Energy Economics* 117: 106442.
44 Zhang, Y., Huang, T., and Bompard, E.F. (2018). Big data analytics in smart grids: a review. *Energy Informatics* 1 (1): 1–24.
45 Barja-Martinez, S., Aragüés-Peñalba, M., Munné-Collado, Í. et al. (2021). Artificial intelligence techniques for enabling Big Data services in distribution networks: a review. *Renewable and Sustainable Energy Reviews* 150: 111459.
46 Li, J., Herdem, M.S., Nathwani, J., and Wen, J.Z. (2023). Methods and applications for Artificial Intelligence, Big Data, Internet of Things, and Blockchain in smart energy management. *Energy and AI* 11: 100208.
47 Colombari, R., Geuna, A., Helper, S. et al. (2023). The interplay between data-driven decision-making and digitalization: a firm-level survey of the Italian and US automotive industries. *International Journal of Production Economics* 255: 108718.
48 Strielkowski, W., Vlasov, A., Selivanov, K. et al. (2023). Prospects and challenges of the machine learning and data-driven methods for the predictive analysis of power systems: a review. *Energies* 16 (10): 4025.
49 Liu, Z., Guo, Z., Chen, Q. et al. (2023). A review of data-driven smart building-integrated photovoltaic systems: challenges and objectives. *Energy* 263: 126082.
50 Entezari, A., Aslani, A., Zahedi, R., and Noorollahi, Y. (2023). Artificial intelligence and machine learning in energy systems: a bibliographic perspective. *Energy Strategy Reviews* 45: 101017.
51 Olaniyi, O.O., Okunleye, O.J., and Olabanji, S.O. (2023). Advancing data-driven decision-making in smart cities through big data analytics: a comprehensive review of existing literature. *Current Journal of Applied Science and Technology* 42 (25): 10–18.
52 Poplavskaya, K., Lago, J., and De Vries, L. (2020). Effect of market design on strategic bidding behavior: model-based analysis of European electricity balancing markets. *Applied Energy* 270: 115130.
53 Sun, Y., Hao, W., Chen, Y., and Liu, B. (2020). Data-driven occupant-behavior analytics for residential buildings. *Energy* 206: 118100.

54 Ableitner, L., Tiefenbeck, V., Meeuw, A. et al. (2020). User behavior in a real-world peer-to-peer electricity market. *Applied Energy* 270: 115061.

55 Hong, Q., Meng, F., Liu, J., and Bo, R. (2023). A bilevel game-theoretic decision-making framework for strategic retailers in both local and wholesale electricity markets. *Applied Energy* 330: 120311.

56 Al-Jumaili, A.H.A., Muniyandi, R.C., Hasan, M.K. et al. (2023). Big data analytics using cloud computing based frameworks for power management systems: status, constraints, and future recommendations. *Sensors* 23 (6): 2952.

57 Fang, Z., Xu, X., Bilal, M., and Jolfaei, A. (2024). Renewable prediction-driven service offloading for IoT-enabled energy systems with edge computing. *Wireless Networks* 30 (5): 3721–3733.

58 Xie, H., Jiang, M., Zhang, D. et al. (2023). IntelliSense technology in the new power systems. *Renewable and Sustainable Energy Reviews* 177: 113229.

2

Advanced Predictive Modeling for Energy Consumption and Demand

Seyed Mohsen Hashemi and Abbas Marini

Power System Operation and Planning Research Group, Niroo Research Institute (NRI), Tehran, Iran

2.1 The Role of Load Forecasting in Power System Planning

Power systems have continuously expanded since their inception, driven by the growing demand for electricity in terms of consumption levels, consumer distribution, and timing of usage. Initially, power generation facilities were typically located near consumers, and the concept of a power system was not well established in the literature on electricity supply. Over time, the rapid escalation in electricity demand necessitated the development of a comprehensive network, or power system, capable of generating electricity at one location and transmitting it to various consumption centers, such as industrial companies and residential areas. The expansion of power generation capacity and transmission lines has been fundamentally linked to the increasing reliance on electrical energy across diverse activities. Without a corresponding growth in electricity demand, there would be little justification for enhancing the capacity of power plants or expanding the distribution and transmission networks. In other words, stakeholders involved in the development of power systems must remain informed about electricity consumption demands, as this information serves as a foundation for determining how to advance different sectors of the power system. Understanding electricity demand encompasses various dimensions, including growth rates, geographical distribution or locations of increased demand, and the temporal aspects of demand growth. Several historical data is needed to determine the mentioned aspects of demand forecast [1], and different AI-based approaches have been developed for that in recent years [2].

The electricity industry can be regarded as a fundamental infrastructure sector in any country. The costs associated with the development of various components of the power system are substantial and typically require significant time periods for implementation. For instance, the development of certain types of power plants, such as nuclear power plants, necessitates a time frame of seven to nine years, while large thermal power plants require a minimum of three years. Additionally, constructing a long transmission line, which may include multiple substations, also demands several years of planning and execution. Consequently, stakeholders involved in power system development must be informed about the anticipated growth in electricity demand over the next several years. Thus, planning issues related to generation and network development

Data-Driven Energy Management and Tariff Optimization in Power Systems:
Shaping the Future of Electricity Distribution through Analytics, First Edition.
Edited by Hamidreza Arasteh, Pierluigi Siano, Niki Moslemi, and Josep M. Guerrero.

encompass a series of models that take into account the forecasted demand for electricity consumption over a long-term horizon, such as the next 20 years. These models propose trends for capacity development in generation or transmission during this period [3]. Typically, these decisions are based on historical time series data reflecting annual demand growth trends [4].

2.2 Need for Short-Term Demand Forecasting

The necessity for annual forecasting of electricity demand growth was discussed in the previous section as the most critical parameter required to determine the annual development trajectory of the power system. A pertinent question arises: is there also a need to understand electricity demand in the short term? While having an overall forecast of demand for each year is beneficial, it does not fulfill the requirement for short-term demand forecasts. The production and consumption of electricity are continuous processes; thus, operators within the power system require relatively accurate information regarding consumption needs in the coming weeks, days, hours, or even minutes.

Within a 24-hour period, electricity consumption demand is continuously fluctuating. Generally, during late-night to early morning hours, there is a significant reduction in electricity demand due to decreased activity in residential and industrial sectors. Conversely, as midday approaches, an increase in electricity consumption is observed, creating a continuous cycle of rising and falling demand. On the other hand, network operators must have relatively precise information about hourly fluctuations in consumption demand over several hours ahead to facilitate effective operation of power generation units. For example, some power plants require several hours to start up and become operational; similarly, adjustments in their generation levels necessitate a specific duration based on the technology employed [5, 6].

For example, if there is a forecast indicating an increase in electricity demand between 10 a.m. and 12 p.m., the system operator must issue commands to bring some power generation units online or increase their output several hours prior to this time frame. Therefore, in addition to the need for long-term electricity demand forecasting to determine trends in generation and network development, access to high-accuracy short-term demand forecasts is one of the most critical requirements for effective power system operation. The importance of short-term demand forecasting may even surpass that of long-term forecasting from certain perspectives. This is because decisions made by network operators are based on near-real-time data [7, 8]; thus, there is not enough time to correct high probable errors in this process. In contrast, if significant errors occur during long-term demand forecasting, there remains an opportunity to update the forecasting process and improve its accuracy over subsequent intervals. In recent years, the proliferation of smart grids has enabled consumers to respond effectively to real-time fluctuations in power market prices and participate in demand response programs [9]. Consequently, utilities must be well-informed about consumers' power consumption patterns and their sensitivities to price changes [10, 11]. This understanding is essential for optimizing demand management strategies and enhancing overall grid efficiency in both the power distribution and transmission levels [12–15].

2.3 Components of Power Demand and Factors Affecting Demand Growth

In this section, we aim to analyze electricity demand from multiple viewpoints, detailing the application of each perspective.

2.3.1 Electricity Demand from the Consumer Type Perspective

Electricity demand can be categorized in various ways based on different perspectives. For instance, from the standpoint of consumer types, electricity demand is divided into residential, commercial, industrial, public, and agricultural categories. In this classification, each consumer type is examined independently based on its specific conditions and the primary factors influencing it. Ultimately, both short-term and long-term behaviors of these categories are predicted independently. For example, in the long term, population growth and the subsequent increase in residential units are among the most significant factors affecting the growth of electricity demand in the residential sector [16]. Conversely, the growth of electricity demand in the industrial sector is somewhat influenced by economic growth and its forecasts for the upcoming period [17]. In the short term, particularly over time frames ranging from several days to a few hours, the factors shaping electricity demand differ significantly. For instance, in areas where cooling systems are electrically powered, daily or hourly temperature fluctuations lead to substantial changes in residential electricity demand [18]. In contrast, temperature has relatively little impact on short-term variations in electricity demand within the industrial sector during brief time intervals.

2.3.2 Electricity Demand from the Supply Perspective

In the previous section, we categorized demand based on the types of consumers. However, this type of classification may not be appropriate in the perspective of how electricity demand is supplied. In practice, important factors for predicting demand include peak load and its average over a long-term horizon. Figure 2.1 illustrates a typical load curve within a power network.

It is evident that there are significant fluctuations in electricity consumption demand over a one-year period. If we arrange the electricity demand in descending order from the highest to the lowest values, we obtain a curve similar to that shown in Figure 2.2.

This shape is called a load duration curve (LDC) and generally serves as the basis for decision-making for generation expansion in power systems [19, 20]. Electricity consumption

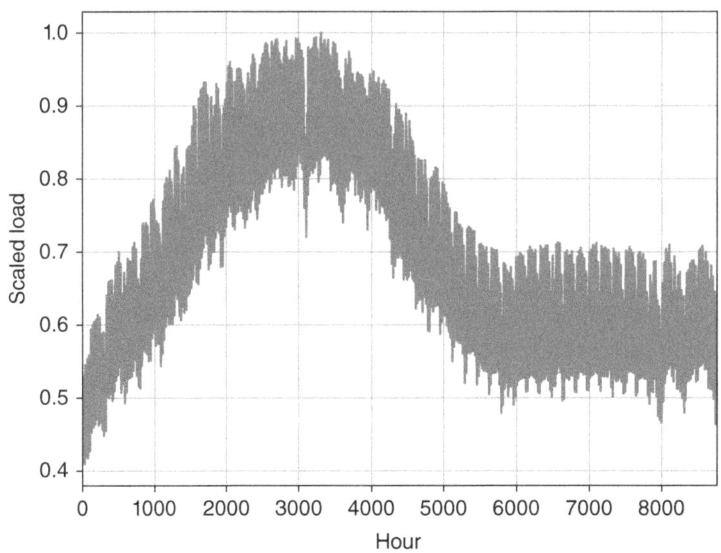

Figure 2.1 Typical electricity load curve.

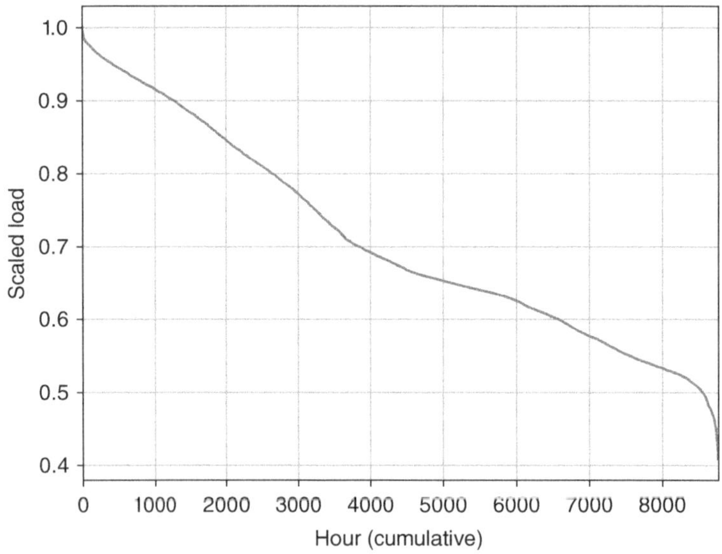

Figure 2.2 Scaled load duration curve.

demand remains above 0.5 p.u. throughout most of the year, representing the base load of the network. Conversely, during about 1200 hours of the year, consumption goes beyond 0.9 p.u., indicating the network's peak load. The average electricity consumption across all hours of the year is around 0.7 p.u. These types of data, along with forecasts of their trends over future planning periods, are among the most critical inputs in generation and transmission planning processes. Among the extractable data from this graph, the area under the curve and peak load are crucial for determining trends in power system development. The area under the curve indicates the volume of energy required to meet electricity demand. For instance, if electricity demand is entirely supplied by thermal power plants, depending on their efficiency and the heating value of various fuels used in these plants, the area under the curve will reflect the amount of fossil fuel needed to meet electrical energy demands. The peak load determines the generating capacity required to ensure that load is met throughout all hours of the year. Typically, the forecasted annual peak load within the study horizon determines the annual generation capacity requirement that has to be provided considering reliability criteria such as reserve capacity.

2.4 Electricity Demand in Networks with High Renewable Energy Sources

As previously mentioned, electricity demand requires to be forecasted, which inherently involves uncertainty. In recent years, the significant growth of renewable energy sources in the energy generation portfolios of various countries has introduced additional uncertainties in predicting the output of these resources. This is primarily due to their high dependence on weather conditions, rendering them as non-dispatchable power plants. In other words, it is assumed that there is no capability to control these plants or adjust their output levels to desired values.

As the renewable power generation and forecasted load are both uncertain, the concept of net load aggregates them into a single random variable, playing a crucial role in operational and planning issues within power systems [21]. Figure 2.3 displays the forecasted load, the predicted output

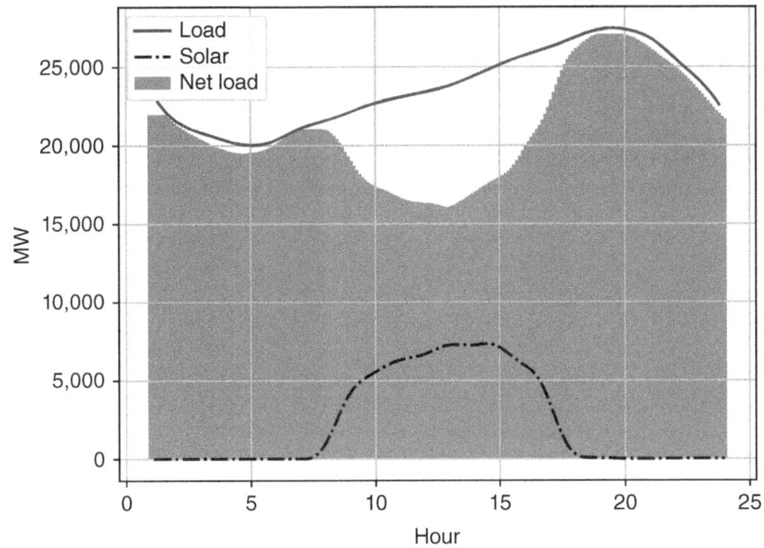

Figure 2.3 Load profile, solar power generation, and net load.

from renewable sources, and the net load concurrently. As can be observed, although the daily fluctuations in load are not significant, the net load exhibits considerable hourly variations during daylight hours, which arise from the variability in generation from renewable power plants. So, system operators must consider the forecasted net load curve when determining how to dispatch the dispatchable power plants.

2.5 Machine Learning and Its Applications in Demand Forecast

Machine learning is a part of artificial intelligence that allows systems to learn from past data and get better at tasks without being specifically coded for them. The key function of machine learning is creating algorithms for recognizing patterns, making decisions, and predicting based on past data [22, 23]. The main parts needed to use and run machine learning models have been depicted in Figure 2.4 and are explained in the following.

Data Collection: It is an important requirement in machine learning. The quality and the number of recorded data directly affect the performance of machine learning algorithms.

Data Analysis: In general, the recorded data of the real-world systems have to be preprocessed before applying the ML algorithms. For example, in the case of metering the power consumption, issues such as missing data in some time periods, incorrect data recording due to improperly calibrated measuring equipment, and similar factors can adversely affect data quality.

Model Selection: Depending on the problem, there are several ML algorithms that can be used. Whether the problem is categorized as classification, clustering, or regression determines how it is approached and suggests multiple algorithmic options.

Model Training: To determine the model parameters, the selected model have to be trained using the recorded data.

Model Assessment: In the previous step a large data set was used to train the model and determine its parameters. Now its performance should be assessed based on a set of unseen data called test data.

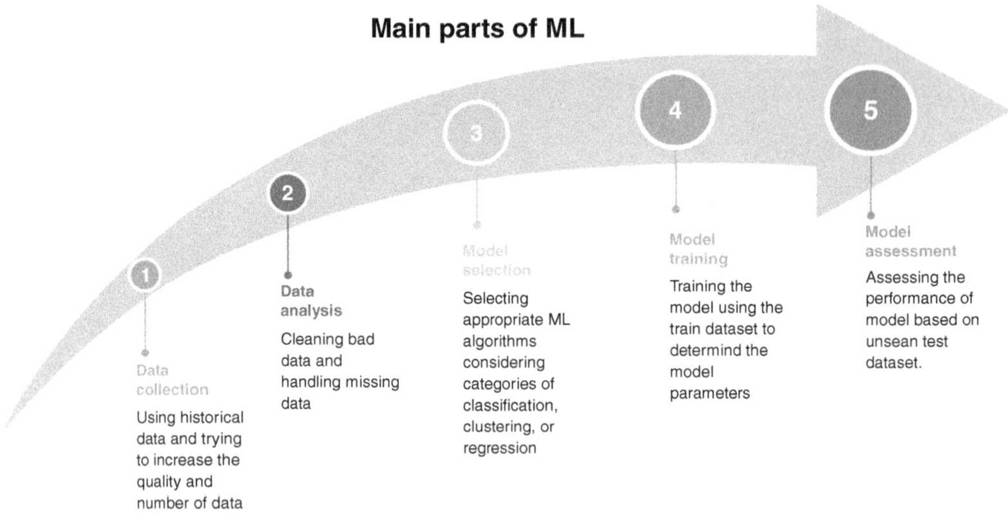

Figure 2.4 Main parts of ML.

As depicted in Figure 2.5, ML algorithms are categorized into the following subsections:

- **Supervised Learning**: These algorithms use the labeled data to create a model. The created model should determine the label of the new data based on its features and the trained model.
- **Unsupervised Learning**: These algorithms use the unlabeled data to recognize the interconnection of the data points based on their similarity and divide them into some clusters.
- **Reinforcement Learning**: these algorithms don't use a large historical dataset to learn the model. They use some penalty and revenue mechanism that is based on model performance in an environment. The model is updated during the time to discover the optimal behavior.

The following subsection provides a clustering algorithm for the demand forecast as one of the applications of ML in power systems.

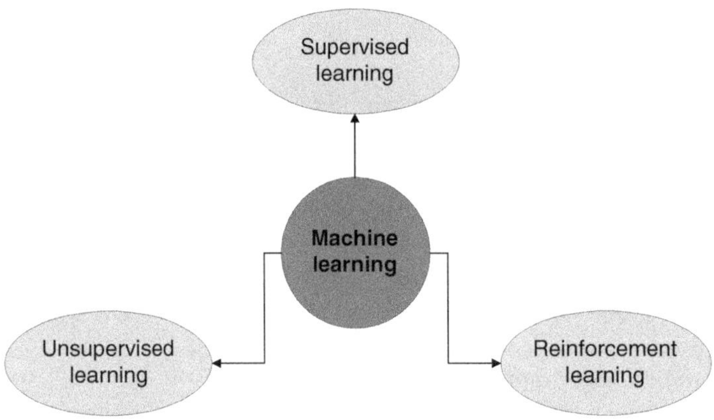

Figure 2.5 Main subsections of ML.

2.5.1 Application of Clustering in Load Forecasting

In previous sections, we discussed the concept of LDC and its application in addressing power system development issues. By determining LDC for all years within the planning horizon, it is possible to ascertain the required electrical energy demand for each year, as well as the minimum generation capacity needed. This provides a significant portion of the data necessary for determining trends in power generation development. However, in recent years, the relevance of this method has somewhat diminished due to the increasing emphasis on flexibility in power systems, which refers to the system's capability to respond to sudden variations in the power demand or supply [24–26]. This limitation arises from the inability of LDC to model the hourly load fluctuations. Figure 2.6 illustrates the actual load shape in a typical power system across different days over four seasons of the year. It is evident that the load curve behavior varies significantly across these four scenarios. Additionally, we observe rapid fluctuations in these graphs, indicating that consumption can change abruptly within short time intervals. Also, in some seasons, such as spring and autumn, the load shape of the days has more variance in comparison to the others. Certainly, overlooking such characteristics inherent in load behavior may lead to the development of power plants that, while capable of meeting peak loads and annual energy requirements, may lack the ability to cope with such load behaviors and consequently result in partial load shedding.

The use of the representative day method is one of the most effective approaches for modeling long-term generation planning issues, enabling the analysis and examination of load behavior over an extended period [27–29]. Ideally, having a separate load forecast for each day within the planning horizon to inform power plant expansion trends would be desirable. However, this is impractical due to the complexity of optimization and planning models. Modeling the forecasted loads for all the days within the planning horizon and analyzing system operations for each would introduce an overwhelming number of variables and constraints, rendering the

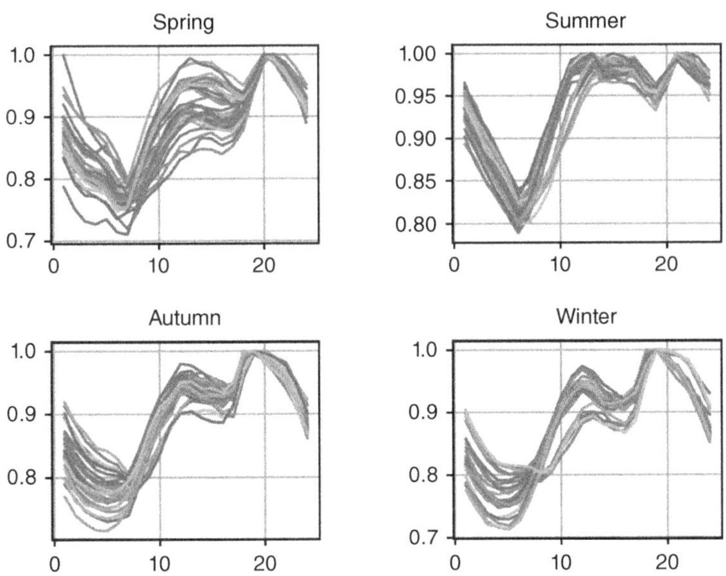

Figure 2.6 Load shape in different seasons of a real power system.

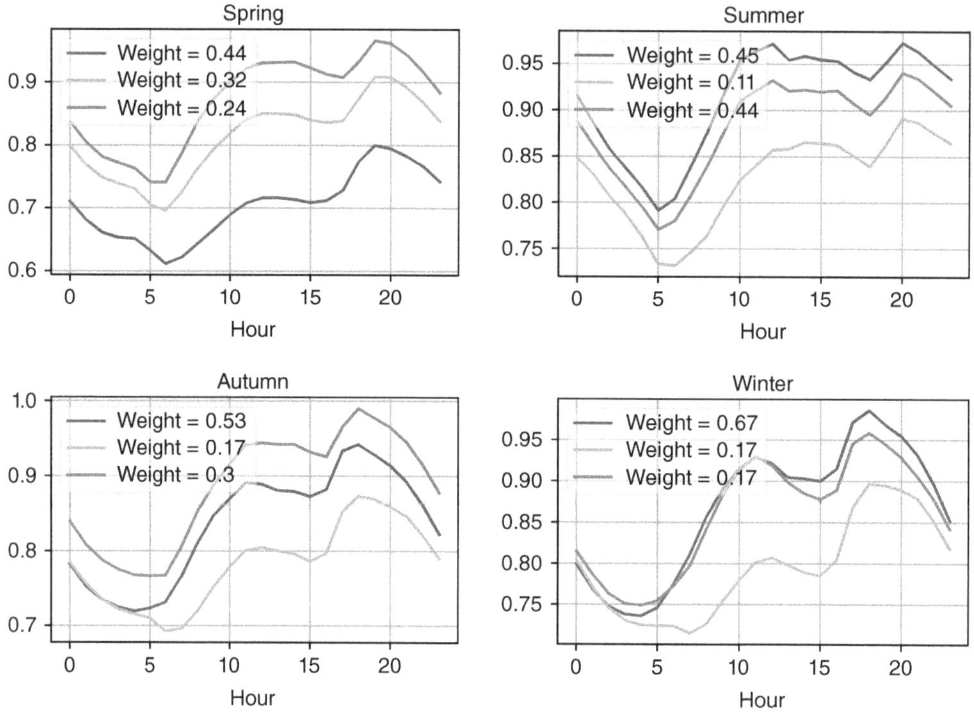

Figure 2.7 Clustering the daily load profile of each season.

problem unsolvable. Therefore, the concept of a representative day is commonly employed in this context. Using machine learning techniques, particularly clustering methods, the hourly load time series data for all days within the planning horizon can be grouped into a set of representative days that closely resemble the overall dataset behavior. Days with similar demand fluctuations are categorized into the same clusters, with each cluster's centroid representing a load curve that typifies the behavior of all its members. Increasing the number of clusters results in cluster members being more similar to their centroid; however, this also limits the applicability of these results in expansion planning due to increased complexity and computational demands in the optimization model. Figure 2.7 illustrates the representative days selected for each season. Clusters having higher weights include more members. These representative days have been created by clustering algorithms such as K-means [30, 31].

2.6 The Impact of Macro-decisions on Long-term Load Forecasting

One of the most significant characteristics of the long-term load forecasting process is the influence of unpredictable factors on it. In short-term load forecasting problems, many of the factors affecting load are generally known, allowing for the application of machine learning methods based on these features, as well as the utilization of recorded data. However, this is not the case for long-term forecasting. Electrification schemes are among the most effective factors influencing electricity demand growth over a long-term horizon [32], an aspect that is not reflected in historical data. More precisely, in addition to recorded electricity consumption data, information from other energy sectors will also be necessary. For instance, when considering the electrification of vehicles, data such as historical consumption of various fossil fuels in the transportation sector (including gas and gasoline), urban and intercity travel demands, and the shares of public versus private transport in travel patterns become essential. One of the most critical energy sectors where electrification has gained attention in recent years is the electrification of heat demand [33, 34]. In many countries, fossil fuels are the primary energy carriers used to meet consumers' heating demands. Given environmental challenges such as pollutant emissions and global warming, replacing these fuels with clean and renewable alternatives has become a priority in the planning horizons of many countries. In other words, meeting heating demand through electricity consumption is being considered. The subsequent section will examine the impact of heat demand electrification on overall electricity demand.

2.6.1 Natural Gas as a Primary Energy Carrier for Heating Demand

Natural gas is one of the most widely used energy carriers for meeting heating demand. In Iran, which is one of the largest producers of gas in the world, natural gas can be considered the primary energy carrier utilized to satisfy heating demand across various consumption sectors, including residential, industrial, and commercial. In industrial applications, the majority of gas consumption is directed toward process heating; additionally, a portion can also be attributed to space heating demand. In the residential and commercial sectors, heating demand can be divided into two categories: space heating demand and demand for hot water and cooking. In other words, these two segments of demand should be analyzed separately, as they exhibit different behaviors. Space heating demand occurs during the cold seasons and intensifies with decreasing temperatures. When modeling the load demand resulting from the electrification of heating systems, several considerations must be taken into account:

- **Temporal Demand Behavior**: Long-term electricity demand forecasting is generally based on a representative day method, which represents hourly demand behavior across different hours of sample days. However, recorded data regarding gas consumption is typically available on a daily basis rather than hourly.
- **Spatial Demand Distribution**: The spatial distribution of electricity consumption demand is a critical factor that influences network development and the location of power generation units. In many cases, historical data on gas consumption is presented at the provincial level, necessitating a more precise determination of spatial distribution.

One approach to increase the temporal and spatial resolution of gas consumption data is to examine its relationship with variables that have been recorded with higher temporal and spatial accuracy. For instance, historical temperature data for various geographical regions is available from NASA, covering a significant time period with appropriate temporal and spatial resolution. This study aims to investigate the correlation between gas consumption in different regions and their respective temperatures. Figure 2.8 displays the hourly average temperature for a winter day across approximately 1200 cities in Iran.

It is observed that the coldest cities are located in the central and northern regions (with the exception of cities along the Caspian Sea), while southern cities exhibit higher average temperatures. As expressed, it can be expected that temperature serves as a significant criterion in the allocation of gas to provinces during the cold months of the year. However, factors such as urban population and the types of heating systems employed also play a crucial role in the accuracy of this assertion. In some vast provinces, such as Kerman and Fars, it is evident that recorded temperatures for various cities exhibit high variance, indicating that the spatial distribution of heating demand

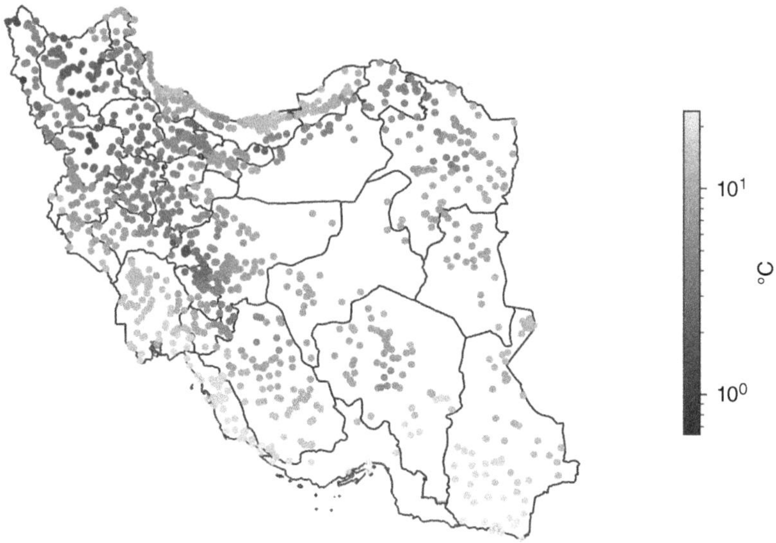

Figure 2.8 Temperature profile in a typical day of winter in Iran.

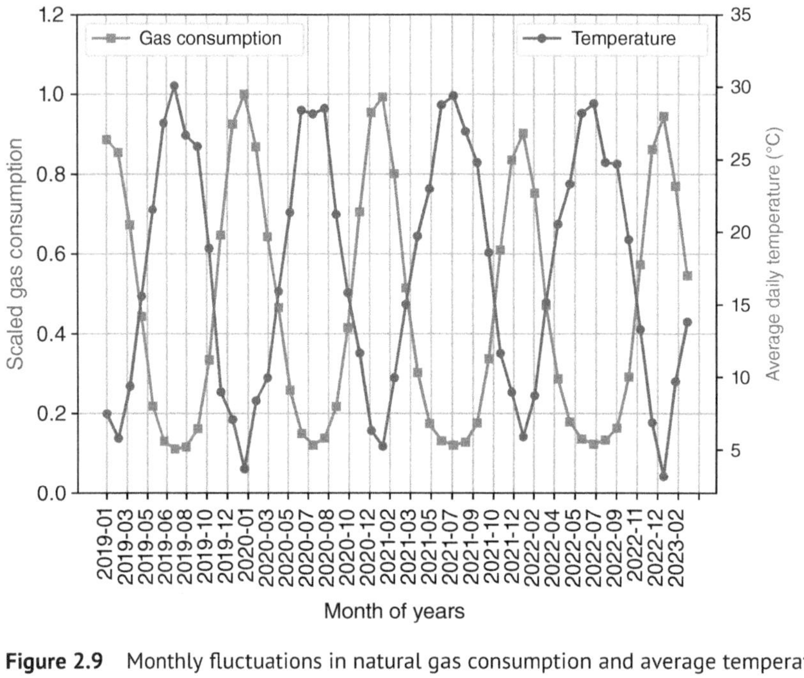

Figure 2.9 Monthly fluctuations in natural gas consumption and average temperature in Kerman Province.

within these provinces is very important. In other words, gas consumption in these provinces cannot be evenly distributed among different cities. Figure 2.9 presents normalized data on residential gas consumption in Kerman province during four years. The maximum annual consumption is consistently increasing, with temperature being the most influential factor affecting monthly fluctuations in gas consumption. Notably, peak natural gas consumption occurs on the coldest days of the year.

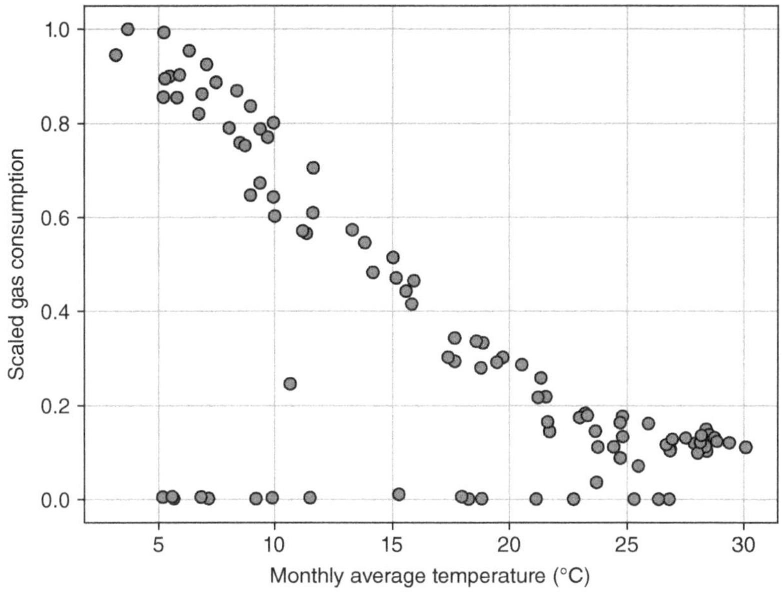

Figure 2.10 Natural gas consumption and temperature date in Kerman province in the past eight years.

When monthly average temperature and natural gas consumption during 2014–2022 are plotted against each other, the resulting figure illustrates the relationship between these two variables, as depicted in Figure 2.10.

We aim to establish a predictive model for natural gas consumption based on average temperature. Utilizing machine learning techniques, we can employ a regression model to facilitate this modeling process.

As previously mentioned, before applying machine learning algorithms to the data, it is essential to preprocess the data by identifying and removing outliers. Based on the analysis, there appear to be significant measurement errors for a considerable number of data points. Specifically, in some instances, regardless of the recorded temperature, the gas consumption is approximately zero, indicating potential inaccuracies in data recording. Since these data points can be visually represented in two dimensions, it is relatively straightforward to identify and eliminate these erroneous entries. However, a systematic anomaly detection process is necessary to identify and exclude outliers effectively. Clustering methods can be employed for this purpose, including the following approaches:

- **K-means**: In this method, data points are grouped into k clusters based on their similarity. The algorithm operates by minimizing the Euclidean distance between each data point assigned to a cluster and the centroid of that cluster while maximizing the distance between the points assigned to different clusters. The number of clusters is provided as an input to the algorithm.
- **DBSCAN**: This method classifies data points based on their density. One of its advantages is that it does not require specifying the number of clusters as an input parameter.
- **Spectral Clustering**: This method operates by forming a graph based on the data points. In this graph, the data points serve as nodes, and edges are defined based on the similarity between them. By representing the graph in a matrix form known as the Laplacian matrix and determining its eigenvalues and eigenvectors, dimensionality reduction is achieved. Subsequently, clustering methods such as K-means can be applied in this reduced-dimensional space. One of the key strengths of this approach is its ability to uncover complex patterns among the data points and effectively classify them.

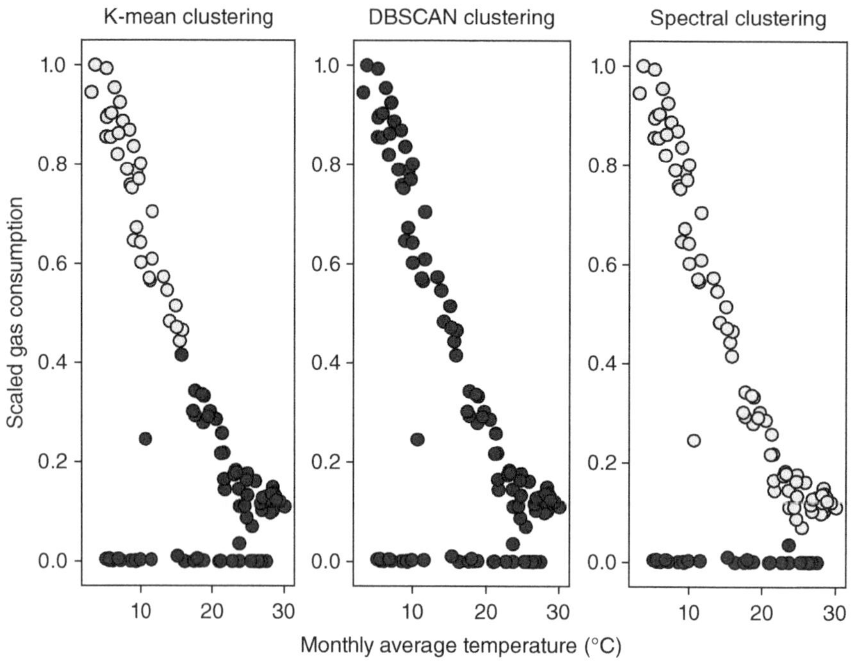

Figure 2.11 Performance of different clustering algorithms used for anomaly detection in data of gas consumption and temperature.

By applying these algorithms to the data, the results will be as in Figure 2.11. The K-means method incorrectly classifies the data, grouping some points that appear valid alongside those that are erroneous within the same cluster. In contrast, the DBSCAN algorithm places all data points into a single cluster. Ultimately, the spectral clustering method effectively identifies and separates nearly all of the erroneous data points into distinct clusters.

By removing the outliers, data will be appropriate to be used in the machine learning algorithms. Given that the target variable for prediction, namely natural gas consumption, is continuous numerical data, regression methods can be employed. Assuming mean squared error (MSE) as the loss function, the regression model will be realized by minimizing Eqs. (2.1) and (2.2).

$$J(w) = \frac{\sum_{i=1}^{N} [y^{(i)} - h(x^{(i)}, w)]^2}{n} \tag{2.1}$$

$$h(x, w) = w_0 + \sum_{i=1}^{d} w_i x_i \tag{2.2}$$

where

h represents the regression model
$y^{(i)}$ denotes the recorded gas consumption values for the ith data sample
$x^{(i)}$ refers to its recorded temperature

Additionally, w represents the coefficients of the regression model, n is the total number of samples, and d is the dimension of each data sample. In case of the polynomial regression, in addition to x_i, as the features of data sample x, the value of $x_i^2, x_i^3, \ldots, x_i^k$ is considered as the data features, too.

Table 2.1 Data features and algorithms used in simulation cases.

Features	A: Monthly gas consumption B: Monthly average temperature C: Maximum annual gas consumption D: Average annual gas consumption
Regression algorithms	L: Linear regression P: Polynomial regression (degree 2)

Table 2.2 Simulation cases.

	Feature 1 (target)	Feature 2	Feature 3	Feature 4	Regression algorithm
Case 1	A	B			L
Case 2	A	B			P
Case 3	A	B	C		L
Case 4	A	B	C		P
Case 5	A	B	C	D	L
Case 6	A	B	C	D	P

For example, for polynomial regression of degree two, the model will be as Eq. (2.3).

$$\hat{y} = h(x, w) = w_0 + \sum_{i=1}^{d} w_i x_i + \sum_{i=1}^{d} w'_i x_i^2 \tag{2.3}$$

The value of parameters w can be calculated by minimizing the loss function using the gradient descent method. In order to analyze the performance of the model, six distinct cases are defined and simulated as indicated in Table 2.1. There are four features and two regression algorithms that are used in these cases and depicted in Table 2.2.

By running the regression model based on the settings of case 1 and case 2 and depicting their forecasted values for the test data, Figure 2.12 is created. As can be observed, case 2 has better performance in terms of forecasting the gas data consumption for the given test temperatures.

R^2 or coefficient of determination is a statistical measure to assess the performance of the regression model. It indicates how much of the target feature variation can be explained by the variation of the model feature. Its formulation is provided in Eq. (2.4).

$$R^2 = 1 - \frac{\sum_{i=1}^{N} [y^{(i)} - \hat{y}^{(i)}]^2}{\sum_{i=1}^{N} [y^{(i)} - \bar{y}]^2} \tag{2.4}$$

Figure 2.13 indicates the performance of the regression models developed in different simulation cases. Models with higher R^2 and lower MSE have better performance. As can be observed, cases that use polynomial regression have better performance in comparison to the others. On the other hand, models having more features have better performance. Cases four and six have similar performance and are the best among all the cases. However, case four is simpler as it considers only

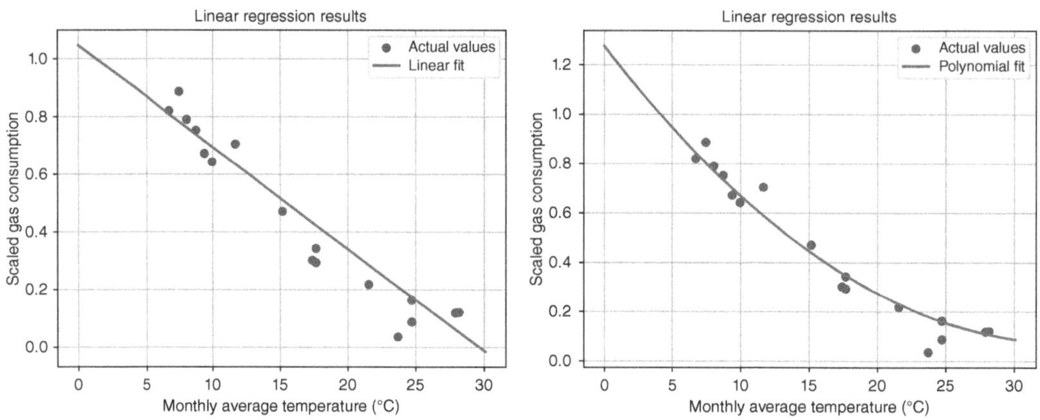

Figure 2.12 Performance of developed regression models in cases 1 and 2 considering the test data.

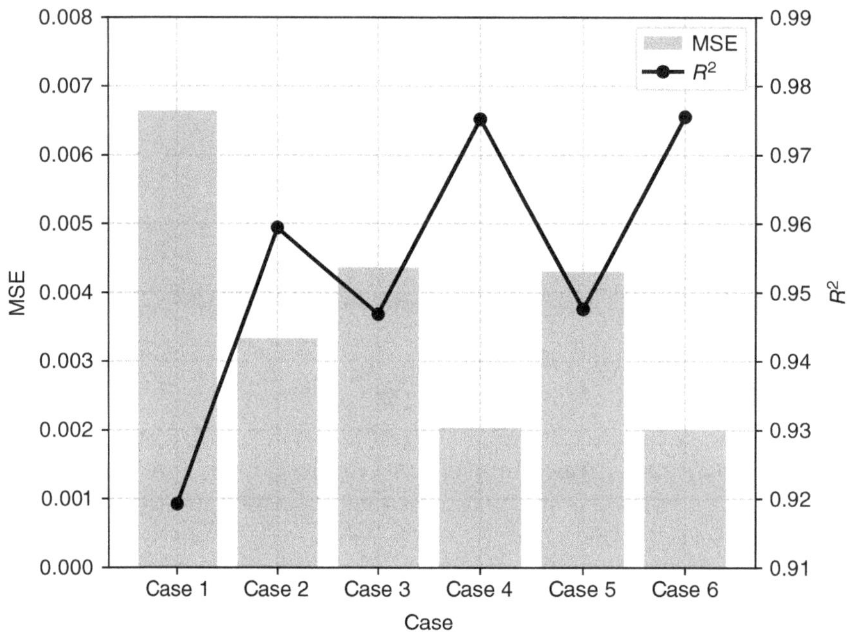

Figure 2.13 Performance of developed regression models in different simulation cases.

two features of the input data to forecast the target feature. These input features are the average temperature of the month and maximum gas consumption in the year.

2.7 Conclusion

This chapter has examined the application of ML techniques in power demand forecasting, a critical component in both long-term and short-term power system planning. We highlighted electrification as a significant factor influencing long-term power demand forecasts, with a particular focus on the electrification of heating demand. Our analysis centered on a case study using natural gas consumption data from an Iranian province, coupled with NASA's temperature records. We employed a regression model to predict gas consumption, incorporating various features such as temperature. Furthermore, we explored anomaly detection using diverse clustering algorithms, including K-means, DBSCAN, and spectral clustering methods. The study underscored the importance of advanced data analysis in extracting meaningful patterns from metered data. We noted that in many instances, particularly in load management programs, raw metered data often fails to provide an accurate representation of actual consumer power demand. This limitation necessitates more sophisticated analytical approaches to uncover relevant patterns and trends. Looking ahead, our research will focus on the application of tailored ML algorithms to raw input data. This approach aims to extract more meaningful insights and enhance prediction accuracy. By doing so, we anticipate making significant strides in improving the reliability and efficiency of power demand forecasting, ultimately contributing to more effective energy management and planning strategies.

References

1 Alsup, M. (2020). Forecasting electricity demand in the data-poor Indian context. PhD thesis. Massachusetts Institute of Technology. https://dspace.mit.edu/handle/1721.1/129084 (accessed 26 October 2024).

2 Aziz, A., Mahmood, D., Qureshi, M.S. et al. (2024). AI-based peak power demand forecasting model focusing on economic and climate features. *Frontiers in Energy Research* 12: 1328891.

3 Hashemi, S.M., Tabarzadi, M., Fallahi, F. et al. (2024). Water and emission constrained generation expansion planning for Iran power system. *Energy* 288: 129821.

4 Gonzato, S., Bruninx, K., and Delarue, E. (2021). Long term storage in generation expansion planning models with a reduced temporal scope. *Applied Energy* 298: 117168.

5 Yuanyuan, Z., Huiru, Z., and Bingkang, L. (2023). Distributionally robust comprehensive declaration strategy of virtual power plant participating in the power market considering flexible ramping product and uncertainties. *Applied Energy* 343: 121133.

6 Zhang, J., Mu, Y., Wu, Z. et al. (2024). Two-stage affine assessment method for flexible ramping capacity: an inverter heat pump virtual power plant case. *Applied Energy* 365: 123254.

7 Wang, Y., Wang, Y., Zhao, Z. et al. (2023). Multi-timescale optimal operation strategy for renewable energy power systems based on inertia evaluation. *Energies* 16 (8): 3577.

8 Ye, Y., Qiu, D., Wu, X. et al. (2020). Model-free real-time autonomous control for a residential multi-energy system using deep reinforcement learning. *IEEE Transactions on Smart Grid* 11 (4): 3068–3082.

9 Arasteh, H., Moslemi, N., and Hashemi, S.M. (2022). Demand response measurement and verification approaches: analyses and guidelines. In: *Industrial Demand Response: Methods, Best Practices, Case Studies, and Applications* (ed. H.H. Alhelou, A. Moreno-Muñoz, and P. Siano), 133. Institution of Engineering and Technology.

10 Hashemi, S.M., Vahidinasab, V., Ghazizadeh, M.S., and Aghaei, J. (2018). Valuing consumer participation in security enhancement of microgrids. *IET Generation, Transmission & Distribution* 13: 595.

11 Hashemi, S.M., Vahidinasab, V., Ghazizadeh, M., and Aghaei, J. (2020). Load control mechanism for operation of microgrids in contingency state. *IET Generation, Transmission & Distribution* 14 (23): 5407.

12 Hashemi, S.M., Vahidinasab, V., Ghazizadeh, M.S., and Aghaei, J. (2019) Reliability-oriented DG allocation in radial microgrids equipped with smart consumer switching capability. *2019 International Conference on Smart Energy Systems and Technologies (SEST)*, Porto, Portugal (9–11 September 2019), pp. 1–6. IEEE.

13 Mirhoseini, S.P., Hashemi, S.M., Alizadeh, B. et al. (2023). A reliability-based strategy for active distribution network expansion planning. *2023 13th Smart Grid Conference (SGC)*, Tehran, Iran (5–06 December 2023), pp. 1–8. IEEE. https://ieeexplore.ieee.org/abstract/document/10459273/ (accessed 8 June 2024).

14 Hashemi, S.M., Mirhoseini, S.P., Alizadeh, B., and Tabarzadi, M. (2024). A comprehensive method for expansion planning of active distribution system considering reliability assessment. *AUT Journal of Electrical Engineering* 57 (1): 113–130.

15 Hashemi, S.M. and Vahidinasab, V. (2021). Energy management systems for microgrids. In: *Microgrids: Advances in Operation, Control, and Protection* (ed. A. Anvari-Moghaddam, H. Abdi, B. Mohammadi-Ivatloo, and N. Hatziargyriou), 61–95. Springer.

16 Ahmad, T. and Zhang, D. (2020). A critical review of comparative global historical energy consumption and future demand: the story told so far. *Energy Reports* 6: 1973–1991.

17 Abbasi, K.R., Hussain, K., Abbas, J. et al. (2021). Analyzing the role of industrial sector's electricity consumption, prices, and GDP: a modified empirical evidence from Pakistan. *AIMS Energy* 9 (1): 29–49.

18 Haben, S., Giasemidis, G., Ziel, F., and Arora, S. (2019). Short term load forecasting and the effect of temperature at the low voltage level. *International Journal of Forecasting* 35 (4): 1469–1484.

19 Lyzwa, W. and Wierzbowski, M. (2016). Load duration curve in the long-term energy mix optimization. *2016 13th International Conference on the European Energy Market (EEM)*, Porto, Portugal (6–9 June 2016), pp. 1–5. IEEE. https://ieeexplore.ieee.org/abstract/document/7521278/ (accessed 26 October 2024).

20 Poulin, A., Dostie, M., Fournier, M., and Sansregret, S. (2008). Load duration curve: a tool for technico-economic analysis of energy solutions. *Energy and Buildings* 40 (1): 29–35.

21 Sreekumar, S., Sharma, K.C., and Bhakar, R. (2020). Grey system theory based net load forecasting for high renewable penetrated power systems. *Technology and Economics of Smart Grids and Sustainable Energy* 5 (1): 21. https://doi.org/10.1007/s40866-020-00094-4.

22 Mahesh, B. (2020). Machine learning algorithms – A review. *International Journal of Science and Research (IJSR)* 9 (1): 381–386.

23 Sarker, I.H. (2021). Machine learning: algorithms, real-world applications and research directions. *SN Computer Science* 2 (3): 160. https://doi.org/10.1007/s42979-021-00592-x.

24 Arasteh, H., Bahramara, S., Kaheh, Z. et al. A system-of-systems planning platform for enabling flexibility provision at distribution level. In: *Flexibility in Electric Power Distribution Networks* (ed. H.H. Alhelou, E. Heydarian-Forushani, and P. Siano), 41–65. CRC Press.

25 Hashemi, S.M., Arasteh, H., Shafiekhani, M. et al. (2023). Multi-objective operation of microgrids based on electrical and thermal flexibility metrics using the NNC and IGDT methods. *International Journal of Electrical Power & Energy Systems* 144: 108617.

26 Hashemi, S.M., Vahidinasab, V., Ghazizadeh, M.S., and Aghaei, J. (2020). Security constrained operation of radial micro grids based on the loads' vulnerabilities and flexibilities. *Tabriz Journal of Electrical Engineering* 50 (3): 1441.

27 Nahmmacher, P., Schmid, E., Hirth, L., and Knopf, B. (2016). Carpe diem: a novel approach to select representative days for long-term power system modeling. *Energy* 112: 430–442.

28 García-Cerezo, Á., Baringo, L., and García-Bertrand, R. (2020). Representative days for expansion decisions in power systems. *Energies* 13 (2): 335.

29 Li, C., Conejo, A.J., Siirola, J.D., and Grossmann, I.E. (2022). On representative day selection for capacity expansion planning of power systems under extreme operating conditions. *International Journal of Electrical Power & Energy Systems* 137: 107697.

30 Pinto, E.S., Serra, L.M., and Lázaro, A. (2020). Evaluation of methods to select representative days for the optimization of polygeneration systems. *Renewable Energy* 151: 488–502.

31 Tambunan, H.B., Barus, D.H., Hartono, J. et al. (2020). Electrical peak load clustering analysis using K-means algorithm and silhouette coefficient. *2020 International Conference on Technology and Policy in Energy and Electric Power (ICT-PEP)*, Bandung, Indonesia (23–24 September 2020), pp. 258–262. IEEE. https://ieeexplore.ieee.org/abstract/document/9249773/ (accessed 26 October 2024).

32 Lewis, J. and Severnini, E. (2020). Short-and long-run impacts of rural electrification: evidence from the historical rollout of the US power grid. *Journal of Development Economics* 143: 102412.

33 Zhang, M., Millar, M.-A., Yu, Z., and Yu, J. (2022). An assessment of the impacts of heat electrification on the electric grid in the UK. *Energy Reports* 8: 14934–14946.

34 Thomaßen, G., Kavvadias, K., and Navarro, J.P.J. (2021). The decarbonisation of the EU heating sector through electrification: a parametric analysis. *Energy Policy* 148: 111929.

3

Demand Response and Customer-Centric Energy Management

Alireza Mansoori, Mohsen Parsa Moghaddam, and Reza Zamani

Faculty of Electrical and Computer Engineering, Tarbiat Modares University, Tehran, Iran

3.1 Introduction

The movement of societies toward sustainable development has led to infrastructures disruption. In this regard, the power system as the infrastructure of infrastructures is undergoing important transformations in various technical, economic, social, and environmental aspects. The requirement for efficient power system transformation is to modify the traditional power system management mechanisms. The main key to this issue includes moving toward a customer-centric concept in such a way that by maximizing the customer-side potentials, higher efficiency will be achieved in realizing the interests of power system stakeholders.

It is noteworthy that, as a catalyst, identifying drivers and enablers can accelerate the achievement process of the customer-centric power system. The drivers include the development of macro policies, which will be discussed in the first and second parts of this chapter, including the background and transforming issue sections. On the contrary, when dealing with enablers, it is necessary to pay attention to more details, including management concepts, technological supports, and data analytics issues. Therefore, the enablers are investigated in the third part of the chapter by introducing the layered structure of the customer-centric power system, including physical, cyber, and social layers. Finally, a summary related to paving the way for achieving the future customer-centric power system and related literature trends is present.

3.2 Background

In recent years, due to stringent environmental regulations, reduced access to fossil fuels, and the need to meet growing electrical energy demands, distributed energy resources (DER), including renewable energy sources (RES), have significantly increased. For instance, one of the goals outlined in China's roadmap is to supply 35% of the country's required electrical energy from renewable sources by the target year of 2030. In 2017, several countries achieved efficient and substantial integration of RES into the main power grid. Leading nations in this field include Denmark, Uruguay, and Germany, with RES penetration rates of 53%, 28%, and 26%, respectively. Other countries, such as India, Ireland, Portugal, Spain, and the United States, recorded a penetration rate of around 20% in 2017. Figure 3.1 illustrates the status of leading countries in integrating renewable

Data-Driven Energy Management and Tariff Optimization in Power Systems: Shaping the Future of Electricity Distribution through Analytics, First Edition.
Edited by Hamidreza Arasteh, Pierluigi Siano, Niki Moslemi, and Josep M. Guerrero.

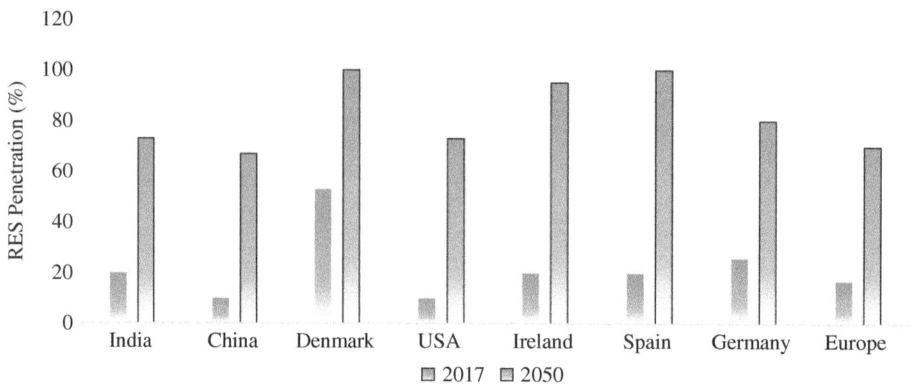

Figure 3.1 Long-term targeting of leading countries in RES integration until 2050. Source: Ratnam et al. [1] / with permission of Elsevier.

energy into the main grid and highlights their targets for 2050. According to this figure, these leading countries aim to supply over 60% of their electrical energy needs from renewable sources in their long-term planning [1]. As discussed, decarbonizing the power system is one of the key goals of governments, which necessitates the increasing penetration of RES.

Another example of DER is electric vehicles (EVs). Given environmental goals, the adoption of EVs can significantly reduce pollution levels. According to the International Energy Agency, the number of EVs reached 4.79 million in 2019, with more than half of these (2.58 million) in China. Moreover, the penetration of EVs is increasing at a remarkable rate of around 36% annually. As a result, it is predicted that by 2030, the number of EVs worldwide will reach 245 million. Assuming the average installed battery capacity per EV is 50 kWh, the total battery capacity of EVs in 2030 could reach 12.25 TWh [2]. Notably, other statistics, as shown in Figure 3.2, indicate that while the share of electricity in the energy mix for the transportation sector is growing, with an expected

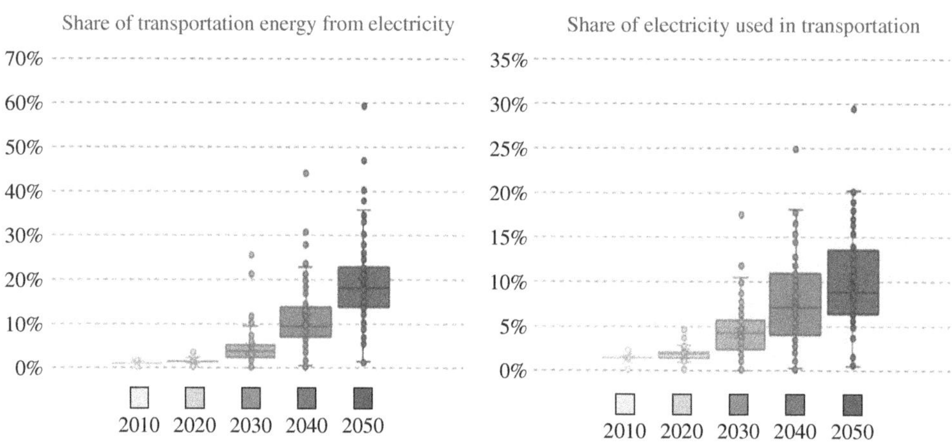

Figure 3.2 Expected electrification in the transportation sector until the horizon of 2050. Source: Muratori and Mai [3] / MDPI / CC BY 4.0.

value of around 20% by 2050, the share of electricity consumption in the transportation sector will also increase significantly, accounting for approximately 10% of total electricity consumption in that year [3].

3.3 Future Power Systems Aspects, Trends, and Challenges

In addition to the advantages outlined in the introduction, the presence of DERs can lead to challenges such as reduced inertia, reverse power flow, congestion, voltage fluctuations, stochastic generation, and increased uncertainty in the system's net load. It is important to note that uncertainty arises due to forecasting errors in the output of RES, which is why RESs are often referred to as non-dispatchable sources. Moreover, net load refers to the load that is not met by renewable energy generation. As a result, the need for greater flexibility in the power system is becoming more pronounced given these challenges [4].

Flexibility in the power system, in general, must ensure the reliable operation of the system while addressing the challenges associated with the integration of DER. Notably, as the penetration of these resources increases, the need for enhanced flexibility to maintain the balance between generation and consumption over short- to long-term time frames (from seconds to years) becomes even more critical. Consequently, one of the long-term impacts of DER penetration is the inadequacy of the conventional approach of demand tracking by generation to provide the necessary flexibility. Thus, there is a need to shift toward demand-side management (DSM) programs, including demand response (DR) and energy efficiency (EE), where customers take on an active role and become generation trackers and new supporters of the grid. However, in the transition toward sustainability, the effectiveness of demand-side programs depends on embracing the customer-centric concept in power systems. Key enablers of such a system include the design of value-based control and economic mechanisms. It is worth noting that the term "value" here refers to economic signals that serve as a common language and a key driver in the win-win management of electricity flow. Through purposeful, fair, and differentiated valuation of the electricity exchanged in the grid, not only is a public interest in meeting the flexibility needs of the power system enhanced, but network stress is reduced, leading to more cost-effective and efficient management. As illustrated in Figure 3.3, among the valuation indicators in this field, the indicators of the full-fledged power system can be mentioned. The perfect power system is defined as one that outperforms traditional power systems in terms of security, quality, reliability, availability, connectivity (both communication and energy networks), efficiency, affordability, and customer-centricity. In such a system, electricity energy management is based on these indicators to meet the flexibility requirements [5–7].

3.4 Transforming to Customer-Centric Era

Involving customers in meeting the power system flexibility needs offers a more efficient solution than traditional demand-tracking by generation. This reduces the need for large investments in network expansion, such as building new lines and power plants. Customers, by actively participating

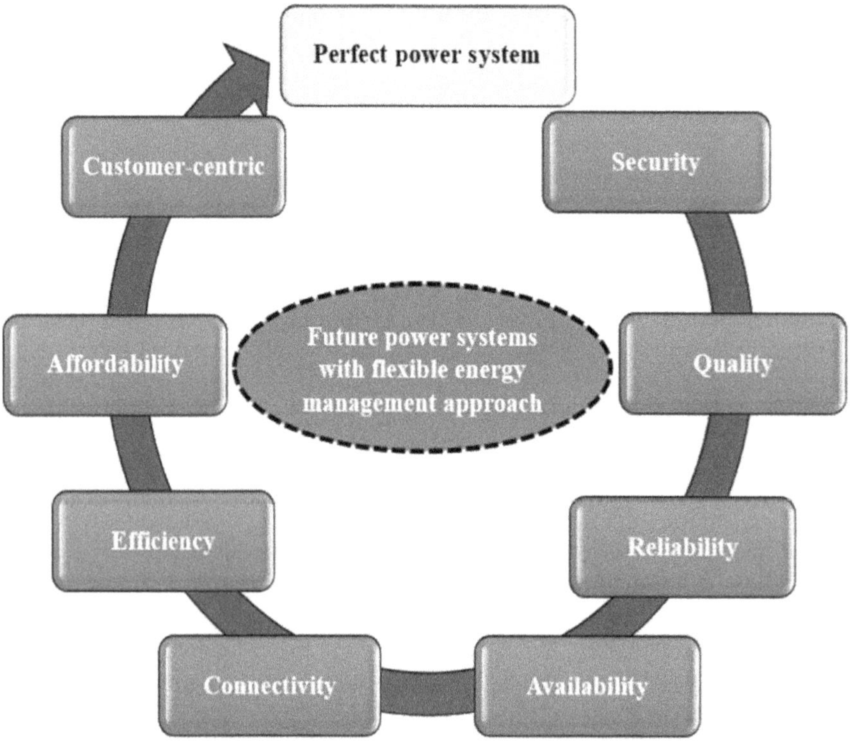

Figure 3.3 Key features of a perfect power system.

and making prices, can significantly lower their energy costs. They can also profit by meeting grid needs through their capabilities. Achieving this shift requires fair and distinct valuation of the electricity exchanged in the grid. This value-based approach can be a key motivator for increasing customer involvement in providing power systems flexible energy management.

3.4.1 Differences Between Customer-Centric DR Solution and Other Ways in the Future Power System

In general, as shown in Table 3.1, the customer-centric power system can be compared to the traditional one. Notably, according to this table, customer-centric solutions offer a key solution for addressing the high penetration of DER in future power systems. This approach proves advantageous from technical, economic, social, and environmental perspectives.

3.4.2 Drivers and Enablers

As illustrated in Figure 3.4, the emergence of such a system requires the sequential development of five disruptive innovations, detailed as follows [8]. It is important to note that there is no universal order of precedence for these innovations, as the existing conditions and policies vary across countries.

- **Decarbonization**: As previously mentioned, the primary driver for customer-centric approaches is the reduction of emissions, which leads to the increased penetration of DER, particularly RES.

Table 3.1 Comparison between customer-centric and conventional frameworks.

Features	Customer-centric framework	Conventional framework
Flexibility of environmental actors	Transmission system operators (TSO), distribution system operators (DSO), conventional generation units (CGU), aggregators, active customers	Transmission system operators (TSO), distribution system operators (DSO), conventional generation units (CGU)
Investment cost	Lower investment due to the utilization of distributed resources from customers	Extensive investments in the network expansion and building expensive dispatchable generation units such as gas units
Business models	Decentralized and bottom-up	Centralized and top-down
The interdependency of cyber-physical-social systems (CPSS)	High	Low
Social welfare	Higher social welfare due to increased power of choice and availability of cheaper options	Lower social welfare due to the presence of limited resources and options
Agile flexibility	High agility in providing both long-term and short-term flexibility is facilitated by the presence of granular resources with rapid response capabilities	Low agility is a consequence of the nature and limitations of the available resources
Facilitation of the demand-side resource activation process	High	Low
Technologies	Supporting technologies at the decentralized level	Supporting technologies at the centralized level

- **Deregulation**: Given the challenges associated with DER integration, new regulations are required to optimally provide the necessary flexibility. These regulations may support the transition toward a democratized grid.
- **Democratization**: A democratized grid enhances public acceptance for both power system operators and customers. The term "democratized" can be interpreted in two ways:
  - Harnessing the power and resources of the public to meet the system needs.
  - Valuing the services provided by the public, which can be a key factor in increasing fairness and public interest in participating in the power system flexibility.
- **Decentralization**: Traditional power systems were designed to provide large-scale centralized generation to customers rather than integrating millions of customer-owned generators. The widespread deployment of DERs creates operational and market challenges, making centralized control of numerous market actors difficult or even impossible for independent system operators (ISOs). From a centralized operational perspective, the following four main challenges arise for future power systems [9, 10]:
  - A centralized DER operation approach must account for a large number of customers. Such large-scale coordination may be impractical due to computational and communication requirements. Specifically, computational demands are tied to the complexity of the optimization problem, while communication needs relate to the maximum number of data exchanged through communication channels, data message size, and network latency.

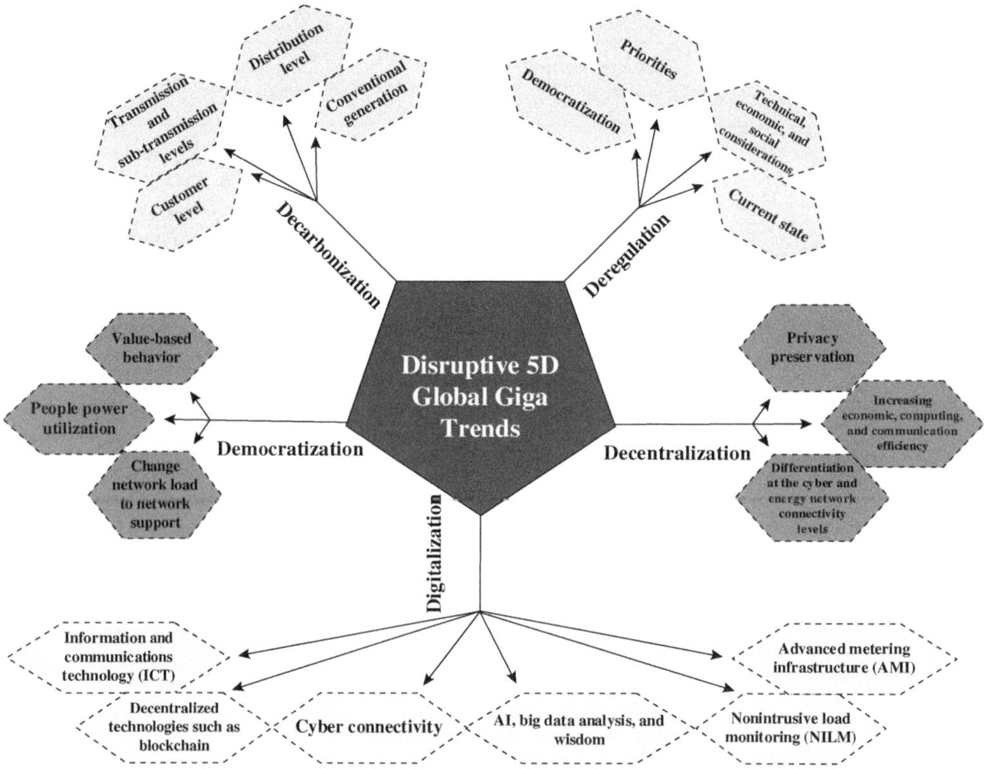

Figure 3.4 Disruptive 5D innovations chain.

- A centralized structure requires all customers (or their representatives) to send their data to a central entity, which could result in communication overhead and significant data storage demands.
- Centralizing data raises privacy concerns, as users may be reluctant to share their private information with an external entity.
- Many DERs, such as energy storage systems (ESS), have discrete operational points that can only be represented using integer variables. Solving the resulting mixed-integer programming problem in a centralized manner would significantly increase computational complexity.
- From an electricity market perspective, top-down centralized control cannot reflect a democratized grid. For example, in centralized structures, feed-in tariffs (FiT) and time-of-use (ToU) tariffs are often used in retail markets. Since there is typically a significant gap between the FiT (energy injected into the grid) and ToU (energy consumed from the grid) prices, the incentives provided may fall short of customers' expectations and fail to cover the investment returns and costs associated with energy generation [11]. In this situation, alternative solutions are needed so that customers can profitably and efficiently use their flexibility compared to current business models. As an alternative, local markets with a bottom-up approach have been proposed as a new business model. These markets can address many of the operational and market challenges related to DER integration in power systems.

- **Digitalization**: Finally, to enable the four stages outlined above, it is necessary to integrate information and communication technologies with real-time monitoring and response capabilities to meet the power system needs. Distributed ledger technologies such as blockchain, customer energy management systems (CEMS), non-intrusive load monitoring, and edge computing are some of the key digital technologies in democratized grids.

3.5 Customer-Centric Power System Structure

A key factor in enhancing flexibility and efficiently managing a customer-centric power system is the identification of its various layers and components to define new incentives. Therefore, in the remainder of this chapter, a comprehensive review of the existing literature is conducted, detailing management strategies for engaging active customers. The advantages and disadvantages of each approach are examined. Finally, based on the presented material, research gaps and future directions in this field are identified and discussed. Unlike the traditional power grid with its hierarchical and unidirectional structure, a customer-centric system involves bidirectional flows of both data and power. In such a system, customers evolve into active participants and collaborators with the grid. As depicted in Figure 3.5, the new power system will consist of a three-layer structure, which can be viewed as a cyber-physical-social system. The cyber layer focuses on communication infrastructure, computational elements, and data exchange among actors. Meanwhile, the social layer addresses the diversity in customer behavior, designing management mechanisms and methods for engaging customers in meeting the system flexibility requirements. Finally, the physical layer deals with the system's physical resources and constraints. The adequacy of the system to accommodate active customers and utilize their flexibility heavily depends on the state of the physical layer. The presence of this layer is a necessary but insufficient condition to meet flexibility demands across various time horizons.

3.5.1 Physical Layer

Broadly speaking, the physical layer of a customer-centric power system can be examined from two perspectives: physical resources and physical constraints that are examined in detail below.

3.5.1.1 Physical Resources

Various resources and technologies contribute to meeting the power system flexibility needs, including:

- **Energy Storage Systems (ESS)**: This technology offers varying degrees of flexibility depending on factors such as response time, storage capacity, charge and discharge rates, and efficiency. ESS can store excess renewable energy and discharge it during periods of shortfall. Electrical energy is now considered a commodity that, with advances in technologies like pumped-storage plants and

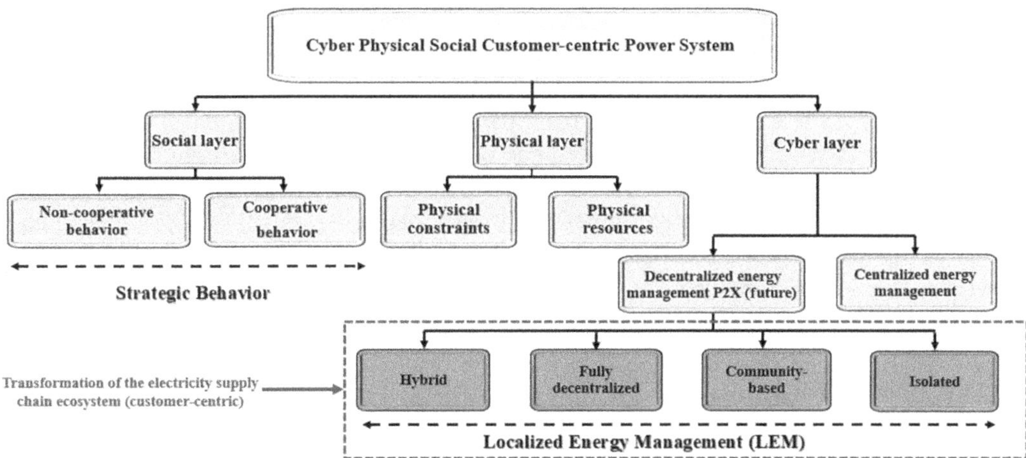

Figure 3.5 The layered structure of customer-centric power system.

electrochemical batteries, can be stored for relatively long periods. This storage capability enables the grid to absorb excess renewable energy, reducing the occurrence of negative electricity prices and enhancing system flexibility. ESS also plays a crucial role in mitigating power generation fluctuations, and its utilization becomes inevitable with the increased penetration of intermittent RES and reduced reliance on fossil fuels. Moreover, with the growth of DER, customers can profit through energy arbitrage – purchasing electricity during low-demand (low-price) periods and selling it during peak-demand (high-price) periods [12]. ESS can provide various flexibility services to the distribution and transmission networks. For instance, ESS has been employed for congestion management in distribution networks and for supporting transmission ramping capacity needs [13, 14]. Mobile storage solutions like EVs can also offer demand flexibility and frequency regulation services [15, 16]. Additionally, thermal storage systems reduce peak demand by storing energy during off-peak hours. In some markets, thermal and electrical ESS participate together to meet energy and reserve requirements [16].

- **Enhanced Control of RES**: Unlocking this flexibility potential requires adequate physical infrastructure to control renewable output in response to issues like excess generation and grid congestion. Controllers, such as proportional-integral-derivative (PID) or fuzzy PID, are essential for this. As previously mentioned, the inherent uncertainty and variability of RES contribute to system flexibility shortages. Control of RES output improves flexibility, particularly when production exceeds demand or during congestion. In these cases, flexibility can be achieved by curtailing RES output. However, this requires mechanisms to incentivize and manage lost energy, as well as sufficient physical capacity to curtail generation. It is worth noting that reducing RES output is considered a last resort for providing flexibility due to the negative impact on revenues for RES owners [17].
- **DSM Infrastructure**: Realizing the flexibility potential of demand-side resources depends on the availability of physical infrastructure, such as communication and smart metering systems. DSM has significant potential for providing flexibility through changes in customer behavior in response to financial incentives. These changes can be either permanent, such as through EE measures, or temporary, such as load shifting during DR programs. DSM is seen as an affordable and effective tool for enhancing system flexibility and facilitating greater penetration of intermittent RES. Programs like DR offer financial incentives for active customers to allocate part of their load as a commodity. Recent developments have seen third-party aggregators enable responsive loads to participate in both local and wholesale markets [18].
- **Smart Grids**: A smart grid integrates sensors, two-way communication, and computational capabilities to improve the delivery of electricity from environmental, technical, and economic perspectives. The physical layer of a smart grid includes meters, communication infrastructure, and electrical connections, which are vital for connecting various components. With smart grids, flexibility can be enhanced through real-time monitoring and forecasting of demand [19].
- **Multi-Energy Carrier Systems**: These systems, including integrated energy hubs such as gas and electricity, are important sources of flexibility that enhance energy security. They enable energy conversion and storage across different carriers, increasing system flexibility. Technologies like combined heat and power and combined cooling, heat, and power are examples [20]. In summary, Figure 3.6 outlines the physical resources that form the backbone of the future customer-centric power system ecosystem.

3.5.1.2 Physical Constraints of the System

In the interactions between agents (an agent is a computer-based system capable of autonomous and independent actions on behalf of its user or owner rather than being explicitly told what to

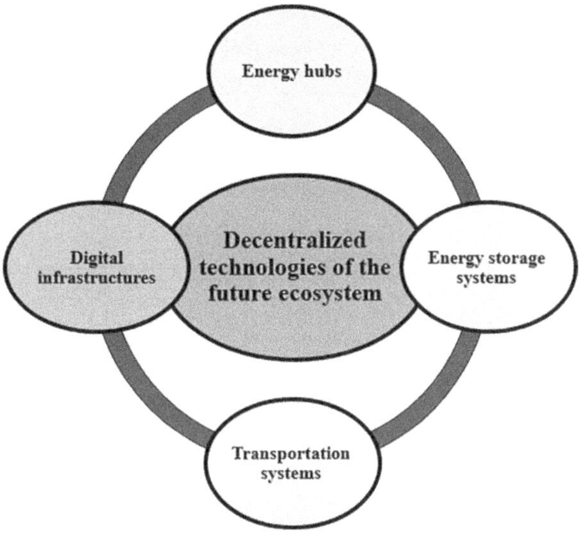

Figure 3.6 Physical resources of future customer-centric power systems.

do at every moment), various limitations exist that prevent the completely free and independent operation of the customers. These constraints, which are inherently tied to the physics of the power system, must be integrated into the decision-making processes at the cyber-social layer. Neglecting these constraints could result in violations of physical system limitations, leading to significant operational challenges, including unscheduled outages. Consequently, there has been increasing attention to the physical layer in the interactions among active customers and with other stakeholders, such as distribution system operators (DSOs).

The challenges posed by the physical constraints of the power system can be summarized as follows:

- **Electrical Power Flow Constraints**: As customers become active participants, their connection to the distribution system through peer-to-peer (P2P) interactions may lead to overvoltage and reverse power flows. New methodologies based on sensitivity analysis are needed to examine the impacts of P2P exchanges on the network and their associated costs [21]. The use of distributed optimization algorithms to support network voltage for smart inverters used by active customers is one potential solution for voltage control [22]. Additionally, operational overhead or congestion may arise from power injections by customers [23].
- **Energy Losses in the Power Supply Chain**: While transactions between sellers and buyers through P2P interactions can increase node voltage and lead to congestion, losses within the network are inevitable. These losses result in energy supply exceeding local demand and increased costs. The allocation of network costs associated with P2P interactions by the system operator can further help reduce network losses. Optimal routing strategies for power flow are also crucial in this context [24].
- **System Inertia and Stability**: Synchronous generators play a critical role in system stability through their inertia, especially in response to disturbances in voltage or frequency. However, the penetration of DER may lead to a reduction in the number of operational synchronous generators, posing significant challenges for maintaining system inertia. A notable instance of system failure due to insufficient inertia occurred during the South Australia blackout in 2016 [1].

As outlined above, the integration of active customers into the power system, while beneficial, affects the physical constraints of the system. Thus, as depicted in Figure 3.7, it is essential to

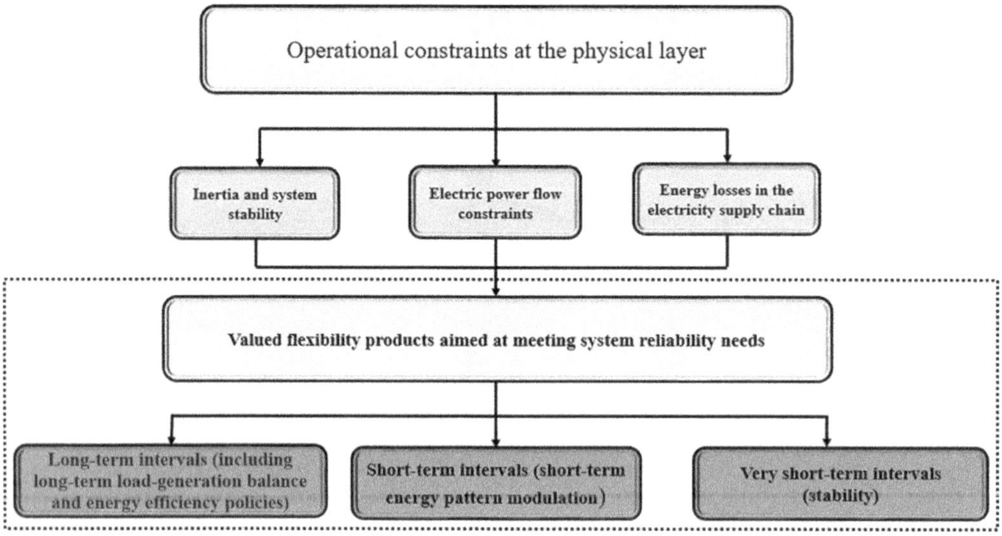

Figure 3.7 Introducing new flexibility products according to the physical constraints of the power system.

define new ancillary services or flexibility products based on response time and the required level of flexibility to address these challenges. The introduction of new ancillary services will naturally incentivize customers to utilize their available resources to meet system requirements while generating profit. Ancillary services include both spinning and non-spinning reserves, as well as regulation services to meet various balance needs of the system in both normal and emergency conditions. However, with the growing penetration of DER, these services may not suffice to provide the necessary flexibility for the system.

Thus, the flexibility of the power system will be enhanced with the introduction of new ancillary services in future power systems. The existing literature identifies the following ancillary services necessary to meet the requirements of the power system in the presence of active customers:

- **Frequency Response**: As previously mentioned, high penetration of DER can adversely affect the system frequency response due to a reduction in the number of rotating components in the grid. The participation of controllable loads in maintaining system frequency is a key strategy in this domain. Additionally, decentralized approaches have been proposed that utilize self-regulating loads to manage power system imbalances, which can help mitigate frequency deviations by providing active power response. Examples of self-regulating loads include temperature-controlled loads such as refrigerators and air conditioners that operate based on system frequency [25].
- **Ramping Capacity**: This product is primarily sought by TSOs to address uncertainties in renewable generation. It has been introduced as a new load-tracking product by ISOs like California ISO and the midcontinent ISO under the title of flexible ramping product [26, 27]. The presence of this product in electricity markets creates opportunities for demand-side resources, such as ESSs and EVs [28].
- **Energy**: In the distribution network, flexibility products are increasingly related to energy modulation to optimize network usage in response to rising demand, reverse power flows, congestion management, system balancing, voltage maintenance, and deferring network investment. The involvement of new players, such as aggregators and prosumers, and their interactions will play a significant role in the intelligent modulation of energy consumption

patterns. For instance, in a local market model operated by a distribution market operator, which can also function as a balancing responsible party, aggregators purchase energy balancing services. Moreover, entities like DSOs can leverage local resource potentials by sending requests for necessary flexibility services to aggregators, addressing issues such as network congestion [29].

3.5.2 Cyber-Social Layers

In the cyber and social layers, the exploitation of flexibility resources occurs through operational directives and market-based procedures. It is crucial to note that the proper functioning of the physical layer of the power system depends on the decisions made in these two layers, highlighting their significant importance in a customer-centric system. Generally, in a customer-centric power system, the type of behavior exhibited toward customers is determined by their interactions with other system players within these layers. Based on the layered structure introduced in Figure 3.5 and illustrated in Figure 3.8, various behavioral approaches exist when interacting with active customers, each of which is examined in detail below. All of these approaches are grounded in game theory:

- **Cooperative Game**: Customers may be inclined to act in ways that maximize shared public objectives, resulting in greater social value compared to individual, separate, and non-cooperative decisions. In specific circumstances, agents may undertake actions that yield significant benefits for the community, even if their gains from generating these positive externalities are limited within a competitive framework. A common public goal could simply be maximizing social welfare. Additionally, managing physical network constraints could be considered a social common goal.
- **Non-Cooperative Game**: In this scenario, customers will act in a self-interested manner, solely seeking to maximize their own efficiency, which could encompass various dimensions such as cost or comfort level. In such cases, the customer-centric system faces a multitude of heterogeneous preferences, as each customer has distinct criteria for decision-making. Employing a non-cooperative approach may not effectively manage flexibility within a customer-centric

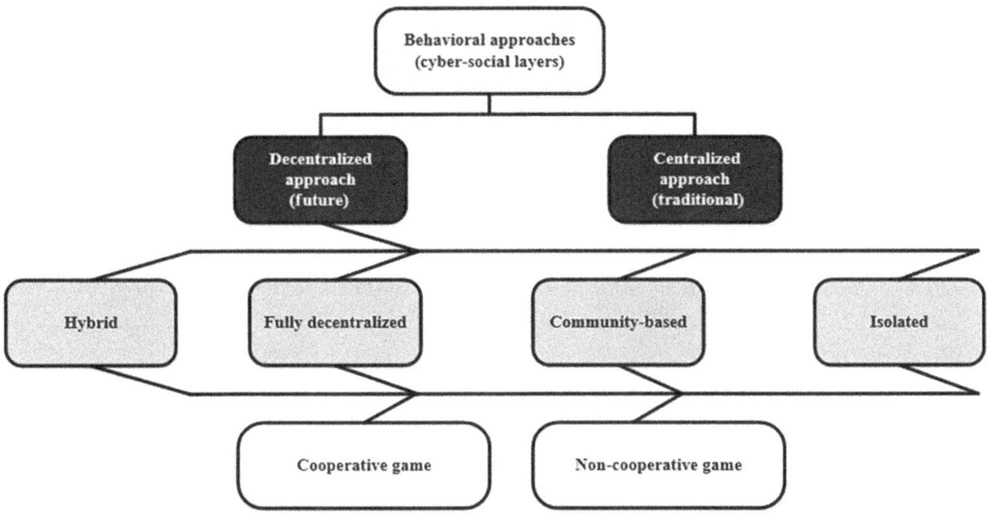

Figure 3.8 Classification of behavioral approaches in dealing with customers.

system and could lead to a lack of overall system flexibility. In such situations, there is a need to design market mechanisms that guide customer decisions to incorporate other objectives, such as addressing system constraints, to enhance overall flexibility. By steering individual decisions toward a cooperative game, it is possible to balance individual interests with the flexibility requirements of the system.

3.5.2.1 Centralized Approach (Traditional)

According to Figure 3.9, this approach refers to a situation in which a central entity has full access to the information of all DER and can make decisions regarding their control actions. In such circumstances, the system flexibility objectives are pursued without considering the customer preferences. A well-known example of this centralized distribution approach is top-down switching and direct load control.

Transitioning toward market-based approaches can enhance customers motivation to align with the system operator objectives. If more distinct metrics are integrated into these market approaches, the level of fairness can be increased, resulting in greater market transparency, higher customer motivation, and more efficient utilization of flexibility resources by the operator. These metrics can be expressed in line with the perfect power system indices. For example, in the day-ahead market, the availability index could be considered when selecting physical flexibility resources to meet the operator reserve requirements, accounting for equipment performance uncertainties [30].

3.5.2.2 Decentralized Approach (Future)

With the increasing number of active customers in power systems, the centralized model becomes excessively large and, due to the growing number of variables, suffers from the curse of dimensionality [31]. In such a scenario, the problem either becomes unsolvable or, if solved, deviates significantly from the optimal solution. Furthermore, in the traditional centralized structure, it is not feasible to account for customers preferences in detail, which prevents the fair utilization of their potential, ultimately leading to a reduction in social welfare. Additionally, having a single central entity responsible for the control and planning of all DER increases the system vulnerability.

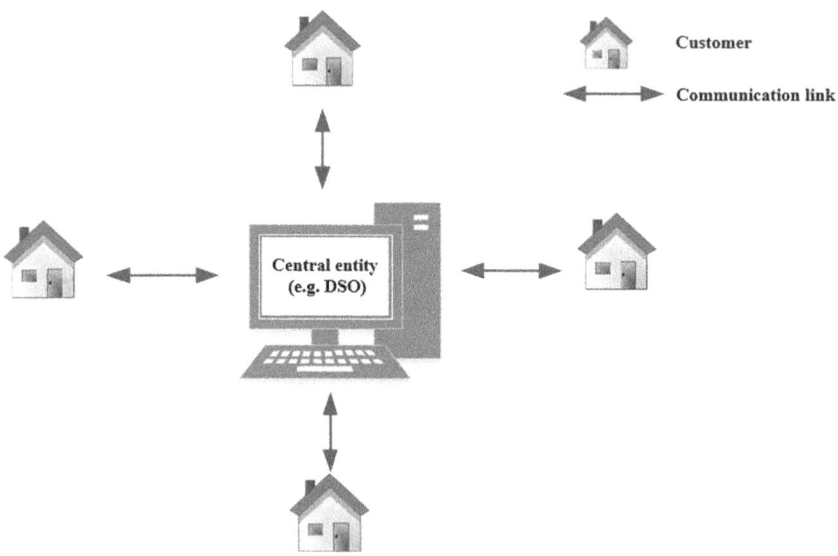

Figure 3.9 Centralized energy management approach.

Table 3.2 The advantages and disadvantages of different isolated approaches.

Isolated approach	Advantages	Disadvantages
CEMS	• Suitable for low penetration level DER • Suitable for a low level of communication infrastructure	• Rewards are limited to retail tariffs • Ignoring network constraints
CEMS-OE	Short-term solutions for addressing network issues in the face of DER growth	Inadequate utilization of the full potential of DERs

To address these challenges, decentralized approaches, which are more compatible with the ecosystem of future power systems, have been introduced. The following sections provide a detailed examination of these approaches.

Isolated Approach In this approach, customers do not share their information with the central entity or their peers and only monitor external environmental information, i.e. prices. The goal of customers in this approach is to maximize their personal benefits in various aspects, such as cost and comfort levels. Therefore, this approach, also known as the CEMS, is considered a non-cooperative approach. To utilize customer flexibility within the isolated approach, the CEMS with operational envelopes (CEMS-OE) has been introduced, where CEMS schedules the customers resources based on operational envelopes provided by the DSO. Thus, it can be said that CEMS-OE plays a cooperative role within the power system [9]. A summary of the advantages and disadvantages of this approach is presented in Table 3.2.

Community-Based Approach As illustrated in Figure 3.10, in this approach, an entity called the community manager (CM) manages the commercial activities within the energy community and serves as an intermediary between the local community and the rest of the system. This flexible market design can be applied to microgrids or groups of customers nearby. It is notable that in the existing literature, the CM is often referred to as an aggregator, who, depending on the available flexibility, can participate in various markets, such as local and wholesale markets, and is capable of providing different flexibility services at both distribution and transmission levels.

The process of local market settlement in the community-based approach can be conducted in both centralized and distributed forms. The results indicate that when implemented with a distributed structure, the community-based approach can more efficiently achieve privacy preservation, security, minimum data exchange and operational costs. One well-known distributed settlement method within community-based approaches is the alternating direction method of multipliers (ADMM) [32].

Based on the evaluation of indicators such as the quality of service (QoS) and the quality of experience (QoE), the policies adopted by the CM can significantly influence the energy exchanges within neighborhoods. It is worth noting that the QoS indicator reflects the volume of energy exchanged within a community, while the QoE indicator represents the total costs and revenues of the community members [33]. One of the policies that the CM can adopt is the institutionalization of a new concept known as energy classes, where electricity becomes a heterogeneous product based on the characteristics of its source. This leads to value creation in the energy domain. For instance, factors such as the type of energy generation technology, distance, or even the customers reputation can contribute to the formation of energy classes [34]. In short, the advantages and disadvantages of this approach are stated in Table 3.3.

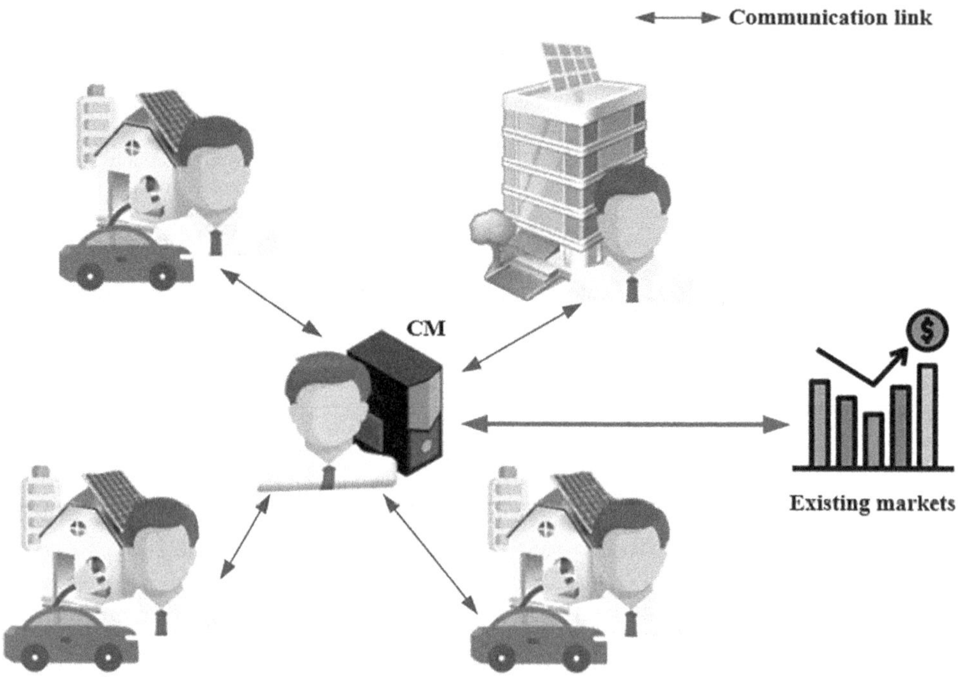

Figure 3.10 Community-based approach with the ability to participate in different markets.

Table 3.3 The advantages and disadvantages of community-based approach.

Behavioral approach	Advantages	Disadvantages
Community-based	• Suitable for high DER integration • Network awareness • Suitable scalability by using distributed methods such as ADMM and RCI	• Achieving preferences for all community members • Collection of all customers data and their management by the community manager • Scalability in centralized structure

Fully Decentralized Approach As shown in Figure 3.11, in this approach, customers share information bilaterally, directly with their peers, without the involvement of a central entity or an organized intermediary. Adopting a fully decentralized approach can further activate customers preferences and enable product differentiation [35]. This also leads to increased privacy, as at any given time, only energy quantities and price offers/bids are exchanged between participants. Comparisons between the approaches discussed thus far indicate that fully decentralized P2P and community-based approaches offer better technical and even economic opportunities than traditional and isolated models [36].

One of the challenges of a fully decentralized approach is the complexity of its management processes compared to other approaches, particularly in ensuring power system flexibility. Generally, the role of the DSO in providing the required flexibility for P2P transactions can be categorized into three main forms:

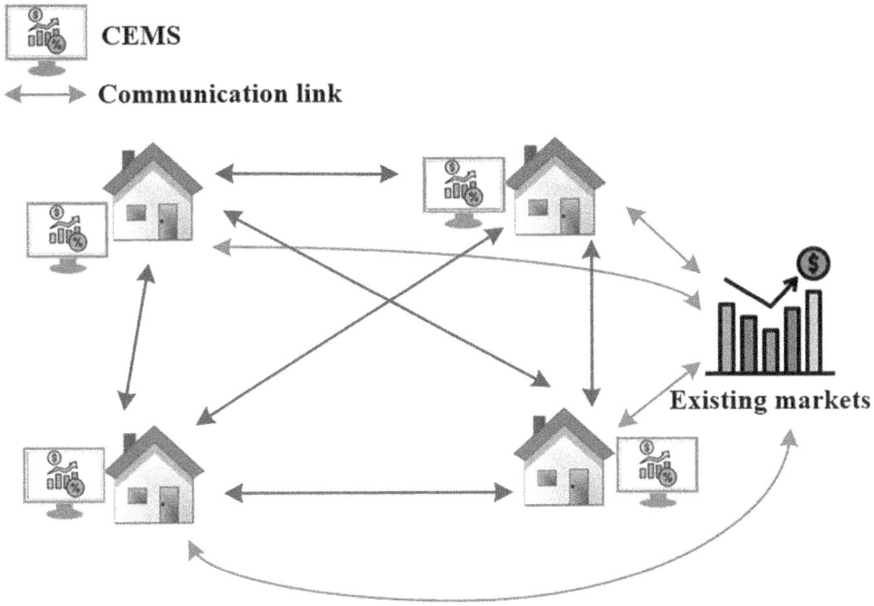

Figure 3.11 Fully decentralized behavioral approach.

- **Real-Time Enforcement of Network Constraints in P2P Interactions**: Under this scenario, the DSO checks network constraints after each agreement between two agents through P2P negotiations. If network violations occur, price signals are issued, and the outcome is communicated to both the buyer and seller of the contract [37, 38].
- **Ignoring Constraints in the P2P Market Process**: In this method, the DSO assesses whether network constraints have been violated after the P2P market is finalized. If any violations of technical or physical constraints are detected, the DSO may undertake corrective actions such as network reinforcement or utilizing on-load tap-changing transformers to resolve the issue [39, 40].
- **Offline Consideration of Constraints in P2P Interactions**: In this case, the system operator considerations regarding network constraints are incorporated into customers objective functions as predefined cost terms. For example, by forecasting energy consumption, the operator predicts the retail prices based on the uncertainties of the upstream network and determines these prices for the next day P2P market [41]. The advantages and disadvantages of this approach are presented in Table 3.4.

Table 3.4 Advantages and disadvantages of fully decentralized approach.

Behavioral approach	Advantages	Disadvantages
Fully decentralized	• High attention to the customer preferences • Privacy preserving • Less vulnerable due to the lower dependency on a central entity	• Network awareness • Scalability

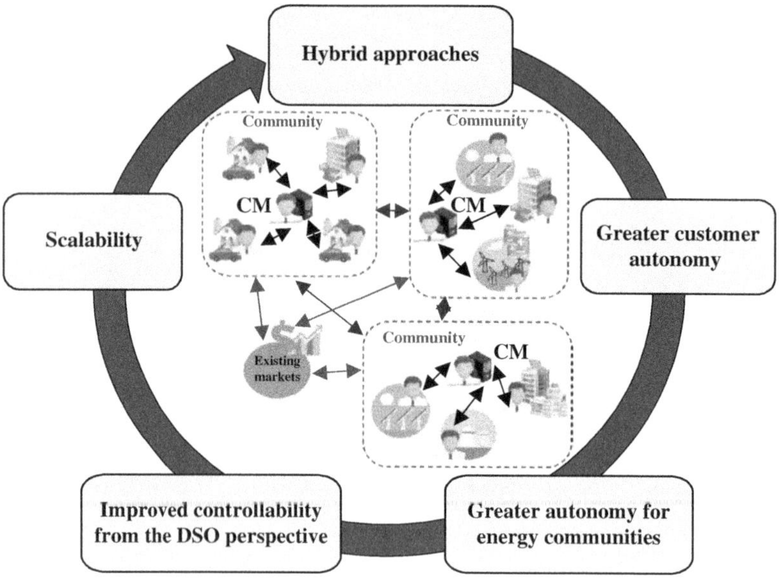

Figure 3.12 Hybrid approach benefits.

Hybrid Approach As discussed, each behavioral approach has its advantages and disadvantages. For instance, the fully decentralized approach offers less control compared to the community-based approach, while the latter places less emphasis on customer preferences than the fully decentralized model. Therefore, as depicted in Figure 3.12, it is necessary to design new behavioral approaches that minimize the drawbacks of the previously discussed models and enhance the flexibility they can offer. One such approach is the hybrid model. In Ref. [42], a hybrid energy exchange approach is utilized. The philosophy behind this approach is to simultaneously leverage the advantages of both fully decentralized and community-based models. The concept is that the energy distribution network is divided into energy communities, with each community forming a local market. These communities can engage in full P2P energy interactions under the supervision of the DSO. Within each energy community, customers exchange energy P2P through the CM. Notably, the reason for employing a fully decentralized approach at the upper layer of the designed mechanism (i.e., interactions between CMs) is that, given the smaller number of CMs compared to individual customers, P2P exchanges in this layer are more feasible and can lead to faster price discovery. In contrast, at the lower layer, due to the large number of customers, the community-based approach is used for better control and to provide more efficient cross-community flexibility services. To ensure privacy, scalability, and ease of accommodating customer preferences, edge and distributed computing methods are employed at the lower layer. By integrating community-based and fully P2P approaches as described, a more efficient customer-centric system can be achieved. This system offers better control for the system operator while granting more freedom of action to customers, who can interact with their local community, the grid, or other energy communities.

3.6 Conclusion and Future Trends

The widespread integration of DER at various transmission and distribution levels offers benefits such as reduced reliance on fossil fuels and lower emissions. However, it also presents challenges like congestion, reverse power flow, increased losses, and voltage limit violations within the power system. From an economic perspective, issues such as the need for substantial investments in expensive fossil fuel units or grid expansion to enhance power system flexibility may arise. In this context, leveraging demand-side potential as an effective solution can outperform traditional approaches to providing flexibility from technical, economic, social, and environmental standpoints by harnessing the power of customers.

Naturally, achieving this goal requires a shift away from the traditional top-down, centralized power system structure, which is ill-suited to fully exploit the potential of active customers (those capable of adjusting their energy production/consumption in response to price or activation signals). The power system must instead evolve into a perfect, customer-centric system with distributed management. This new customer-centric system, given the large number of actors involved, requires the efficient utilization of robust information and communication infrastructures to engage more stakeholders in meeting its needs. Thus, compared to the traditional approach, the customer-centric power system exhibits a higher degree of physical-cyber-social integration due to the active, two-way interaction of customers. The physical layer of this system deals with physical resources and constraints, which, while necessary, are not sufficient to provide the required flexibility. Physical constraints, in particular, are the primary drivers of the need for flexibility in the power system. On the other hand, the cyber and social layers necessitate the design of behavioral approaches and decision-making mechanisms to engage customers, ensuring that adequate flexibility is provided in both short- and long-term horizons. The importance of the cyber-social layer can be highlighted by the fact that the reliable utilization of physical resources to address system constraints depends on the mechanisms designed in this layer. In this chapter, a layer-by-layer assessment of the customer-centric power system has been conducted. Given the importance of this subject, future trends in this field can be identified in the following three key areas:

- **Behavioral Analyses**: In the field of customer behavior modeling, most existing studies employ a concave utility function. In this function, the horizontal axis represents the amount of energy, while the vertical axis reflects the value customers derive from a given amount of energy. This utility function clearly represents the nonlinear behavior of customers. As a result, recent research in customer-centric interactive energy mechanisms has increasingly focused on the behavioral uncertainties and bounded rationality of customers. Bounded rationality refers to deviations from the optimal customer response, which can be attributed to cognitive limitations [43]. One of the key approaches to modeling behavioral uncertainty is prospect theory (PT). According to PT, the subjective value function is convex for losses and concave for gains, as human behavior tends to be more radical (with a steeper slope) when facing losses, while more conservative when realizing gains. As illustrated in Figure 3.13, customers can be categorized into three groups: risk-takers, risk-averse individuals, and customers with rational behavior. The distinguishing factor between these groups lies in the coefficients of their behavioral curves (i.e., their sensitivity to gains and losses relative to a reference point) [44]. It is worth noting that, in addition to the model-based approach of PT, data-driven approaches such as spatiotemporal neural networks can also be utilized for modeling behavioral uncertainties [45].

- **Reinforcement Learning (RL)**: In recent years, the topic of RL has received much attention. One of the notable features of RL is its data-driven performance, which offers advantages over model-based approaches, such as adaptability, a model-free nature, the ability to learn from historical data, and the capability to identify optimal behavioral policies in highly complex and dynamic environments [46]. As illustrated in Figure 3.14, a comparison between the functionalities of RL algorithms and the characteristics of a customer-centric power system. It is evident

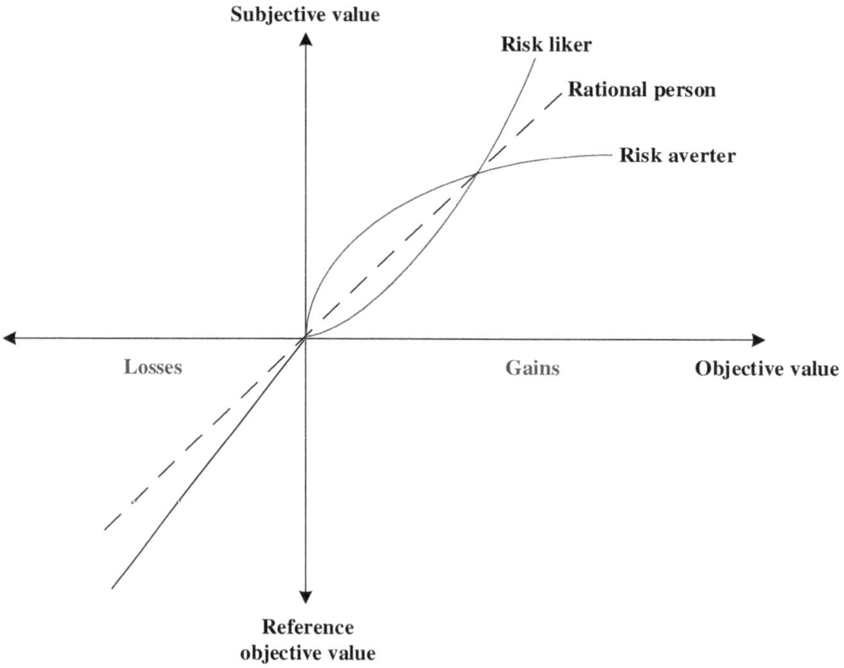

Figure 3.13 Different customer behavior categories. Source: Chen et al. [45] / with permission of IEEE.

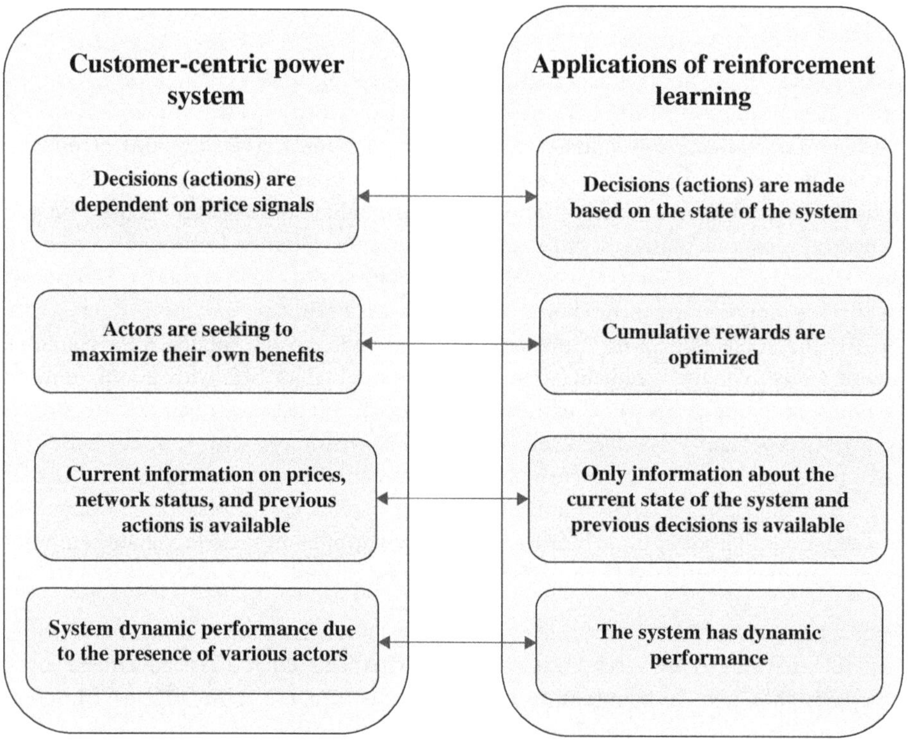

Figure 3.14 One to one correspondence RL and customer-centric power system features.

from this figure that all functionalities of RL algorithms correspond one-to-one with the features of a customer-centric power system. Given this alignment, applying RL can facilitate the extraction of behavioral models for agents. This, in turn, allows for more rapid and efficient provision of the necessary flexibility based on the insights gained. For instance, customers interacting with their external environment can predict the impacts of their decisions on that environment and avoid choices that negatively affect the performance of the power system. In such scenarios, the concept of a digital twin becomes more tangible. A digital twin is a virtual representation of a real system that leverages extensive data acquired from the actual system to extract its behavioral patterns and predict future behaviors. Essentially, a digital twin is a data-driven simulation that integrates artificial intelligence. Thus, within a customer-centric power system, RL can be utilized to extract environmental behavioral patterns and develop a corresponding digital twin. Based on the resulting digital twin, agents can anticipate their impact on the environment and optimize their actions accordingly.

- **Cybersecurity**: Customer-centric systems establish control and economic processes to continuously balance supply and demand. However, the flexibility inherent in these systems necessitates the utilization of Internet of Things technologies, which significantly increases the vulnerability of cyber assets. The bottom-up structure of customer-centric systems is particularly susceptible to cyberattacks, especially those with economic motivations targeting stakeholders. Therefore, new analytical approaches for continuous system monitoring and the detection of malicious activities are essential. Among the cyberattacks that can impact this domain are manipulations of quantity-price offers from peers, which can adversely affect both the technical and economic aspects of the customer-centric power system. The technical implications include violations of operational constraints and a lack of awareness among network operators. Economically, such manipulations can lead to a divergence of market prices from competitive levels and undermine market efficiency [47].

References

1 Ratnam, K.S., Palanisamy, K., and Yang, G. (2020). Future low-inertia power systems: requirements, issues, and solutions-A review. *Renewable and Sustainable Energy Reviews* 124: 109773.

2 Ravi, S.S. and Aziz, M. (2022). Utilization of electric vehicles for vehicle-to-grid services: progress and perspectives. *Energies* 15 (2): 589.

3 Muratori, M. and Mai, T. (2020). The shape of electrified transportation. *Environmental Research Letters* 16 (1): 011003.

4 Impram, S., Nese, S.V., and Oral, B. (2020). Challenges of renewable energy penetration on power system flexibility: a survey. *Energy Strategy Reviews* 31: 100539.

5 Shahidehpour, M. (2008). Energy efficiency perfect power system techniques & applications. *IEEE Power & Energy Magazine* 6 (6): 25–26.

6 Shahidehpour, M. (2010). Role of smart microgrid in a perfect power system. *IEEE PES General Meeting*, Minneapolis, MN (25–29 July 2010), p. 1. IEEE.

7 Kelly, J., Von Dollen, D., and Brook, I.O. (2007). The Illinois Institute of technology's perfect power system prototype. *Grid-Interop Forum* 137: 1–4.

8 Moghaddam, M.P., Zamani, R., Alhelou, H.H., and Siano, P. (2022). *Decentralized Frameworks for Future Power Systems: Operation, Planning and Control Perspectives*. Elsevier.

9 Guerrero Orbe, J.E. (2020). *Transactive Energy in Low-Voltage Networks*. University of Sydney.

10 Khorasany, M., Mishra, Y., and Ledwich, G. (2018). Market framework for local energy trading: a review of potential designs and market clearing approaches. *IET Generation, Transmission & Distribution* 12 (22): 5899–5908.

11 Nelson, T., Simshauser, P., and Nelson, J. (2012). Queensland solar feed-in tariffs and the merit-order effect: economic benefit, or regressive taxation and wealth transfers? *Economic Analysis and Policy* 42 (3): 257.

12 Cavanagh, K., Ward, J., Behrens, S. et al. (2015). Electrical energy storage: technology overview and applications. EP154168. Australia.

13 Prajapati, V.K. and Mahajan, V. (2021). Reliability assessment and congestion management of power system with energy storage system and uncertain renewable resources. *Energy* 215: 119134.

14 Hu, J., Sarker, M.R., Wang, J. et al. (2018). Provision of flexible ramping product by battery energy storage in day-ahead energy and reserve markets. *IET Generation, Transmission & Distribution* 12 (10): 2256–2264.

15 Zhou, Y. and Cao, S. (2019). Energy flexibility investigation of advanced grid-responsive energy control strategies with the static battery and electric vehicles: a case study of a high-rise office building in Hong Kong. *Energy Conversion and Management* 199: 111888.

16 Peng, C., Zou, J., and Lian, L. (2017). Dispatching strategies of electric vehicles participating in frequency regulation on power grid: a review. *Renewable and Sustainable Energy Reviews* 68: 147–152.

17 Brandstätt, C., Brunekreeft, G., and Jahnke, K. (2011). How to deal with negative power price spikes? – flexible voluntary curtailment agreements for large-scale integration of wind. *Energy Policy* 39 (6): 3732–3740.

18 Cruz, M.R., Fitiwi, D.Z., Santos, S.F., and Catalão, J.P. (2018). A comprehensive survey of flexibility options for supporting the low-carbon energy future. *Renewable and Sustainable Energy Reviews* 97: 338–353.

19 Gellings, C.W. (2020). *The Smart Grid: Enabling Energy Efficiency and Demand Response.* River Publishers.

20 Akrami, A., Doostizadeh, M., and Aminifar, F. (2019). Power system flexibility: an overview of emergence to evolution. *Journal of Modern Power Systems and Clean Energy* 7 (5): 987–1007.

21 Guerrero, J., Chapman, A.C., and Verbič, G. (2018). Decentralized P2P energy trading under network constraints in a low-voltage network. *IEEE Transactions on Smart Grid* 10 (5): 5163–5173.

22 Almasalma, H., Claeys, S., and Deconinck, G. (2019). Peer-to-peer-based integrated grid voltage support function for smart photovoltaic inverters. *Applied Energy* 239: 1037–1048.

23 Tushar, W., Saha, T.K., Yuen, C. et al. (2020). Peer-to-peer trading in electricity networks: an overview. *IEEE Transactions on Smart Grid* 11 (4): 3185–3200.

24 Wang, R., Wu, J., Qian, Z. et al. (2017). A graph theory based energy routing algorithm in energy local area network. *IEEE Transactions on Industrial Informatics* 13 (6): 3275–3285.

25 Bian, Y., Wyman-Pain, H., Li, F. et al. (2017). Demand side contributions for system inertia in the GB power system. *IEEE Transactions on Power Systems* 33 (4): 3521–3530.

26 Xu, L. and Tretheway, D. (2012). Flexible ramping products. CAISO Proposal. https://www.caiso.com.

27 Navid, N. and Rosenwald, G. (2013). Ramp capability product design for MISO markets. Market development and analysis.

28 Zhang, X., Hu, J., Wang, H. et al. (2020). Electric vehicle participated electricity market model considering flexible ramping product provisions. *IEEE Transactions on Industry Applications* 56 (5): 5868–5879.

29 Olivella-Rosell, P., Lloret-Gallego, P., Munné-Collado, Í. et al. (2018). Local flexibility market design for aggregators providing multiple flexibility services at distribution network level. *Energies* 11 (4): 822.

30 Mansoori, A., Moghaddam, M.P., and Delkhosh, H. (2023). A hybrid stochastic-robust approach for power system security-constrained scheduling in the presence of flexibility facilities. *IEEE Transactions on Power Systems* 39 (2): 4064–4076.

31 Zamani, R., Moghaddam, M.P., and Haghifam, M.-R. (2022). Dynamic characteristics preserving data compressing algorithm for transactive energy management frameworks. *IEEE Transactions on Industrial Informatics* 18 (11): 7587–7596.

32 Javadi, M.S., Nezhad, A.E., Jordehi, A.R. et al. (2022). Transactive energy framework in multi-carrier energy hubs: a fully decentralized model. *Energy* 238: 121717.

33 Moret, F. and Pinson, P. (2018). Energy collectives: a community and fairness based approach to future electricity markets. *IEEE Transactions on Power Systems* 34 (5): 3994–4004.

34 Morstyn, T. and McCulloch, M.D. (2018). Multiclass energy management for peer-to-peer energy trading driven by prosumer preferences. *IEEE Transactions on Power Systems* 34 (5): 4005–4014.

35 Le Cadre, H., Jacquot, P., Wan, C., and Alasseur, C. (2020). Peer-to-peer electricity market analysis: from variational to generalized Nash equilibrium. *European Journal of Operational Research* 282 (2): 753–771.

36 Guerrero, J., Gebbran, D., Mhanna, S. et al. (2020). Towards a transactive energy system for integration of distributed energy resources: home energy management, distributed optimal power flow, and peer-to-peer energy trading. *Renewable and Sustainable Energy Reviews* 132: 110000.

37 Zamani, R., Moghaddam, M.P., and Haghifam, M.-R. (2024). Network sustainability enhancement through optimal fairness management for peer-to-peer energy trading. *IEEE Transactions on Smart Grid* 16 (1): 288–300.

38 Ullah, M.H. and Park, J.-D. (2021). Peer-to-peer energy trading in transactive markets considering physical network constraints. *IEEE Transactions on Smart Grid* 12 (4): 3390–3403.

39 Davoudi, M., Moeini-Aghtaie, M., and Ghorani, R. (2021). Developing a new framework for transactive peer-to-peer thermal energy market. *IET Generation, Transmission & Distribution* 15 (13): 1984–1995.

40 Wang, Y., Huang, Z., Shahidehpour, M. et al. (2019). Reconfigurable distribution network for managing transactive energy in a multi-microgrid system. *IEEE Transactions on Smart Grid* 11 (2): 1286–1295.

41 Morstyn, T., Teytelboym, A., Hepburn, C., and McCulloch, M.D. (2019). Integrating P2P energy trading with probabilistic distribution locational marginal pricing. *IEEE Transactions on Smart Grid* 11 (4): 3095–3106.

42 Khorasany, M., Mishra, Y., and Ledwich, G. (2020). Hybrid trading scheme for peer-to-peer energy trading in transactive energy markets. *IET Generation, Transmission & Distribution* 14 (2): 245–253.

43 Radoszynski, A.M., Dvorkin, V., and Pinson, P. (2019). Accommodating bounded rationality in pricing demand response. *2019 IEEE Milan PowerTech*, Milan, Italy (23–27 June 2019), pp. 1–6. IEEE.

44 Xia, Y., Xu, Q., Huang, Y. et al. (2022). Preserving privacy in nested peer-to-peer energy trading in networked microgrids considering incomplete rationality. *IEEE Transactions on Smart Grid* 14 (1): 606–622.
45 Chen, L., Liu, N., Liu, L. et al. (2021). Data-driven stochastic game with social attributes for peer-to-peer energy sharing. *IEEE Transactions on Smart Grid* 12 (6): 5158–5171.
46 Ye, Y., Qiu, D., Sun, M. et al. (2019). Deep reinforcement learning for strategic bidding in electricity markets. *IEEE Transactions on Smart Grid* 11 (2): 1343–1355.
47 Zhang, Y., Krishnan, V.V.G., Pi, J. et al. (2019). Cyber physical security analytics for transactive energy systems. *IEEE Transactions on Smart Grid* 11 (2): 931–941.

4

Applications of Data Mining in Industrial Tariff Design and Energy Management: Concepts and Practical Insights

Hamidreza Arasteh[*,1,2], *Niki Moslemi*[1], *Majid Miri Larimi*[3], *Pierluigi Siano*[4], *Sobhan Naderian*[5], *and Josep M. Guerrero*[2,6]

[1] *Power Systems Operation and Planning Research Department, Niroo Research Institute, Tehran, Iran*
[2] *Center for Research on Microgrids (CROM), Huanjiang Laboratory, Zhuji, Shaoxing, Zhejiang, 311800, China*
[3] *Faculty of Electrical and Computer Engineering, Tarbiat Modares University, Tehran, Iran*
[4] *Department of Management and Innovation Systems, University of Salerno, Fisciano SA, Italy*
[5] *James Watt School of Engineering, University of Glasgow, Glasgow, Scotland*
[6] *Zhejiang University, Hangzhou, Zhejiang, China*

4.1 Introduction

4.1.1 Data Mining: Concepts, Procedures, and Tools

Data analysis and data mining are two interrelated fields that leverage vast datasets to extract valuable insights, enabling informed decision-making, knowledge discovery, and prediction. Especially in the era of big data, where high amounts of structured and unstructured data are generated and stored, these disciplines play a pivotal role [1].

Data analysis encompasses a set of techniques and processes aimed at examining, cleaning, transforming, and modeling data to uncover meaningful patterns and relationships. This process involves various steps, including exploratory data analysis, statistical analysis, data visualization, and hypothesis testing. Its primary objective is to gain a comprehensive understanding of the data and identify significant trends and correlations.

Also, data mining employs algorithms and computational techniques to discover patterns, relationships, and latent information within datasets. This domain encompasses methods and algorithms such as classification, clustering, association rule (AR) mining, and anomaly detection. The goal of data mining is to identify patterns and extract knowledge from the data [2].

Data mining refers to the systematic process of uncovering patterns, relationships, and insights from extensive datasets. It entails extracting valuable information and knowledge from intricate and frequently unstructured data sources, including databases, data warehouses, text documents, social media feeds, etc. Leveraging diverse statistical and computational algorithms, data mining methods aim to reveal hidden patterns, trends, and correlations that facilitate informed decision-making, prediction, and the identification of valuable insights [1, 2].

Data mining typically encompasses several key steps, including data preparation, data exploration, model building, and result validation. This intricate process integrates methodologies from diverse fields such as machine learning, statistics, pattern recognition, and database systems.

* Corresponding Author

Data-Driven Energy Management and Tariff Optimization in Power Systems: Shaping the Future of Electricity Distribution through Analytics, First Edition.
Edited by Hamidreza Arasteh, Pierluigi Siano, Niki Moslemi, and Josep M. Guerrero.

The primary objective of data mining is to extract valuable and actionable insights from extensive datasets. These insights are instrumental in enhancing decision-making processes, refining business strategies, enhancing customer experiences, and gaining competitive advantages across various domains [3].

Data mining holds a pivotal role in today's data-driven landscape, offering important benefits. Below are several key reasons highlighting the significance of data mining [1, 2].

1) **Knowledge Discovery**: Data mining facilitates the exploration of patterns, trends, and relationships concealed within vast datasets. Through comprehensive analysis, organizations can discover valuable insights and extract knowledge that are not possible using the traditional analytical methods.
2) **Decision-Making and Strategy**: Data mining enables informed decision-making by offering pertinent information and insights. Through an understanding of customer behavior, market trends, and business patterns, companies can use data-driven strategies to optimize performance, enhance strategies, and detect new opportunities.
3) **Predictive Analysis**: Data mining techniques, including predictive modeling, empower companies to forecast future outcomes based on historical data. Through these predictions, organizations can anticipate market trends, customer preferences, demand fluctuations, and potential risks. This capability enables them to proactively take preventive measures and adapt their strategies.
4) **Customer Relationship Management**: Data mining enhances customer relationship management by enabling organizations to comprehend customer preferences, behavior, and needs. Through the analysis of customer data, companies can personalize marketing campaigns, elevate customer satisfaction levels, and formulate targeted strategies aimed at attracting and retaining customers.
5) **Fraud Detection and Prevention**: Data mining methods play a pivotal role in detecting and preventing fraudulent activities. Through the analysis of patterns and identification of anomalies or outliers in data, organizations can pinpoint potentially fraudulent transactions, activities, or behaviors. This approach helps to mitigate financial losses, preserve customer benefits, and keep the integrity of business operations.
6) **Improved Operations and Efficiency**: Data mining optimizes business processes and enhances operational efficiency by analyzing operational data. Through this analysis, organizations can identify bottlenecks, inefficiencies, and areas for improvement. The results facilitate workflows, reduce costs, enhance efficiency, and improve overall performance.
7) **Market Intelligence and Competitive Advantage**: Data mining furnishes organizations with valuable market insights, empowering them to stay ahead in the competitive landscape. Through the analysis of market trends, consumer behavior, competitor strategies, and emerging patterns, companies can pinpoint market gaps, innovate products, refine marketing strategies, and gain a competitive edge.
8) **Product Development and Innovation**: Data mining plays a crucial role in product development and innovation. Through the analysis of customer feedback, market trends, and competitor insights, organizations can determine areas for product enhancement, new opportunities, and innovative features. By aligning their product offerings with customer needs and preferences, businesses can enhance customer satisfaction and gain a competitive edge in the market.
9) **Resource Optimization**: Data mining enables organizations to effectively optimize their resources. By analyzing data related to resource allocation and identifying patterns, businesses

can determine areas of waste and inefficiency and identify cost-saving opportunities. This optimization leads to improved resource management, increased productivity, and cost savings.

10) **Healthcare and Medical Research**: Data mining significantly impacts the healthcare sector by enhancing patient care, disease diagnosis, and medical research. Through the analysis of healthcare data, patterns and trends can be discerned to enhance the effectiveness of diagnosis and treatment. Additionally, data mining aids medical research by identifying correlations, risk factors, and potential treatments or interventions.
11) **Personalization and Targeted Marketing**: Data mining empowers organizations to personalize their marketing efforts and deliver targeted campaigns. Through the analysis of customer data, including demographics, behavior, and preferences, businesses can customize their marketing messages, offers, and recommendations for individual customers. This approach enhances the effectiveness of marketing initiatives and fosters stronger customer engagement and interaction.
12) **Supply Chain Management**: Data mining is pivotal in optimizing supply chain management processes. Through the analysis of supply chain data, organizations can identify inefficiencies, bottlenecks, and delays. This enables streamlined operations, optimized inventory levels, improved logistics, and overall improvement in supply chain performance.
13) **Risk Assessment and Management**: Data mining aids in identifying and mitigating various risks associated with business operations. Through the analysis of historical data, market trends, and external factors, organizations can assess and mitigate risks such as financial, operational, and market risks. This empowers businesses to make informed decisions, devise risk mitigation strategies, and preserve their benefits.

In essence, data mining empowers organizations to harness the potential of their data, extracting valuable insights to drive informed decision-making. This capability allows them to uncover hidden opportunities, streamline operations, enhance customer experiences, mitigate risks, and maintain competitiveness in today's data-centric landscape.

The process of uncovering relationships and patterns within data is accomplished through a series of steps known as the data mining process. Typically, this process comprises the following stages [4]:

1) **Understanding the Problem**: Identifying the objective or issue to be addressed through data mining.
2) **Preparing the Data Samples**: Gathering and preprocessing the data samples to ensure they are suitable for analysis.
3) **Model Development**: Building models using various data mining techniques.
4) **Model Evaluation**: Applying the developed model to the dataset to assess its effectiveness and performance.
5) **Production Deployment**: Evaluating the performance of the model and refining it as needed to improve accuracy and efficiency.

As the utilization of data mining broadens, numerous frameworks for conducting the data mining process have been developed by both academic and business researchers. One notable framework widely adopted in the field is CRISP-DM[1], established by a consortium of companies specializing in data mining [5]. CRISP-DM stands as one of the most commonly employed frameworks for data mining endeavors. Figure 4.1 provides an overview of this framework [4].

1 Cross Industry Standard Process for Data Mining

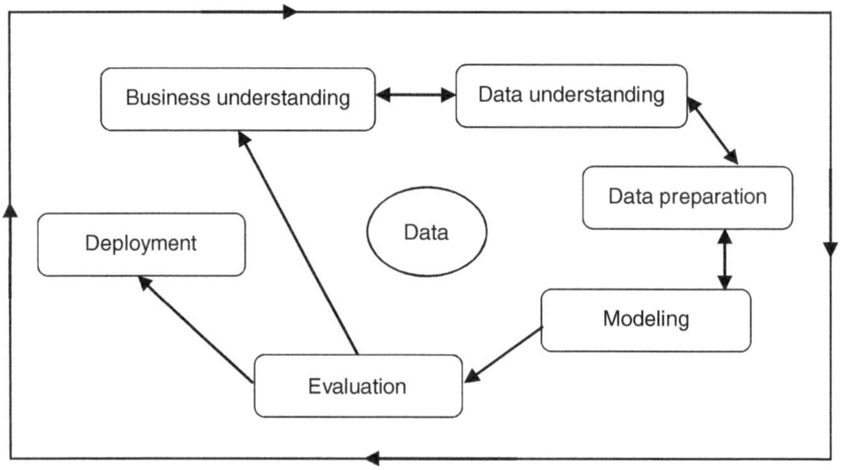

Figure 4.1 Overview of the CRISP-DM framework. Source: Kotu [4].

Other data mining frameworks include SEMMA[2] (developed by SAS Institute [6]), DMAIC[3] used in Six Sigma practice [7], and the Selection, Preprocessing, Transformation, Data Mining, Interpretation, and Evaluation framework, which has been utilized in KDD[4] analysis [8]. Each of these frameworks proposes the execution of a series of tasks to attain optimal outputs. The process of extracting information follows an iterative procedure. The steps in data mining are nonlinear and involve numerous loops, allowing for backward, forward, and lateral movement between different stages, sometimes even returning to the initial step to redefine the data mining problem.

Figure 4.2 illustrates the general steps involved in performing data mining, encompassing business considerations, algorithm selection, and the utilization of data mining tools [4]. The primary objective of any data mining process is to address the central question of analysis. This may entail tasks such as customer segmentation, weather forecasting, or pattern recognition within the data.

Among the various stages of data mining, the modeling phase, the third stage, stands out as a focal point. At this stage, a model is constructed based on a set of sample data, with the aim of making predictions (predictive modeling) or elucidating the underlying patterns within the data (descriptive or explanatory modeling) [4]. However, experienced data mining professionals often discover that the most time-consuming aspect of the data mining process is not model building, but rather data preparation and extracting business insights.

There are numerous data mining tools available, both open-source and commercial, that automate the model-building process. Some of the widely used tools include RapidMiner, R, Weka, SAS, SPSS, Oracle Data Miner, Salford, and Statistica [9].

As mentioned, data mining involves the exploration of large datasets to uncover potentially valuable patterns. This process employs a range of techniques, including machine learning, statistics, and artificial intelligence, to extract insights regarding the probability of future events. Data mining encompasses various tasks such as data classification, data integration, data transformation, data discretization, and pattern evaluation. It employs diverse methods to unveil hidden relationships

2 Sample, Explore, Modify, Model, and Assess (SEMMA)
3 Define, Measure, Analyze, Improve and Control (DMAIC)
4 Knowledge discovery in databases (KDD)

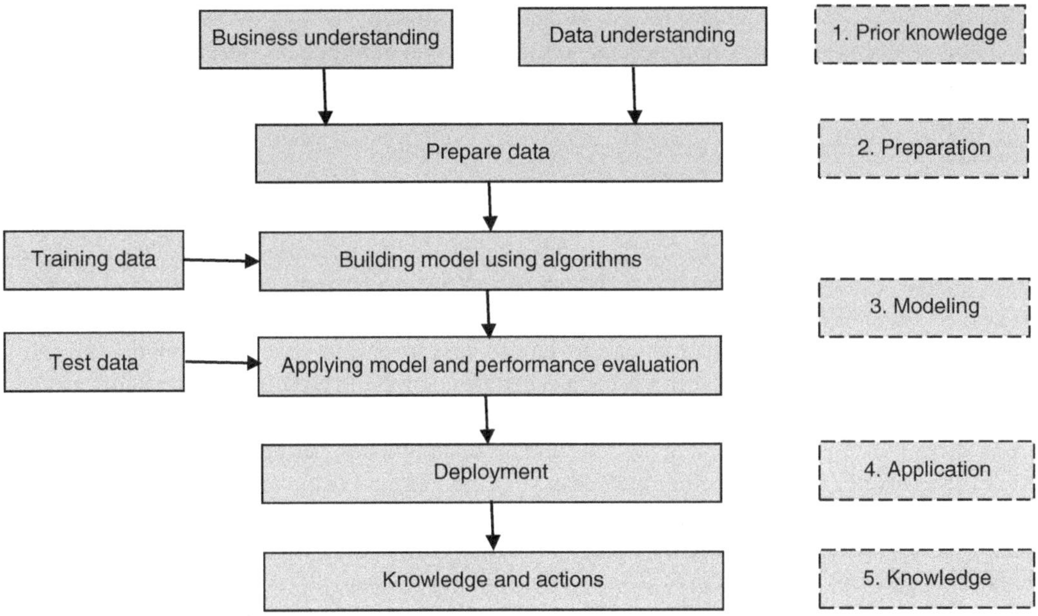

Figure 4.2 Data mining process. Source: Kotu [4].

within data, with applications spanning wide fields such as marketing, sales, fraud detection, scientific research, product development, healthcare, and education. Common data mining techniques include AR learning, classification, regression analysis, outlier detection, predictive modeling, and sequential pattern analysis [10].

4.1.2 Energy Management and the Role of Data Mining

Energy management, encompassing energy efficiency and demand response, constitutes the backbone of modern electric power systems [11]. It plays a pivotal role in the efficient operation and sustainability of electric power systems, encompassing a range of strategies aimed at optimizing energy generation, transmission, distribution, and consumption [12].

In broad terms, energy efficiency programs focus on initiatives aimed at improving the efficiency of appliances and systems, thereby reducing overall energy consumption [13]. Conversely, DR programs are defined as altering consumption patterns in response to pricing signals or system requirements [14, 15]. DR programs can typically be categorized into three main groups: time-based programs, incentive-based programs, and market-based programs [16]. DR can also be considered as a resource capable of enhancing the system's flexibility, alongside other distributed generation resources [17].

Taking into account that the actual participation of customers in DR programs isn't directly measurable and necessitates evaluation, various methodologies have been suggested to estimate customers' baseline and evaluate the real potential of DR. Additional details on the measurement and verification procedures for DR are available in Ref. [18].

Tariff design involves adjusting electricity prices across different time periods to influence consumption patterns [19]. By offering varying prices throughout the day, customers are incentivized to adjust their usage accordingly, potentially shifting load from peak to off-peak hours [20]. This optimization of consumption patterns can lead to more efficient utilization of electricity resources.

Ref. [21] provides a comprehensive review of recent literature pertaining to customer segmentation, demand and price forecasting, and customer engagement strategies. The aim is to identify suitable methods tailored to specific input attributes within the electricity market for dynamic tariff design in the power industry. Ref. [22] introduces novel approaches to community tariff design and addresses response fatigue concerns in energy communities. Firstly, it proposes designing tariffs on an annual basis rather than daily spot prices, ensuring stability akin to traditional retailer tariffs. Additionally, the design considers response fatigue by limiting the number of pricing signals sent to consumers, thereby maintaining their engagement. The proposed design approach is framed as a data-driven framework, leveraging real databases that collect information on demand, renewable generation, and retailer prices that could be used as inputs for a bi-level Stackelberg-based problem, implicitly considering prosumer reactions. A case study on a benchmark energy community evaluates various tariff mechanisms, such as flat, time-of-use (TOU), and happy hours tariffs, validating the proposed approach and analyzing local market effects in energy communities.

In electric power systems, energy management encompasses various activities, including demand forecasting, load balancing, and peak load management, all of which are crucial for ensuring the reliable and cost-effective operation of the grid. By accurately predicting energy demand patterns and strategically adjusting generation and distribution resources accordingly, energy managers can mitigate the risk of supply–demand imbalances, voltage fluctuations, and blackouts. Moreover, energy management initiatives often prioritize the integration of renewable energy sources, energy storage systems, and DR programs to promote sustainability and reduce carbon emissions. Overall, effective energy management not only enhances the operational efficiency and reliability of electric power systems but also contributes to broader goals of environmental conservation and energy sustainability.

The study conducted in Ref. [23] delves into the data-centric approach to energy management of virtual power plants (VPPs). This reference provides an in-depth review of the evolution of VPP technologies, emphasizing their reliance on data. Furthermore, it thoroughly examines the pivotal role data plays in guiding decisions across all phases of VPP operations. Numerous studies have delved into the significance and diverse applications of energy management across various domains within electric power systems [24–27].

4.1.3 Aims and Scope

In this chapter, we delve into the application of data mining techniques within the realm of energy management for the industrial sector. We shed light on the pivotal role of data mining in comprehending consumption patterns and optimizing energy utilization across industrial sectors. Through the analysis of hourly data spanning a year from two representative industries, we illustrate the practical application of data mining in industrial energy management.

Several criteria for analyzing the data from these industries are proposed, aiming to establish meaningful classifications based on pertinent factors. Subsequently, we explore various tariff design mechanisms tailored to address the specific needs of different industry categories, all while carefully considering the requirements of the power system. Throughout, the importance of informed decision-making and strategic planning in the implementation of data-driven energy management solutions is underscored.

Furthermore, our discussion extends to encompass a comprehensive examination of the multifaceted nature of tariff design for the energy management of the industries. We consider both technical and nontechnical factors, recognizing their potential impact not only within power systems but also on broader societal levels. It is emphasized that a thorough understanding of all aspects and potential side effects is essential for devising comprehensive tariff design procedures, ensuring that all foreseeable outcomes are effectively managed, and that the path forward aligns with optimal strategies. The conclusion emphasizes the necessity of collaborating with diverse entities and regulators across various domains to comprehend the multitude of constraints from different perspectives.

In summary, the chapter makes several key contributions, such as:

- Integration of data mining methodologies into industrial energy management, presenting innovative approaches to optimize energy consumption.
- Introduction of systematic approaches for the data mining of industry load.
- Proposal and utilization of diverse criteria for data mining and the classification of industries.
- Real-world application and case study analysis showcasing the efficacy of data mining in identifying consumption patterns.
- Classification of industrial customers for tariff design and load management purposes.
- Comprehensive exploration of various factors influencing tariff design and load management, underscoring the importance of considering multifaceted constraints across domains to achieve optimal outcomes while managing potential side effects.

4.2 Investigating Industrial Load Data: Analysis Through Various Indexes

In this section, hourly load data from two sample industries over the period of one year is utilized. Herein, some of the analyses conducted on this dataset are presented. It is crucial to note that for a thorough analysis of industrial customers, it's imperative to identify and incorporate any factors that could influence the actual load of the industry during specific time periods. For instance, having data on industry participation in load management programs during the study period is essential for calculating the net and unmanipulated amount of load.

It's worth noting that each analysis requires the preparation of available data in a format suitable for analysis, followed by the execution of the necessary analytical procedures. Throughout this project, Python has been employed to organize the data into a suitable format and subsequently conduct the necessary analyses. Some outputs and the resulting analyses are presented below.

1) **Annual Load Curves of Industries for Estimating an Overview of Their Consumption Patterns**: This section displays the annual load curves of two sample industries (Figures 4.3 and 4.4). As depicted in Figures 4.3 and 4.4, Industry #1 exhibits highly variable behavior across different hours throughout the year, whereas Industry #2 demonstrates a nearly constant pattern over the same annual period.
2) **Daily Consumption Curves for Specific Days to Examine Consumption Patterns and Daily Variations**: This section presents the daily consumption curves for some selected days (Figures 4.5–4.10).

   Upon examination of Figures 4.5–4.10, it becomes apparent that a load reduction program has been implemented during summer hours, leading to significant changes in the load behavior of industries.

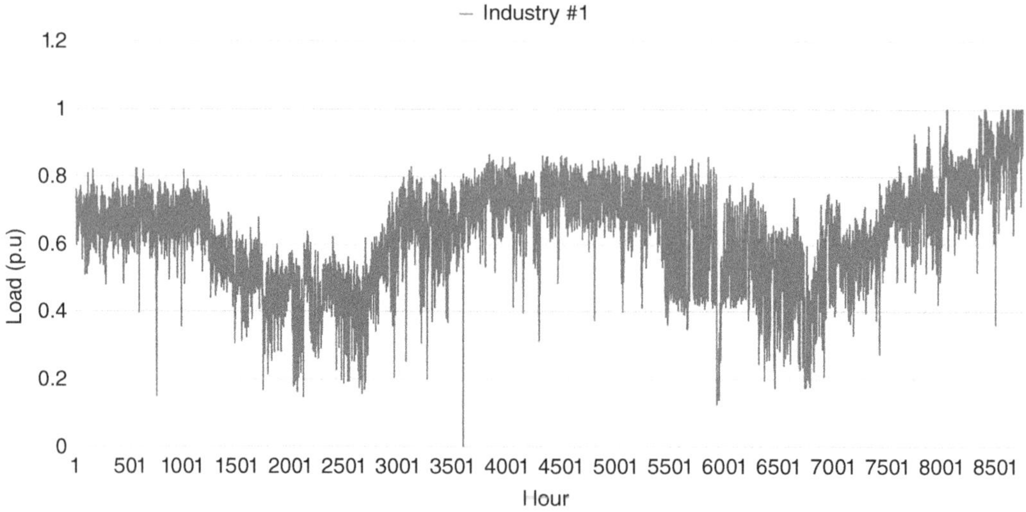

Figure 4.3 Annual load curve of Industry #1.

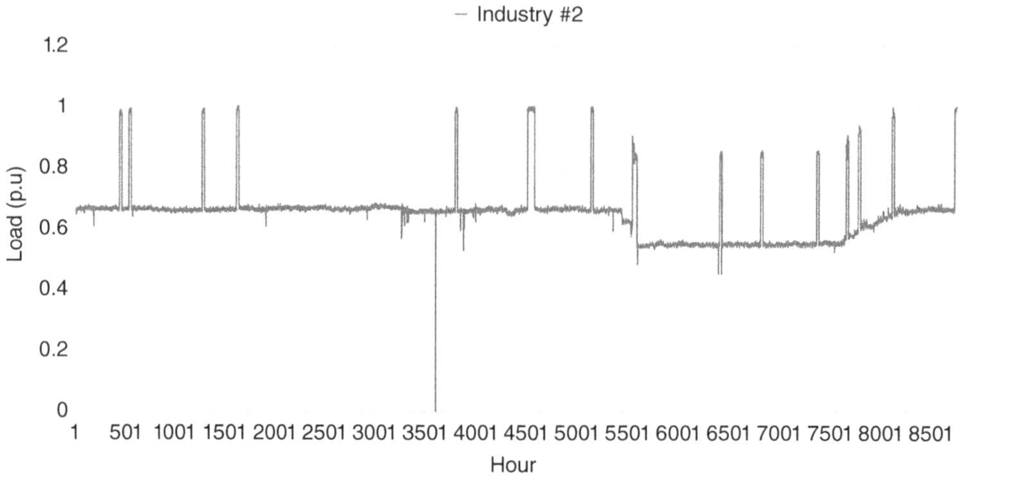

Figure 4.4 Annual load curve of Industry #2.

3) **Creating Separate Consumption Curves for Each Day of the Week Allows for the Examination of Consumption Patterns on Different Days and Facilitates Comparison**: Here, load curves for each specific day of the week are presented, juxtaposed to explore potential analytical insights. Figures 4.11–4.14 depict the load curves for multiple days of the week for Industries #1 and #2.

   If necessary, the load curves of different days of the week can be superimposed, and other types of curves can be generated.

4) **Creating a Load Curve for Each Hour Throughout the Year to Investigate the Load Value at Each Hour and Compare It with the Load Behavior at Other Hours**: Here, load

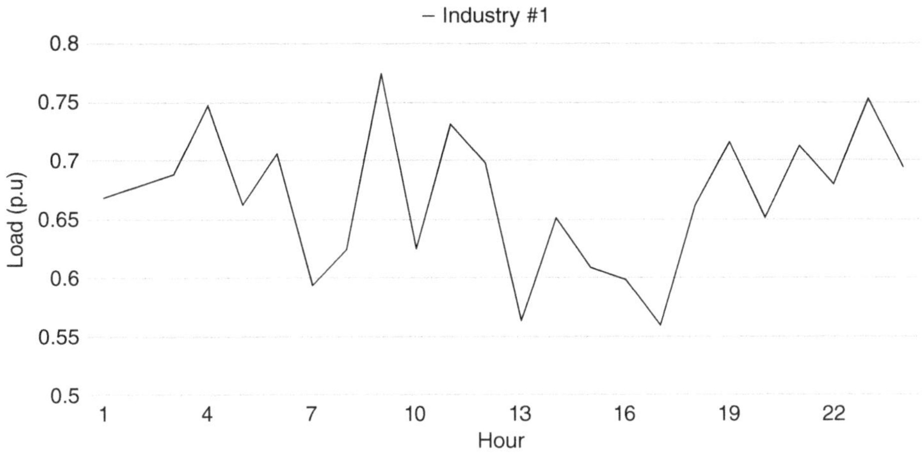

Figure 4.5 Consumption curve of Industry #1 on a Sunday during the fall.

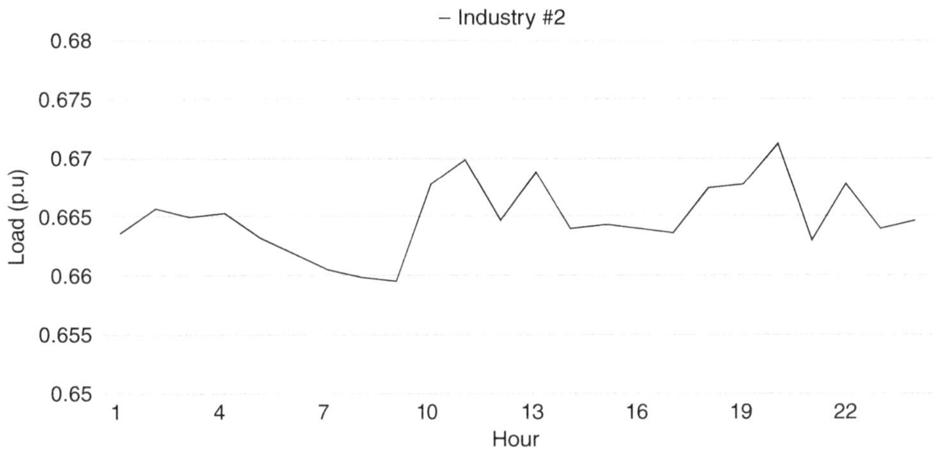

Figure 4.6 Consumption curve of Industry #2 on a Sunday during the fall.

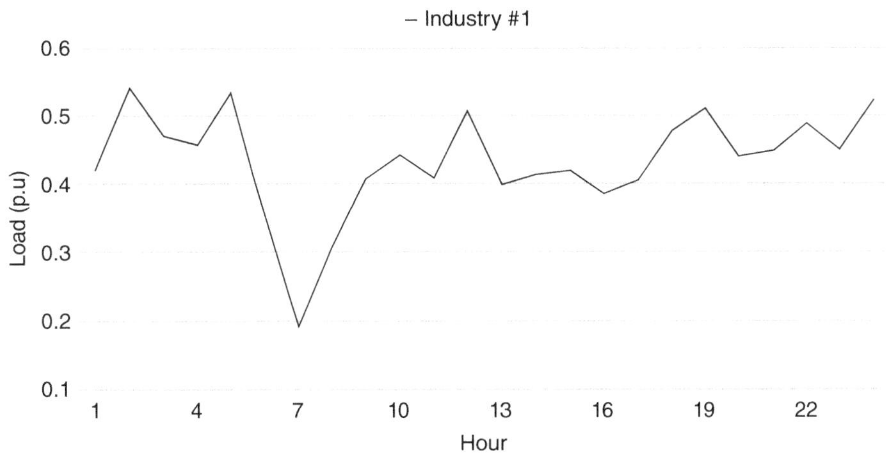

Figure 4.7 Consumption curve of Industry #1 on a Sunday during the winter.

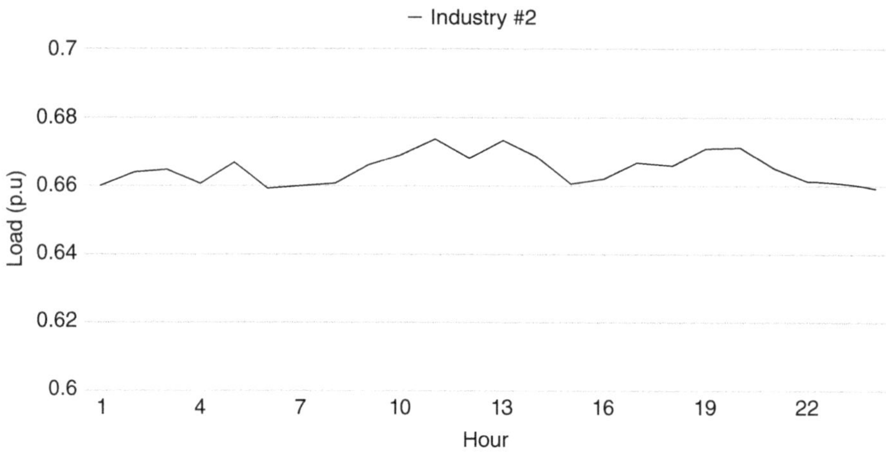

Figure 4.8 Consumption curve of Industry #2 on a Sunday during the winter.

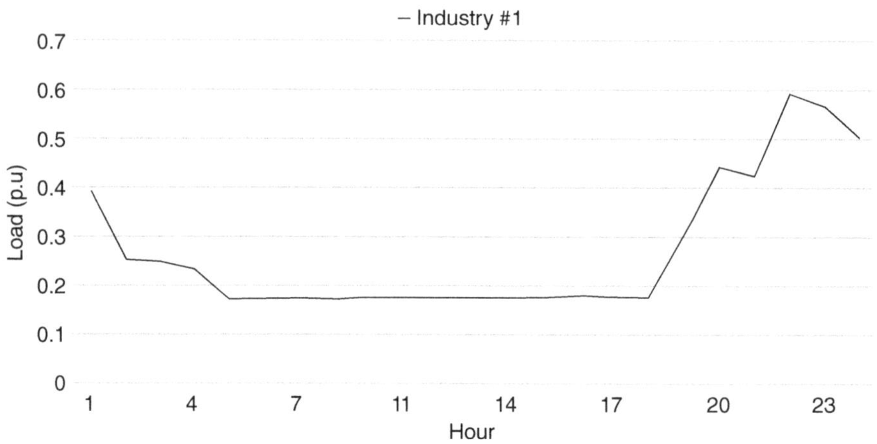

Figure 4.9 Consumption curve of Industry #1 on a Sunday during the summer.

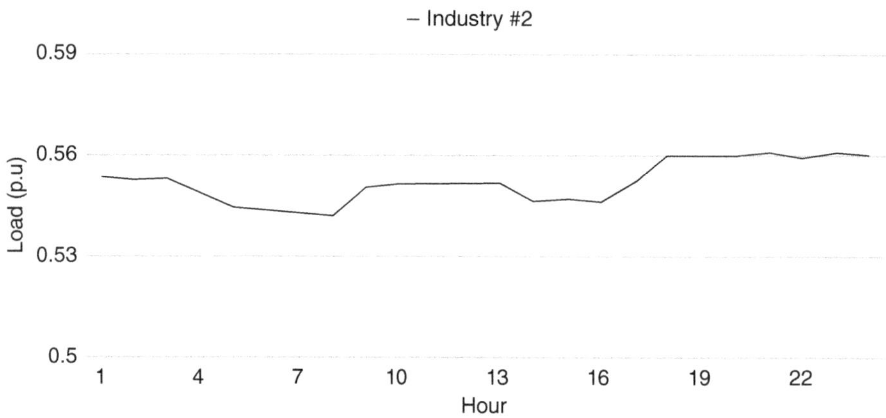

Figure 4.10 Consumption curve of Industry #2 on a Sunday during the summer.

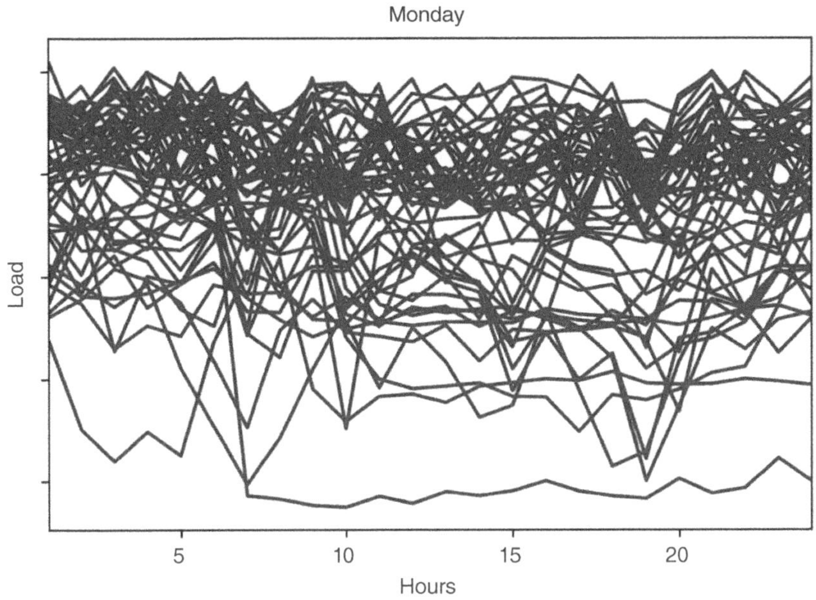

Figure 4.11 Hourly load curve for Monday for Industry #1.

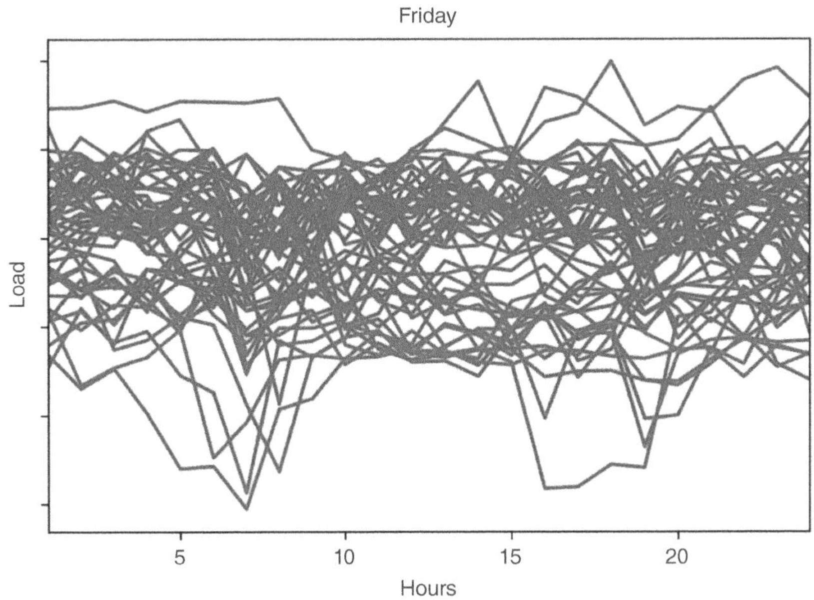

Figure 4.12 Hourly load curve for Friday for Industry #1.

curves are individually plotted for each hour to examine their behavior and make comparisons with other hours, such as Figures 4.15 and 4.16 for Industries #1 and #2.

5) **Calculating the Average Load Value (Both Over the Entire Year and for Each Season Separately) and Comparing It with the Maximum Load Values Within That Period**: This analysis utilizes both annual and seasonal data, facilitating the classification of industries

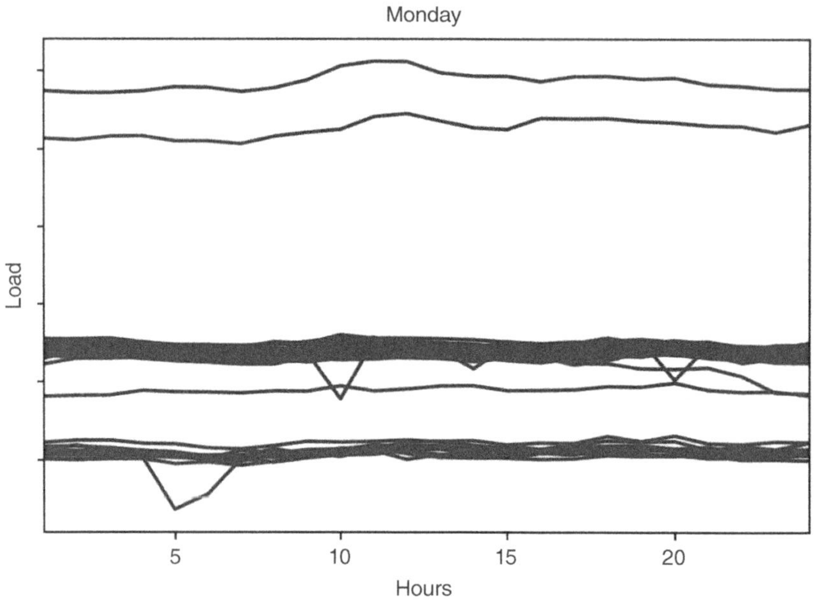

Figure 4.13 Hourly load curve for Monday for Industry #2.

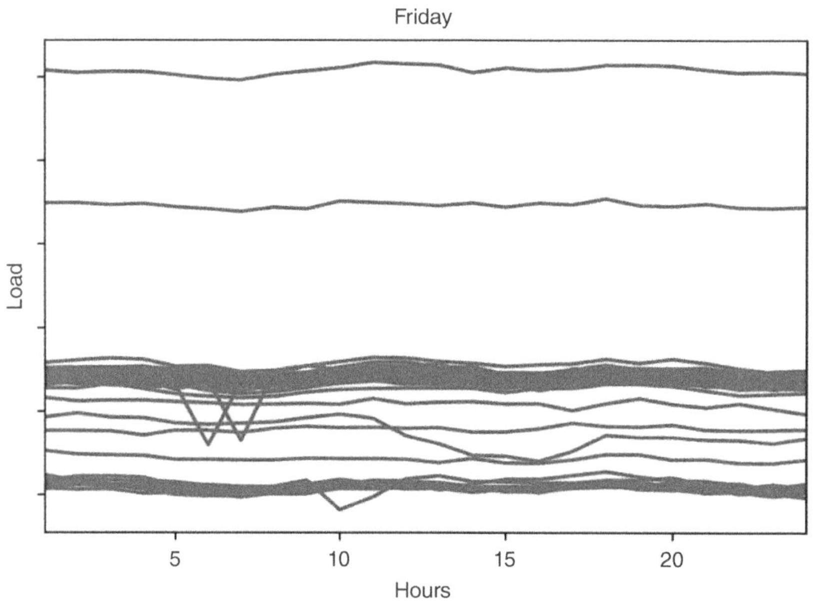

Figure 4.14 Hourly load curve for Friday for Industry #2.

based on the outcomes. In this section, first, the average and maximum load values are computed for the entire annual period. Then, the same calculations are performed separately for each season of the year. Comparing the average load value in each interval with the maximum load value of that interval indicates the deviation of the consumed load in each interval. Subsequently, the load management during the summer period is focused, which holds greater significance based on the requirements of the system under study. A comparison

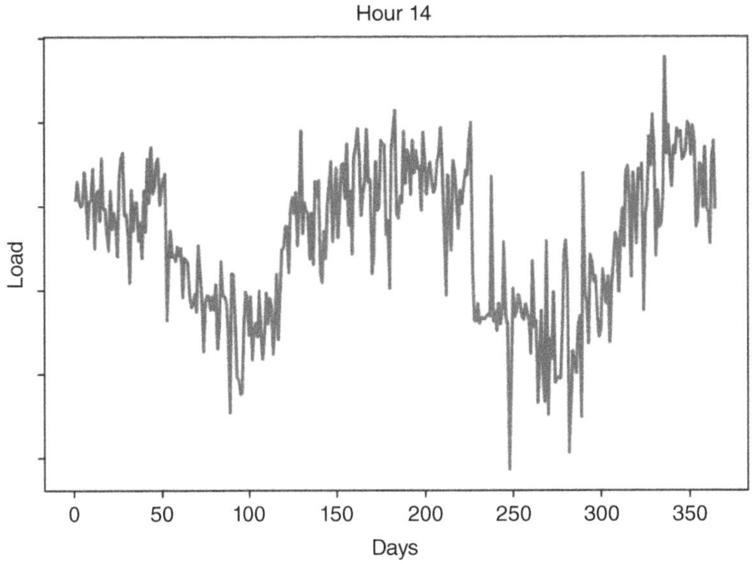

Figure 4.15 Load level at hour 14 throughout the year for Industry #1.

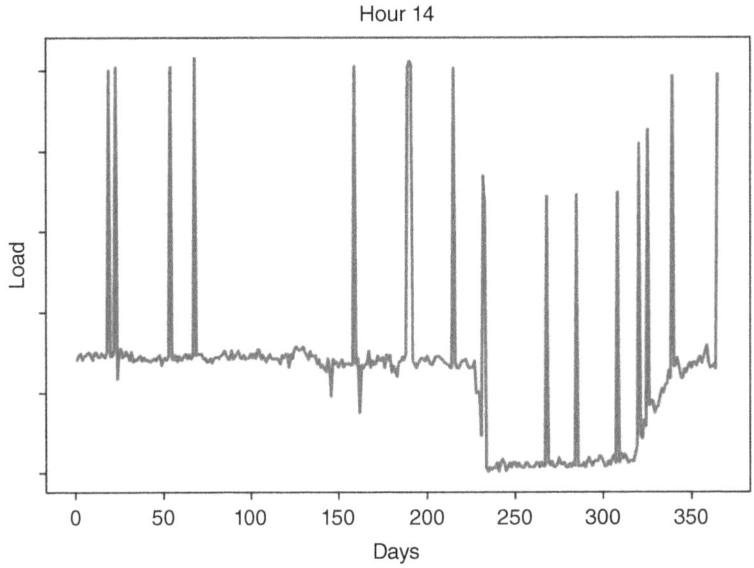

Figure 4.16 Load level at hour 14 throughout the year for Industry #2.

is made between the average load in the summer and the average load throughout the year, excluding the summer. Based on this comparison, industries can be categorized. Industries whose average load during summer surpasses their average load during non-summer periods are placed in one category. For example, neither of the investigated industries, Industry #1 and Industry #2, falls into this category, indicating that their average load during summer is lower than during other seasons, implying no significant challenge from this perspective. However, it's important to note that, based on prior analyses, these industries seem to have participated in other load management programs during the summer, which could influence the data by causing a reduction in load during the summer period. For a more comprehensive investigation, it would be necessary to separately estimate the participation of the industries in the summer load management programs, enabling the calculation of the actual and unaltered load behavior patterns.

6) **Utilizing the Correlation Index to Assess the Correlation Between Hourly Load Values of Each Industry Throughout the Day, Both with Annual and Seasonal Data, and Drawing the Graphs of Correlation Values for Specific Hours to Offer an Averview of Hourly Correlations**: Here, the hourly correlation indexes are initially computed using data from the entire year. The outcome reveals the correlation value of the load at each hour in relation to other hours. Figures 4.17 and 4.18 illustrate the correlations for hours 1, 8, 14, and 20 compared to other hours for Industries #1 and #2, respectively. Furthermore, the calculation of hourly correlation values using seasonal data has been conducted separately. For instance, Figures 4.19 and 4.20 illustrate the values of this index for the summer season.

Upon examining these figures, it becomes apparent that in Industry #1, there is likely no significant dependence between the consumption in each hour and other hours. Conversely, in the case of Industry #2, a high dependence is observed among all hours, suggesting nearly uniform consumption throughout different hours for this industry.

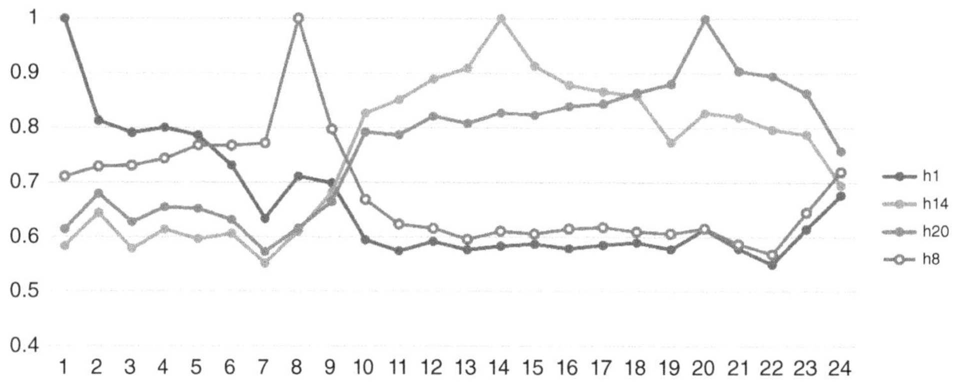

Figure 4.17 Hourly correlation indexes for Industry #1 utilizing data from the entire year (hours 1, 8, 14, and 20).

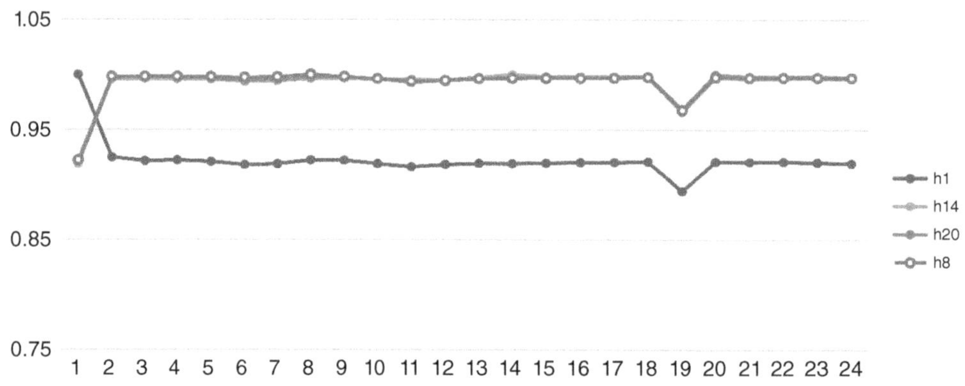

Figure 4.18 Hourly correlation indexes for Industry #2 utilizing data from the entire year (hours 1, 8, 14, and 20).

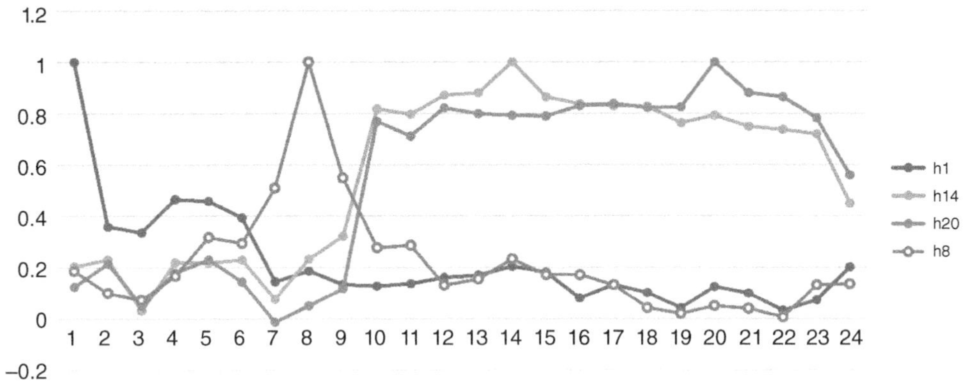

Figure 4.19 Hourly correlation indexes for Industry #1 utilizing data from the summer season (hours 1, 8, 14, and 20).

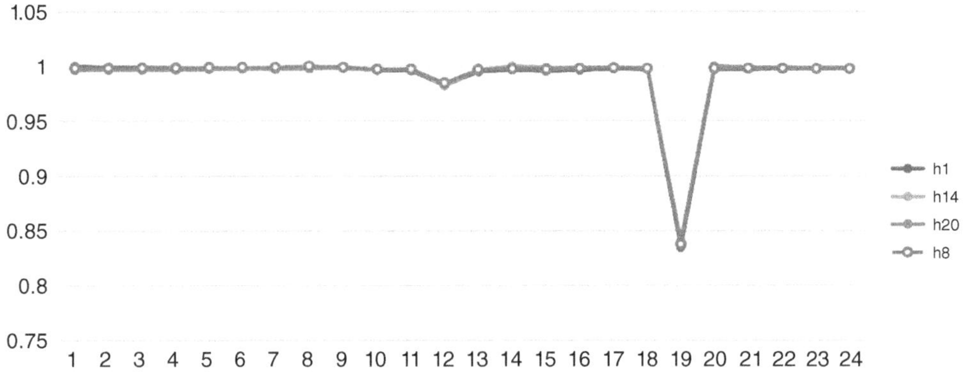

Figure 4.20 Hourly correlation indexes for Industry #2 utilizing data from the summer season (hours 1, 8, 14, and 20).

7) **Utilizing the Correlation Index to Examine the Daily Correlation of Each Industry's Load Values Using Annual Data and Generating a Correlation Chart to Estimate an Overview of the Daily Correlation**: In this section, first, the correlation values between different days over a year, spanning 365 days, are calculated. Figures 4.21 and 4.22 depict the daily correlations for Industries #1 and #2 based on one year's data.

   Based on these figures, it appears that there is no discernible correlation between consumption on different days throughout the year.
8) **Calculating the Average Value of the Hourly Correlation Indexes for Each Industry Separately Based on Annual and Seasonal Data and Categorizing Them Accordingly**: In this section, the average value of the hourly correlation indexes is computed using both annual and seasonal data. Various classifications can be established based on these average hourly correlation values. For instance, if the average value of the hourly correlation index exceeds 0.9, the industries could be placed in the first category; if it is greater than 0.7 and less than or equal to 0.9, they fall into the second category; and if it is less than or equal to 0.7, they are classified into the third category. Thus, industries in the first category exhibit the highest hourly correlation values, while those in the third category have the lowest

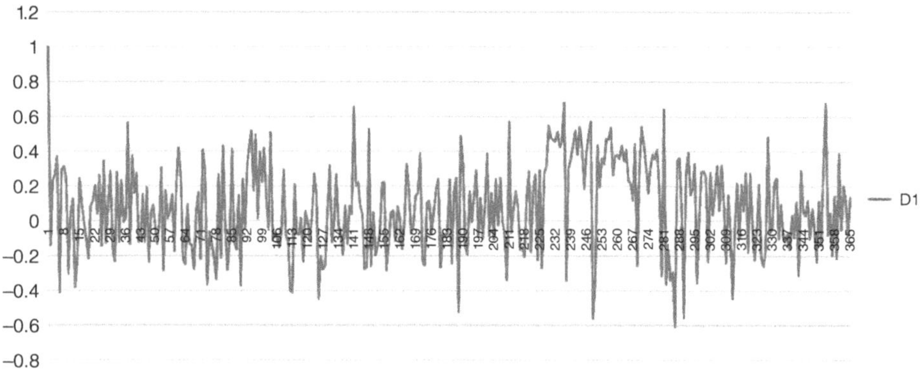

Figure 4.21 Daily correlation index of Industry #1 utilizing data from 365 days of the year (for Day 1 compared to other days).

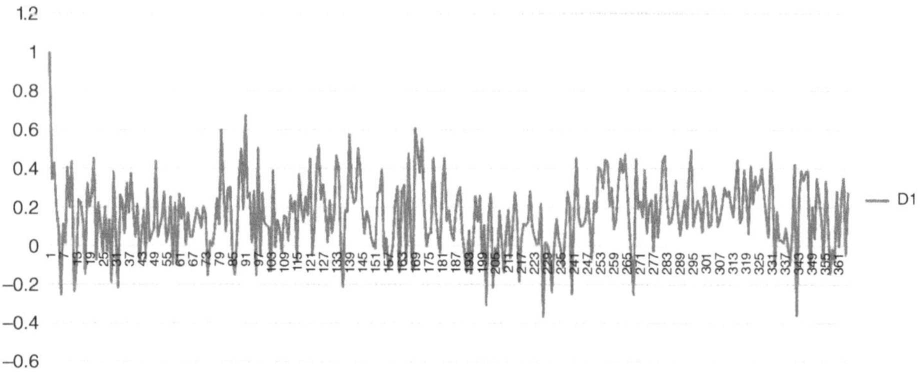

Figure 4.22 Daily correlation index of Industry #2 utilizing data from 365 days of the year (for Day 1 compared to other days).

hourly correlation values. For example, the average hourly correlation index for Industry #1 is calculated as 0.71 and for Industry #2 as 0.99. Consequently, Industry #1 falls into the second category, while Industry #2 is placed in the first category. This classification can also be conducted using hourly data separately for each season. For instance, using summer data, the average value of the correlation index for Industry #1 is calculated as 0.62 and for Industry #2 as 0.98. Following this procedure, Industry #1 would be classified into the third category, while Industry #2 would remain in the first category. For further investigation, Table 4.1 displays the hourly correlation values for Industry #1.

Based on the data in Table 4.1, the following results can be observed:

- **Symmetry**: The correlation matrix exhibits symmetry along the main diagonal since the correlation between time X and time Y is equal to the correlation between time Y and time X.
- **Diagonal Correlation**: Values along the main diagonal (i.e., 1-1, 2-2, 3-3, etc.) are all equal to 1. This signifies that the correlation of an hour with itself is perfect (correlation coefficient is 1), as expected.
- **Correlations of Successive Hours**: As one moves away from the main diagonal, correlation values tend to decrease. Hours further apart demonstrate weaker correlations. For instance, hour 1 exhibits a lower correlation with hour 12 compared to its correlation

Table 4.1 Hourly correlation values of Industry #1.

	1	2	3	4	5	6	7	8	9	10	11	12	13	14	15	16	17	18	19	20	21	22	23	24
1	1.00	0.81	0.79	0.80	0.79	0.73	0.63	0.71	0.70	0.59	0.57	0.59	0.58	0.58	0.59	0.58	0.59	0.59	0.58	0.61	0.58	0.55	0.61	0.68
2	0.81	1.00	0.88	0.86	0.85	0.79	0.68	0.73	0.74	0.65	0.63	0.66	0.63	0.64	0.64	0.63	0.65	0.67	0.65	0.68	0.63	0.61	0.67	0.74
3	0.79	0.88	1.00	0.89	0.86	0.81	0.70	0.73	0.73	0.61	0.59	0.60	0.58	0.58	0.60	0.58	0.61	0.62	0.61	0.63	0.57	0.56	0.63	0.71
4	0.80	0.86	0.89	1.00	0.87	0.81	0.72	0.74	0.76	0.63	0.63	0.63	0.61	0.61	0.62	0.62	0.63	0.64	0.65	0.65	0.61	0.58	0.65	0.72
5	0.79	0.85	0.86	0.87	1.00	0.86	0.74	0.77	0.75	0.63	0.61	0.61	0.58	0.60	0.61	0.60	0.63	0.63	0.65	0.65	0.61	0.58	0.63	0.72
6	0.73	0.79	0.81	0.81	0.86	1.00	0.80	0.77	0.77	0.62	0.60	0.60	0.58	0.61	0.61	0.60	0.62	0.63	0.65	0.63	0.60	0.55	0.63	0.71
7	0.63	0.68	0.70	0.72	0.74	0.80	1.00	0.77	0.76	0.60	0.56	0.56	0.54	0.55	0.53	0.56	0.57	0.58	0.62	0.57	0.53	0.51	0.58	0.66
8	0.71	0.73	0.73	0.74	0.77	0.77	0.77	1.00	0.80	0.67	0.62	0.62	0.60	0.61	0.61	0.61	0.62	0.61	0.61	0.62	0.59	0.57	0.64	0.72
9	0.70	0.74	0.73	0.76	0.75	0.77	0.76	0.80	1.00	0.74	0.72	0.71	0.67	0.68	0.67	0.66	0.66	0.65	0.64	0.66	0.63	0.61	0.69	0.74
10	0.59	0.65	0.61	0.63	0.63	0.62	0.60	0.67	0.74	1.00	0.87	0.87	0.83	0.83	0.81	0.80	0.80	0.79	0.73	0.79	0.79	0.76	0.78	0.70
11	0.57	0.63	0.59	0.63	0.61	0.60	0.56	0.62	0.72	0.87	1.00	0.89	0.86	0.85	0.82	0.81	0.80	0.80	0.74	0.79	0.79	0.76	0.75	0.68
12	0.59	0.66	0.60	0.63	0.61	0.60	0.56	0.62	0.71	0.87	0.89	1.00	0.91	0.89	0.86	0.85	0.84	0.85	0.77	0.82	0.82	0.80	0.78	0.69
13	0.58	0.63	0.58	0.61	0.58	0.58	0.54	0.60	0.67	0.83	0.86	0.91	1.00	0.91	0.88	0.85	0.83	0.84	0.77	0.81	0.81	0.79	0.77	0.67
14	0.58	0.64	0.58	0.61	0.60	0.61	0.55	0.61	0.68	0.83	0.85	0.89	0.91	1.00	0.91	0.88	0.87	0.86	0.77	0.83	0.82	0.80	0.79	0.69
15	0.59	0.64	0.60	0.62	0.61	0.61	0.53	0.61	0.67	0.81	0.82	0.86	0.88	0.91	1.00	0.90	0.88	0.86	0.78	0.82	0.83	0.79	0.77	0.68
16	0.58	0.63	0.58	0.62	0.60	0.60	0.56	0.61	0.66	0.80	0.81	0.85	0.85	0.88	0.90	1.00	0.91	0.89	0.82	0.84	0.85	0.81	0.79	0.70
17	0.59	0.65	0.61	0.63	0.63	0.62	0.57	0.62	0.66	0.80	0.80	0.84	0.83	0.87	0.88	0.91	1.00	0.92	0.83	0.84	0.83	0.81	0.79	0.70
18	0.59	0.67	0.62	0.64	0.63	0.63	0.58	0.61	0.65	0.79	0.80	0.85	0.84	0.86	0.86	0.89	0.92	1.00	0.88	0.86	0.86	0.83	0.80	0.71
19	0.58	0.65	0.61	0.65	0.65	0.65	0.62	0.61	0.64	0.73	0.74	0.77	0.77	0.77	0.78	0.82	0.83	0.88	1.00	0.88	0.85	0.84	0.81	0.73
20	0.61	0.68	0.63	0.65	0.65	0.63	0.57	0.62	0.66	0.79	0.79	0.82	0.81	0.83	0.82	0.84	0.84	0.86	0.88	1.00	0.90	0.89	0.86	0.76
21	0.58	0.63	0.57	0.61	0.61	0.60	0.53	0.59	0.63	0.79	0.79	0.82	0.81	0.82	0.83	0.85	0.83	0.86	0.85	0.90	1.00	0.90	0.86	0.73
22	0.55	0.61	0.56	0.58	0.58	0.55	0.51	0.57	0.61	0.76	0.76	0.80	0.79	0.80	0.79	0.81	0.81	0.83	0.84	0.89	0.90	1.00	0.88	0.75
23	0.61	0.67	0.63	0.65	0.63	0.63	0.58	0.64	0.69	0.78	0.75	0.78	0.77	0.79	0.77	0.79	0.79	0.80	0.81	0.86	0.86	0.88	1.00	0.83
24	0.68	0.74	0.71	0.72	0.72	0.71	0.66	0.72	0.74	0.70	0.68	0.69	0.67	0.69	0.68	0.70	0.70	0.71	0.73	0.76	0.73	0.75	0.83	1.00

Table 4.2 Average correlation coefficients for different hours for Industry #1.

1	0.64	7	0.62	13	0.73	19	0.73
2	0.70	8	0.66	14	0.74	20	0.75
3	0.67	9	0.70	15	0.74	21	0.73
4	0.69	10	0.73	16	0.74	22	0.71
5	0.68	11	0.72	17	0.74	23	0.73
6	0.67	12	0.74	18	0.75	24	0.71

with hour 2. While exceptions may arise, this pattern of decreasing correlation is generally observed in the data. High correlation values (close to 1) are typically found near the main diagonal (top left to bottom right); for example, hour 1 is highly correlated with hour 2, and hour 23 is highly correlated with hour 24. Generally, consecutive hours (i.e., 1-2, 2-3, 3-4, etc.) exhibit high correlation, indicating a strong positive relationship between these hours.

- **Peak Hours**: Certain hours may display higher correlations with multiple other hours; that might indicate peak hours or critical intervals in which some activities or events repetitively occur across various days during these hours.

  Table 4.2 presents the average values of hourly correlation coefficients for Industry #1. Based on Table 4.2, the highest average correlation values are observed around and after hour 10, indicating potential peak hours.
- **Low Correlation**: Certain hours may exhibit relatively low correlation with other hours, suggesting periods of reduced activity or less stability in patterns during those hours. For instance, in Table 4.2, the average correlation is notably low at times, such as 1 a.m. Additionally, correlations with very small values indicate minimal relationships between specific hours. This suggests that these hours are relatively independent of each other regarding their impact on the industry. To examine the behavior of Industry #2, the hourly correlation values for this industry are presented in Table 4.3.

  Based on Table 4.3, the following observations can be made:
- **Symmetry**: Similar to Industry #1, the correlation matrix for Industry #2 is symmetrical along the main diagonal.
- **Diagonal Correlation**: Like Industry #1, the diagonal values (correlation of each hour with itself) are all 1, indicating complete correlation of each hour with itself.
- **Correlations of Consecutive Hours**: Similar to Industry #1, the correlation between closely related hours is generally higher. However, for Industry #2, the correlation between closely related hours (i.e., 1 and 2, 2 and 3, etc.) is notably high, typically exceeding 0.9. This suggests a strong positive relationship between consecutive hours. As the lag of times increases (i.e., 1 and 10, 2 and 12, etc.), correlation values begin to decrease, but they still remain relatively high. It should be noted that this may not always occur, and in some cases, the decrease in correlation with increasing lag times may not be observed. Nonetheless, this trend is generally observable in most instances.
- **Lagged Correlations**: Correlations between nonconsecutive hours (i.e., 1 and 3, 2 and 4, etc.) are also very high, indicating continuity or similarity in industry performance between these hours.
- **Strong Positive Correlation**: Generally, correlation values between hours of the day are very high, indicating a strong positive correlation. This implies that an increase in load at one hour corresponds to an increase in load at another hour, suggesting consistent underlying patterns or trends throughout the day.

Table 4.3 Hourly correlation values of Industry #2.

	1	2	3	4	5	6	7	8	9	10	11	12	13	14	15	16	17	18	19	20	21	22	23	24
1	1.00	0.92	0.92	0.92	0.92	0.92	0.92	0.92	0.92	0.92	0.92	0.92	0.92	0.92	0.92	0.92	0.92	0.92	0.89	0.92	0.92	0.92	0.92	0.92
2	0.92	1.00	1.00	1.00	1.00	1.00	1.00	1.00	1.00	1.00	0.99	0.99	1.00	1.00	1.00	1.00	1.00	1.00	0.97	1.00	1.00	1.00	1.00	1.00
3	0.92	1.00	1.00	1.00	1.00	1.00	1.00	1.00	1.00	1.00	0.99	0.99	1.00	1.00	1.00	1.00	1.00	1.00	0.97	1.00	1.00	1.00	1.00	1.00
4	0.92	1.00	1.00	1.00	1.00	1.00	1.00	1.00	1.00	1.00	0.99	0.99	1.00	1.00	1.00	1.00	1.00	1.00	0.97	1.00	1.00	1.00	1.00	1.00
5	0.92	1.00	1.00	1.00	1.00	1.00	1.00	1.00	1.00	1.00	0.99	0.99	1.00	1.00	1.00	1.00	1.00	1.00	0.97	1.00	1.00	1.00	1.00	1.00
6	0.92	1.00	1.00	1.00	1.00	1.00	1.00	1.00	0.99	0.99	0.99	0.99	0.99	0.99	0.99	0.99	0.99	0.99	0.97	0.99	0.99	0.99	0.99	0.99
7	0.92	1.00	1.00	1.00	1.00	1.00	1.00	1.00	1.00	0.99	0.99	0.99	0.99	0.99	0.99	0.99	0.99	1.00	0.97	1.00	1.00	1.00	1.00	1.00
8	0.92	1.00	1.00	1.00	1.00	1.00	1.00	1.00	1.00	1.00	0.99	0.99	1.00	1.00	1.00	1.00	1.00	1.00	0.97	1.00	1.00	1.00	1.00	1.00
9	0.92	1.00	1.00	1.00	1.00	0.99	1.00	1.00	1.00	1.00	0.99	0.99	1.00	1.00	1.00	1.00	1.00	1.00	0.97	1.00	1.00	1.00	1.00	1.00
10	0.92	1.00	1.00	1.00	1.00	0.99	0.99	1.00	1.00	1.00	0.99	0.99	0.99	1.00	1.00	1.00	1.00	1.00	0.96	1.00	1.00	1.00	1.00	1.00
11	0.92	0.99	0.99	0.99	0.99	0.99	0.99	0.99	0.99	0.99	1.00	0.99	0.99	1.00	1.00	0.99	1.00	0.99	0.96	0.99	0.99	0.99	0.99	0.99
12	0.92	0.99	0.99	0.99	0.99	0.99	0.99	0.99	0.99	0.99	0.99	1.00	1.00	0.99	0.99	0.99	0.99	0.99	0.96	0.99	0.99	0.99	0.99	0.99
13	0.92	1.00	1.00	1.00	1.00	0.99	0.99	1.00	1.00	0.99	0.99	1.00	1.00	1.00	1.00	1.00	1.00	1.00	0.97	1.00	1.00	1.00	1.00	1.00
14	0.92	1.00	1.00	1.00	1.00	0.99	0.99	1.00	1.00	1.00	1.00	0.99	1.00	1.00	1.00	1.00	1.00	1.00	0.97	1.00	1.00	1.00	1.00	1.00
15	0.92	1.00	1.00	1.00	1.00	0.99	0.99	1.00	1.00	1.00	1.00	0.99	1.00	1.00	1.00	1.00	1.00	1.00	0.97	1.00	1.00	1.00	1.00	1.00
16	0.92	1.00	1.00	1.00	1.00	0.99	0.99	1.00	1.00	1.00	0.99	0.99	1.00	1.00	1.00	1.00	1.00	1.00	0.97	1.00	1.00	1.00	1.00	1.00
17	0.92	1.00	1.00	1.00	1.00	0.99	0.99	1.00	1.00	1.00	1.00	0.99	1.00	1.00	1.00	1.00	1.00	1.00	0.97	1.00	1.00	1.00	1.00	1.00
18	0.92	1.00	1.00	1.00	1.00	0.99	1.00	1.00	1.00	1.00	0.99	0.99	1.00	1.00	1.00	1.00	1.00	1.00	0.97	1.00	1.00	1.00	1.00	1.00
19	0.89	0.97	0.97	0.97	0.97	0.97	0.97	0.97	0.97	0.96	0.96	0.96	0.97	0.97	0.97	0.97	0.97	0.97	1.00	0.97	0.97	0.97	0.97	0.97
20	0.92	1.00	1.00	1.00	1.00	0.99	1.00	1.00	1.00	1.00	0.99	0.99	1.00	1.00	1.00	1.00	1.00	1.00	0.97	1.00	1.00	1.00	1.00	1.00
21	0.92	1.00	1.00	1.00	1.00	0.99	1.00	1.00	1.00	1.00	0.99	0.99	1.00	1.00	1.00	1.00	1.00	1.00	0.97	1.00	1.00	1.00	1.00	1.00
22	0.92	1.00	1.00	1.00	1.00	0.99	1.00	1.00	1.00	1.00	0.99	0.99	1.00	1.00	1.00	1.00	1.00	1.00	0.97	1.00	1.00	1.00	1.00	1.00
23	0.92	1.00	1.00	1.00	1.00	0.99	1.00	1.00	1.00	1.00	0.99	0.99	1.00	1.00	1.00	1.00	1.00	1.00	0.97	1.00	1.00	1.00	1.00	1.00
24	0.92	1.00	1.00	1.00	1.00	0.99	1.00	1.00	1.00	1.00	0.99	0.99	1.00	1.00	1.00	1.00	1.00	1.00	0.97	1.00	1.00	1.00	1.00	1.00

- **Potential for Time Series Analysis**: The high correlation between consecutive hours suggests that time series analysis methods could be beneficial for modeling and predicting the behavior of this industry.

  In general, the following points should be considered:
- **Potential Trends**: Depending on the industry and the nature of the data, there may be potential trends or patterns worthy of further investigation. For instance, consistently increasing or decreasing correlation values throughout the day could indicate certain time-dependent trends.
- **Seasonal Variation**: If the data spans multiple days or seasons, it is possible to detect seasonal variation in correlations, indicating that certain hours are highly related to each other at specific periods.
- **Use of Existing Knowledge**: Generally, the correlation values can provide insights into daily consumption patterns, peak hours identification, and informed decision-making. However, for further analysis, it is imperative to use domain knowledge and conduct additional statistical tests to draw meaningful conclusions and make informed decisions based on the results of data mining. It is crucial to note that the correlations presented are derived from historical data, and other factors such as industry type, seasonal patterns, and potential external influences should be taken into account for any meaningful analysis or conclusions. Furthermore, without prior information about an industry and its specific data, drawing definitive conclusions may be challenging. In essence, for further analysis, sufficient and specific relevant knowledge is necessary to fully interpret the correlations and extract practical insights.

9) **Calculation of the Average Value of the Daily Correlation Index**: Here, for each company, the average value of the daily correlation index (between different days) is computed based on both annual and seasonal data. In this analysis, the average value of the correlation indexes for these industries is found to be less than 0.7 in all cases, indicating low correlations.
10) **Calculation of Seasonal Correlation Index**: In this section, the correlation value between the load data of each season is computed. These values determine the degree to which the load pattern of each industry in each season correlates with the load data of other seasons. The values of this index are small for both industries, indicating that no significant correlation can be observed between different seasons.
11) **Calculation of Count, Mean, Standard Deviation (Std), 25th Percentile, 50th Percentile (Median), 75th Percentile, Minimum (Min), and Maximum (Max) Indicators**: In this section, the values of these indicators are calculated separately based on hourly data related to 365 days of the year, as well as separately for different seasons. Each of these indicators provides insights into the distribution and variation of the data, as outlined below.
    - **Count**: Indicates the number of data points in each survey. For instance, in the yearly interval, there are 365 counts/h.
    - **Mean**: Represents the average load per hour.
    - **Std**: Reflects the standard deviation value for each hour, measuring the variability or spread of the load values around the mean. Higher standard deviation values signify more variability or dispersion of the load. For example, in the first industry, hour 6 exhibits a relatively high standard deviation of 166.81, indicating significant load fluctuations during that hour on different days. However, for the second company, standard deviation values range from approximately 18.48 (lowest) to 19.94 (highest).
    - **Min and Max**: Denote the minimum and maximum load values, indicating the lowest and highest observed load values per hour throughout the intervals. As an instance, examining

the minimum and maximum load at different hours in the first industry reveals a wide range of load fluctuations.

- **Quartiles (25% (Q1), 50% (Q2), and 75% (Q3))**: Quartiles provide insights into the distribution of load values. The 50th percentile (median) represents the load value at which half of the data points are below and half are above. The 25th and 75th percentiles represent the values that 25% and 75% of the data fall below, respectively. Quartile analysis aids in understanding the tendency and spread of the load data distribution. A large difference between Q1 and Q3 indicates a broad load distribution with significant variation between low and high load periods. Conversely, a small difference between Q1 and Q3 suggests a more concentrated load distribution around the median. Indeed, if the 25th and 75th percentiles are close, it indicates a constant and stable load throughout the interval, while a significant difference between Q1 and Q3 may suggest fluctuating load patterns. Also, a wide distribution between Q1 and Q3 may imply more outliers. Additionally, a relatively small difference between Q1 and Q2 (25th–50th percentile) indicates a compact lower half of the load data, while a small difference between Q2 and Q3 (50th–75th percentile) indicates a compact upper half of the load data. Comparing medians at different hours provides insights into load levels at those hours. For example, if the median at 8 o'clock is lower than the median at 18 o'clock, it could indicate higher load values in the afternoon compared to the early morning. As an instance, in Industry #2, Q3 values for each hour range from approximately 148.30 to 149.92, indicating that in 75% of the hours of the year, the load is lower than these values, and only in 25% of the hours is the load higher than these values. Overall, analyzing these indicators reveals patterns in the industry's load throughout the year.

  In Industry #2, the following observations can be made:

- Standard deviation values indicate slight variations in load values for each hour throughout the year.
- Load values exhibit a daily pattern.
- Load values appear to remain within a consistent and reasonable range.

  This information is valuable for load forecasting, capacity planning, and understanding peak demand periods in the industry. Each of these indicators provides unique insights for data analysts and aids in different analyses.

12) **Calculation of Coefficient of Variation (CV) Index**: The CV is a measure of variability, calculated as the ratio of standard deviation to the mean. This index enables comparison of load variability across different hours relative to their average load levels. A higher CV indicates relatively higher variability compared to the mean, while a lower CV suggests more stable load patterns. Therefore, in this section, the CV index is calculated for each case. A smaller CV index implies less scattered data and potentially a more uniform load pattern diagram.

    It is noteworthy that while hourly CV values can be calculated individually, here, for the purpose of industry classification, a single CV value is computed for all hours. This approach yields a single number for each industry that can be compared across different industries. The CV values are calculated separately using data from the entire year and different seasons. For Industry #1, the CV value calculated using data from the entire year is 0.23, while for Industry #2, it is 0.13. As expected, the CV value is smaller for Industry #2, indicating a more uniform load pattern. Based on this index, a classification can be made. For example, industries with a CV value less than 0.15 could be assigned to the first category. According to this classification, Industry #2 falls into the first category, while Industry #1 falls into the second category.

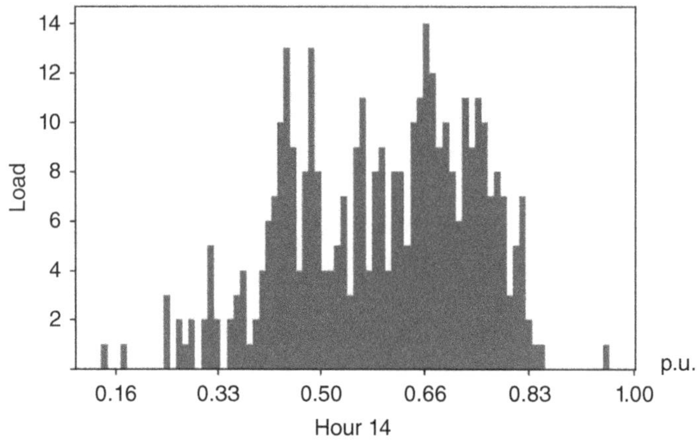

Figure 4.23 Histogram of Industry #1 at 2 p.m.

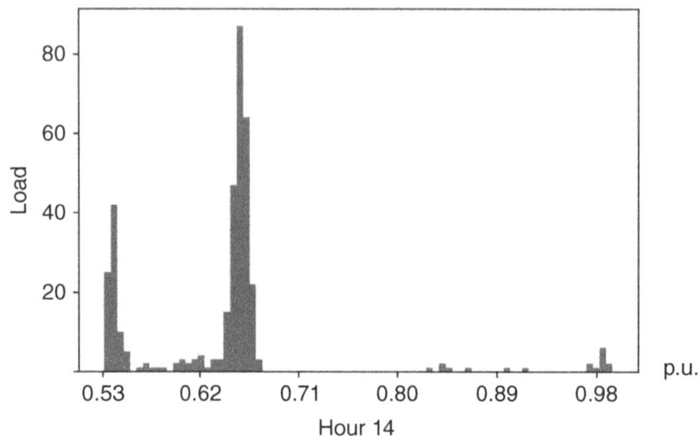

Figure 4.24 Histogram of Industry #2 at 2 p.m.

13) **Draw the Load Distribution Diagram for Each Industry**: In this section, histogram graphs can be plotted to visualize the repetition of load levels. Figures 4.23 and 4.24 display the load histograms of Industries #1 and #2 at 2 p.m., respectively. These diagrams offer a general overview of load level repetition for each industry.
14) **Calculation of Load Regression and Drawing Regression Diagrams**: This section involves computing the regression of the load values at different hours using annual data, followed by the creation of corresponding regression diagrams. This analysis effectively demonstrates the dispersion or uniformity of the data. Figures 4.25 and 4.26 depict the regression values of the load at hour 1 compared to the load at hour 15 for Industries #1 and #2. As observed from the figures, the load data for Industry #1 appears relatively scattered, whereas for Industry #2, the load data is closely aligned with the regression line, indicating a certain degree of uniformity.
15) **Plotting Load Values of One Hour Compared to Load Values of Other Hours**: In this section, the load values in each hour are plotted against the load values in other hours to visualize their distribution and uniformity. Figures 4.27 and 4.28 depict these graphs for Industries #1 and #2, respectively. These plots provide insights into the dispersion or uniformity of the load data.

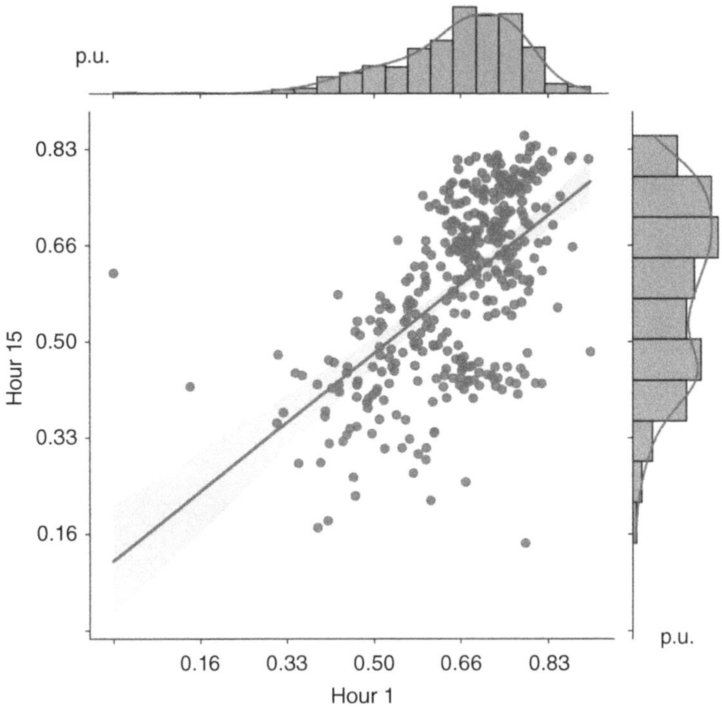

Figure 4.25 Regression plot of load levels at hour 1 compared to load levels at hour 15 for Industry #1.

16) **Clustering of Load Values per Hour and Drawing Corresponding Graphs**: In this section, load values per hour are clustered into specific categories to provide an overall insight into load concentration levels. In this chapter, the K-means method is employed for clustering the data. K-means is a well-known approach in machine learning for clustering and data classification. It partitions data points into K clusters based on similarity, with the goal of minimizing within-cluster variation. The algorithm iteratively assigns data points to the nearest cluster centroid and updates centroids based on the mean of the points in each cluster. Figures 4.29 and 4.30 illustrate two examples of load clustering in Industries #1 and #2 based on load values at 1 a.m. and 3 p.m., respectively. These graphs help visualize the distribution of load values and the concentration levels within each hour.

    Other analyses that can be conducted in this area include:

- **Skewness**: Skewness measures the asymmetry in the distribution of load values. Positive skewness indicates a longer right tail (more load points with high values), while negative skewness indicates a longer left tail (more load points with low values). A skewness of zero indicates a perfectly symmetrical distribution.
- **Kurtosis**: Kurtosis measures the degree of peakedness or flatness of the distribution of load values. Positive kurtosis indicates a more peaked distribution, while negative kurtosis indicates a flatter distribution compared to a normal distribution.
- **Seasonal Decomposition**: Seasonal decomposition can be employed to extract seasonal patterns and trends in load data. This decomposition offers valuable insights into industry load patterns and aids in identifying any recurring seasonal trends.

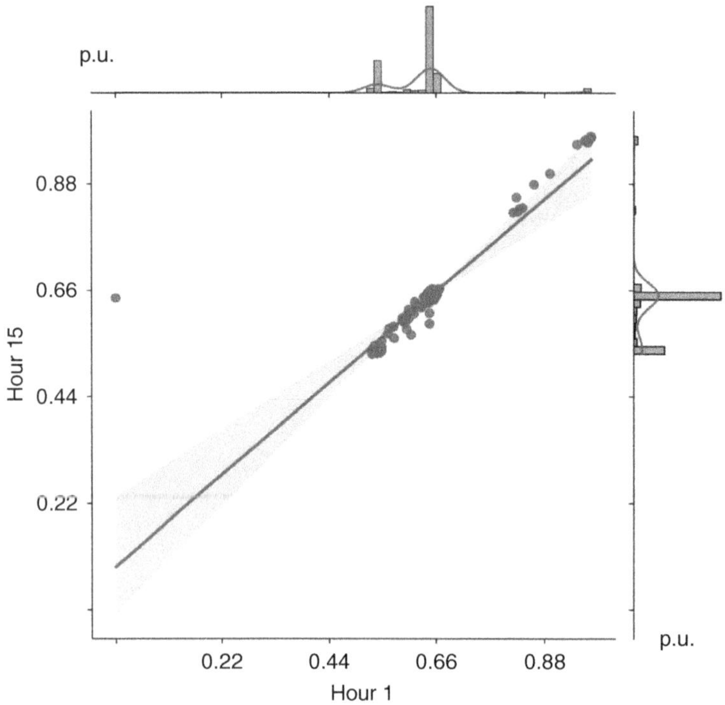

Figure 4.26 Regression plot of load levels at hour 1 compared to load levels at hour 15 for Industry #2.

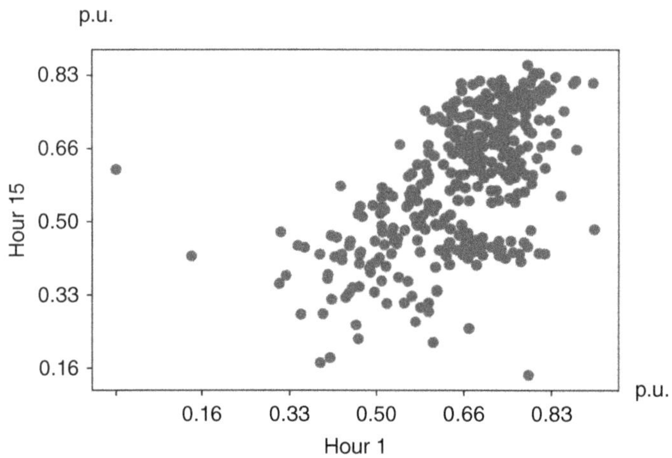

Figure 4.27 Load values at hour 1 compared to load values at hour 15 for Industry #1.

- **Outliers**: Identifying and analyzing outliers in hourly load data is crucial. Outliers are values that deviate significantly from the rest of the data and can indicate unusual load events or system failures. Analyzing outliers can provide insights into potential anomalies in the system or abnormal load behavior, which may require further investigation or corrective action.
- **Coefficient of Quartile Variation (CQV)**: CQV represents another measure of relative dispersion, denoting the interquartile range (IQR) as a percentage of the median. CQV is computed by dividing the IQR (Q3 – Q1) by the median (Q2) and subsequently multiplying by 100 to express it as a percentage.

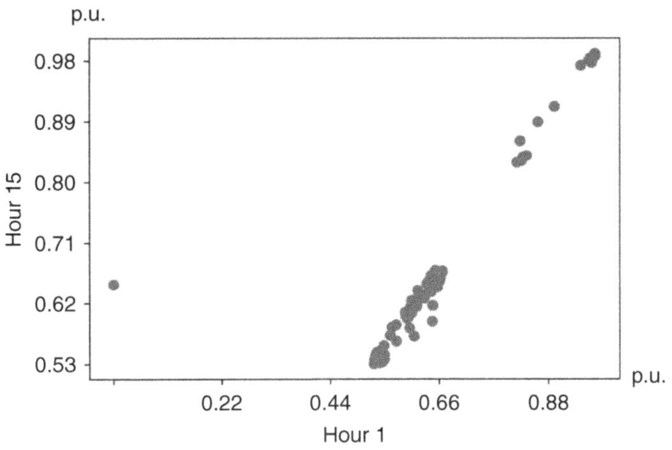

Figure 4.28 Load values at hour 1 compared to load values at hour 15 for Industry #2.

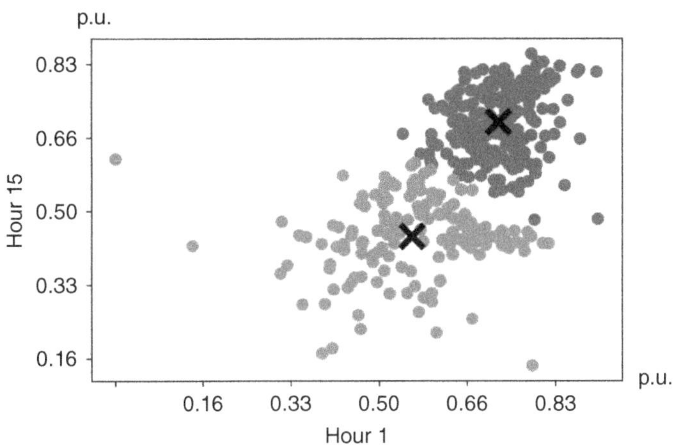

Figure 4.29 Load clustering for Industry #1 based on load values at hours 1 a.m. and 3 p.m.

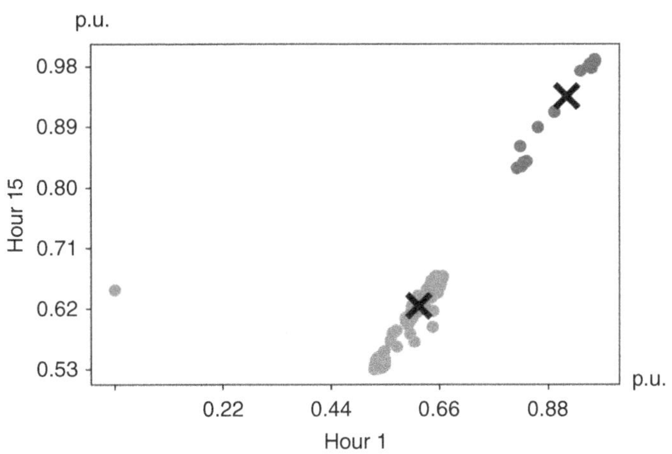

Figure 4.30 Load clustering for Industry #2 based on load values at hours 1 a.m. and 3 p.m.

- **Mean Absolute Deviation (MAD)**: MAD quantifies the average of absolute differences between each data point and the mean of the entire dataset. To calculate the MAD for each hour, sum the absolute differences between each load value and the overall mean, then divide by the number of data points.

In the subsequent section, various potential categorizations will be outlined based on the analysis conducted.

4.3 Classification of Industries

In the preceding section, a comprehensive analysis was undertaken to examine the load patterns across different hours, days, weeks, or seasons for each industry, utilizing various tools and methodologies. The objective was to ascertain suitable strategies for optimizing consumption patterns. Through these analyses, efforts were made to discern distinctive load shapes and consumption patterns among different industries, facilitating the provision of tailored classifications according to each industry's circumstances. Such categorizations and decisions can prove instrumental in formulating optimal energy management policies for industrial sectors. It is imperative to emphasize that a comprehensive understanding of relevant issues and the diverse characteristics of industries is essential to reach definitive conclusions. Additionally, comprehensive analyses, encompassing technical and nontechnical constraints, should be considered. Moreover, the broader implications of proposed policies, including potential side effects such as production reduction or inflation, must be carefully considered. Nevertheless, this section offers a foundational analysis, laying the groundwork for future studies and further exploration of these topics.

Categorizing industries for electricity tariff design necessitates identifying features or characteristics that distinguish their electricity consumption patterns. The following criteria are presented to define various categories for designing electricity tariffs.

1) **Electricity Consumption Patterns**: Industries exhibiting high peaks in electricity consumption may require specific tariff structures to facilitate participation in load management programs. Additionally, industries with steady electricity usage necessitate different tariff schemes compared to those with fluctuating consumption patterns.
2) **Working Hours**: Industries operating round the clock (24/7 operations) may have distinct tariff needs compared to those with fixed working hours. Differentiating between industries operating continuously and those with specific business hours is crucial for tariff design.
3) **Energy Intensity**:
   - **High Energy Intensity**: Industries requiring high energy for their processes may require tailored tariffs for effective cost management. For instance, the steel industry exemplifies high energy intensity, characterized by consistently high energy consumption. In such cases, designing specialized tariffs aimed at load management can be beneficial.
   - **Industries with Average Load Levels**: Similarly, tariffs can be customized to incentivize improvements in load patterns for this type of load.
     Note: Analyzing the correlation between industrial processes and electricity consumption within specific time intervals with the network load during these intervals can aid the classification process, potentially facilitating effective load management during peak periods. Additionally, data pertaining to the industry's load intensity during specific periods, such as peak hours of the network, can offer valuable insights for the classifications.

Note: Industries with different installed capacities may require specific tariff structures to address their capacity-related expenses.

4) **Seasonal Variability**: Industries with consumption fluctuations across different seasons can benefit from seasonal tariff adjustments. Classification based on this feature allows for the implementation of tailored seasonal tariff adjustments.
5) **Energy Efficiency**: Industries that have invested in energy-efficient technologies may require tariffs that reward their efforts. By reflecting these practices in industry classification and tariff regulation, tariff structures can incentivize efficiency improvements and reward such initiatives.
6) **Flexibility and DR Capability**:
   - Industries capable of shifting their operations to off-peak hours can benefit from time-based tariffs. Analyzing load data and its variations across different hours can assess an industry's flexibility in load management. Industries showcasing the ability to shift electricity consumption to off-peak hours can be categorized as having flexible loads. For instance, industries capable of adjusting and reducing energy consumption during peak hours may find value in tariffs with higher prices during these periods.
   - For industries able to swiftly adapt their electricity demand in response to grid signals, dynamic tariff structures may be more suitable. Additionally, correlation values during specific hours can indicate an industry's potential for responding to hourly load variations. These values can help to determine if an industry is suitable for dynamic tariff structures and can adjust its electricity consumption in alignment with grid signals.
7) **Load Factor**: Industries may require tariffs tailored to their specific load factor (the ratio of average load to peak load). This characteristic can significantly impact tariff design, as industries with different load factors may necessitate customized tariff structures to effectively manage their load. Industries with dispersed load levels may benefit from tariffs designed to mitigate energy consumption during peak periods.
8) **Grid Interactions**: Industries equipped with distributed generation resources or energy storage units may require specific tariff considerations. A classification system can facilitate the development of tariffs that promote grid stability or offer incentives for utilizing energy storage, considering the potential for industry interaction with the grid or utilization of local energy storage solutions.
9) **Industry Type**: Various industries possess distinct load profiles based on their operations. Analyzing load data according to specific industry types (e.g., manufacturing, commercial, and healthcare) facilitates understanding of their load patterns. Classification based on industry type enables tailored tariff designs to accommodate the specific characteristics of each industry.
10) **Geographic Location**: Industries located in different regions may require different tariff structures due to variations in regional energy supply and demand. Geographic location plays a crucial role in determining the appropriate tariff structures for industries, ensuring alignment with the energy landscape specific to each region.
11) **Environmental Impact**: Industries prioritizing sustainability may need tariffs that support their environmental objectives.
12) **Regulatory Requirements**: Industries mandated to adhere to energy efficiency or load management regulations may necessitate tailored tariffs to ensure compliance with regulatory standards.
13) **Economic Factors**: Industries with different profit margins may benefit from tariff structures that optimize cost savings while accommodating operational requirements.

By integrating these key metrics with industry-specific insights derived from data and analytics, meaningful classifications can be developed for electricity tariff design that align with distinct consumption patterns and operational characteristics of each industry. The combination and analysis of these criteria enable the creation of tailored tariff categories, addressing the unique needs of various industries. The objective is to design tariffs aimed at promoting energy efficiency, effective load management, and sustainable energy consumption practices. These definitions and examples illustrate the potential for designing appropriate tariffs based on energy consumption patterns and the diverse requirements of industries.

Therefore, using the data and analysis conducted for each industry, specific criteria can be identified to categorize industries effectively. Based on load consumption data over specific time intervals and comprehensive analyses, alongside general knowledge about electricity tariff design and load management, several key criteria or characteristics emerge for classifying industries for the purpose of electricity tariff design. Below, several specific criteria are introduced.

- Maximum demand
  - **Criterion**: The peak load observed during the highest demand period.
  - **Analysis**: Industries with higher peak demand may require tariff structures aimed at reducing load during peak network hours.
- Load variability
  - **Criterion**: Standard deviation of the load, indicating the variability of load values relative to the mean.
  - **Analysis**: Industries exhibiting higher variability may benefit from flexible tariff options to manage their fluctuating energy consumption patterns effectively. Such industries can leverage flexible and time-based tariffs to optimize energy consumption across different intervals.
- Distribution of energy consumption
  - **Criterion**: The spread between the first quartile and the third quartile reflects the dispersion of the load distribution.
  - **Analysis**: A wider distribution, as indicated by the difference between the first quartile and the third quartile, may suggest industries with varied load patterns necessitating relevant tariff plans. Industries exhibiting high load dispersion could benefit from the implementation of diverse tariff structures across various time frames to effectively manage load distribution.
- Sudden load increase during specific time periods (e.g., morning, noon, or night)
  - **Criterion**: The percentage of increase in load from the minimum load within that interval.
  - **Analysis**: Industries experiencing abrupt load spikes during specific time periods may require tariff structures to effectively manage these fluctuations.
- Midday peak ratio
  - **Criterion**: The ratio of the average load value during the midday peak hours (e.g., 12–14) to the average total load consumption.
  - **Analysis**: A higher midday peak ratio suggests the need for tariff structures that accommodate peak load periods during these hours.
- Median load
  - **Criterion**: The median load value observed throughout the time interval.
  - **Analysis**: Median values offer insights into load consumption patterns and aid in designing tariffs relevant to the specific load characteristics of industries.
- Average load
  - **Criterion**: Average of load consumption.
  - **Analysis**: Industries with higher average loads may benefit from tariffs designed to incentivize energy consumption efficiency.

By incorporating such criteria, industries can be categorized into relevant classes for electricity tariff design. Once industries are classified, tariff structures can be designed based on the characteristics of each category to achieve specific objectives, such as enhancing energy efficiency, optimizing load management, and ensuring overall grid stability.

In the subsequent sections, a more detailed examination and analysis of the two industries outlined in this chapter will be attempted. Based on the findings presented in the preceding section, it appears that an industry such as Industry #1, which exhibits no significant dependence between various hours and days, could be categorized to potentially benefit from a range of tariff options. These could include hourly, daily, monthly, or seasonal tariffs, each of which may elicit a distinct response from the industry. However, implementing such tariffs requires a thorough examination of the industry's operational processes to determine its responsiveness and its behavioral modifications.

In the analysis of the load pattern of Industry #2, it is evident that this industry consistently exhibits uniform consumption behavior. Policies aimed at altering the load volume on an hourly basis may not be suitable for this industry. Therefore, industries displaying similar behavior to Industry #2 may benefit more from initiatives targeting changes in daily, weekly, monthly, or seasonal consumption patterns. Consequently, these industries fall into the second category, necessitating a policy distinct from that of the first category to modify their load patterns. However, arriving at a final decision requires a more comprehensive examination of the production process, as well as an assessment of the industry's responsiveness and its approach to reaction and modifications.

For instance, some recommendations for Industries #1 and #2 are outlined below.

- Suggestions for Industry #1
  - **TOU Tariff**: Given the fluctuation in load throughout the day, implementing a TOU tariff could be beneficial. TOU pricing involves applying different rates during peak, shoulder, and off-peak hours. This incentivizes the industry to move energy-intensive activities to off-peak hours, when electricity costs are lower, resulting in potential cost savings.
  - **Energy Efficiency Measures**: Conducting an energy audit to pinpoint areas for improvement and implementing energy efficiency measures can result in long-term cost savings and environmental benefits.
  - **Development and Utilization of Renewable Energy Sources**: Investing in and harnessing renewable energy sources like solar or wind energy can yield long-term cost savings and environmental advantages, including reduced carbon emissions.
  - **Deployment of Electrical Energy Storage Resources**: Implementing energy storage systems to store surplus energy during off-peak periods and utilize it during peak hours can aid in load management.
  - **Adoption of Smart Grid and Energy Management Systems**: Deploying smart grid technologies and energy management systems that offer real-time energy consumption data and enable automated load management can optimize energy usage.

    Overall, a holistic approach encompassing tariff design, load management strategies, and energy efficiency initiatives can facilitate optimization of energy consumption. Nevertheless, conducting a comprehensive analysis of industry-specific requirements, objectives, and limitations is imperative to devise optimal solutions.
- Suggestions for Industry #2
  - **DR Programs**: Financial incentives could be implemented to encourage the industry to reduce electricity consumption during peak periods through various DR programs. Automatic systems could be utilized to notify the industry about peak load periods and offer corresponding incentives for load reduction. By incentivizing load shifting and optimizing the timing

of power-intensive processes, the industry can engage in energy-intensive operations during off-peak periods.
- **Energy Efficiency**: Conducting energy audits and implementing policies to enhance efficiency can effectively reduce overall electrical energy consumption.
- **Development and Use of Renewable Energy Sources**: Investing in and utilizing renewable energy sources can not only reduce energy costs but also offer environmental benefits.
- **Energy Storage Utilization**: Investment in energy storage equipment, such as batteries, enables the storage of electrical energy during off-peak hours for later use during peak hours, facilitating peak load reduction and consumption management.
- **Smart Grid and Energy Management Systems**: Leveraging smart grid infrastructure and energy management systems can optimize load management.

In addition to the aforementioned measures, collaborative initiatives, such as partnerships between industries or regional utilities, aimed at sharing information and implementing load management strategies, can significantly enhance the effectiveness of load management practices and overall energy conservation efforts.

The ultimate goal of these recommendations is to optimize energy consumption within the industry, mitigate peak loads, and enhance energy efficiency. Through the implementation of these tariff design and load management programs, the industry can play a vital role in fostering a more sustainable and resilient energy system while potentially realizing reductions in operating costs.

4.4 Discussion and Conclusions

In recent years, the rapid growth in electric energy consumption has led to an imbalance between production and consumption in some electric power systems, highlighting the urgent need for demand-side management measures. Besides infrastructure development for energy generation, transmission, and distribution progress, demand-side management offers a more immediate solution to control consumption and manage peak loads. This chapter focuses on energy management for industrial customers, aiming to analyze available data to understand consumption patterns and propose efficient programs to modify these patterns.

It's crucial to approach this task with careful examination and strategic decision-making, encompassing accurate problem modeling, definition of objective functions, and consideration of technical and non-technical limitations. Beyond technical aspects, different aspects such as social, legal, and economic factors must also be taken into account, along with the potential side effects of decisions. For instance, changes in electricity tariffs of industries could impact production levels, potentially leading to market shortages, social discontent, or threats to public health if vital products are affected.

To ensure that modifying a customer's electricity consumption pattern doesn't inadvertently harm other areas, the problem and its consequences should be comprehensively understood and modeled as constraints. This necessitates leveraging the knowledge and expertise of experts across various fields.

Furthermore, the question arises whether uniform decisions can be made for all industrial customers or if different decisions and tariffs should be applied based on diverse consumption patterns. Given variations in production processes and other influencing factors, implementing various programs to modify consumption on hourly, daily, weekly, monthly, or seasonal bases for different customer types may be more effective.

Similarly, it is essential to consider whether decisions for each industry should remain consistent across all companies within the same sector or if variations are necessary for different companies within that sector. For instance, this could be particularly relevant for diverse companies within the steel production industry. The legality and social acceptability of making and implementing different decisions for each company or industry should also be examined, alongside potential objections from customers.

In addition to the aforementioned considerations, it is crucial to carefully plan how decisions will be implemented. Will the findings from initial studies be applied to all industrial customers, or will specific industries be selected as case studies initially? If the latter, what criteria will be used to select pilot industries? One approach could involve selecting industries with high load intensity, as they would likely have a significant impact on the national level. While this choice may yield observable impacts and facilitate impact assessments, it is essential to pay attention to the sensitivity and importance of these industries. Improper implementation or errors in analysis could have far-reaching consequences across technical and non-technical dimensions in a society. Alternatively, opting for smaller industries with lower energy intensity may entail less sensitivity to results in terms of technical and non-technical aspects. Additionally, initial implementation steps may have a smaller impact, but potential errors or deficiencies in analysis or policy formulation may also have fewer adverse effects. Ultimately, the decision on whether to focus on larger or smaller industries as case studies will depend on balancing the need for impactful results (with the risks associated with sensitive industries) and the potential consequences of errors.

A data-driven approach is essential for gaining insights into diverse companies and industries, forming the basis for developing suitable categorizations. Subsequently, a thorough examination of the operational processes within each company or industry is imperative to refine these categories and ensure alignment with previous analyses and findings.

Moreover, a meticulous modeling of price elasticity is essential to accurately capture the responsiveness of energy consumers in response to tariffs. Furthermore, a holistic modeling of the problem, encompassing both technical and non-technical constraints, is essential to form strategic decisions regarding tariff modifications for different industries. The goal is to guide consumption patterns to harmonize with the fundamental objectives of the national power system, all while ensuring that any potential consequences in other areas remain manageable and within control.

This chapter outlines fundamental procedures and innovative approaches of data mining as foundational steps in tariff-based energy management. It begins with an exploration of the concepts and importance of data mining, followed by an overview of the data mining process, encompassing tools and methods. Additionally, it delves into energy management and the pivotal role of data mining within this domain.

Furthermore, fundamental steps toward addressing the complexities of industrial energy management and tariff regulation are explained while pinpointing forthcoming challenges and inquiries. Understanding the distinction between companies and industries is crucial, as it provides the cornerstone for constructing effective solutions.

Consequently, the chapter also delves into the analysis of load values from two industries, offering insights into their energy consumption patterns. Several indicators and methodologies are introduced to enable comparison and categorization of energy consumption. The analysis reveals potential categories applicable to different industries. Additionally, the aim of these analyses is to explore consumption patterns and assess the feasibility of modifying these behaviors across various timeframes, including hourly, daily, weekly, monthly, or seasonal. For instance, the analysis revealed that the consumption pattern of Industry #1 exhibits some dispersion and lacks consistency across different hours or days. Conversely, Industry #2 demonstrates a relatively uniform consumption pattern. Accordingly, strategies for modifying consumption patterns for Industry #2 may need to extend beyond hourly adjustments, potentially incorporating programs with longer-term impacts, such as weekly or monthly initiatives.

Based on the discussions presented in this chapter, it is evident that modifying the tariff structures for each industry requires a comprehensive understanding of their consumption and demand levels. This involves utilizing various analytical techniques outlined herein to categorize industries effectively. Subsequently, it is essential to pinpoint the overarching requirements for each category, guiding the formulation of policies tailored to address them effectively. However, final decisions should consider a multitude of technical and non-technical dimensions, including social, economic, and legal factors, as well as restrictions imposed by other entities to prevent adverse market impacts such as inflation or product shortages.

Ultimately, such a process leads to the formulation of an optimization problem, wherein the objective function aligns with the desired goals of consumption pattern modification and tariff design. This optimization problem encompasses a wide array of constraints, covering diverse technical and non-technical dimensions. With such a thorough analysis, it becomes viable to enact measures aimed at tariff modification, consumption pattern adjustments, and risk mitigation across diverse technical and non-technical dimensions.

References

1 Han, J., Kamber, M., and Pei, J. (2011). *Data Mining: Concepts and Techniques.* Morgan Kaufmann.

2 Tan, P.N., Steinbach, M., and Kumar, V. (2014). *Introduction to Data Mining*. Pearson.

3 Aggarwal, C.C. (2015). *Data Mining: The Textbook*. Springer.

4 Kotu, V. (2015). Chapter 2: Data mining process. In: *Predictive Analytics and Data Mining, Concepts and Practice with Rapidminer* (ed. V. Kotu and B. Deshpande), 17–36. Elsevier https://doi.org/10.1016/B978-0-12-801460-8.00002-1.

5 Chapman, P., Clinton, J., Kerber, R. et al. (2000). CRISP-DM 1.0: step-by-step data mining guide. SPSS Inc. https://www.kde.cs.uni-kassel.de/wp-content/uploads/lehre/ws2012-13/kdd/files/CRISPWP-0800.pdf.

6 SAS Institute. (2013). *Getting Started with SAS Enterprise Miner 12.3.* (pp. 1–3) SAS Institute. Shearer, C. (2000). The CRISP-DM model: the new blueprint for data mining. *Journal of Data Warehousing*, 5(4), 13–22.

7 Kubiak, T. and Benbow, D.W. (2005). *The Certified Six Sigma Black Belt Handbook*. Milwaukee, WI: ASQ, Quality Press.

8 Fayyad, U., Piatetsky-Shapiro, G., and Smyth, P. (1996). From data mining to knowledge discovery in databases. *AI Magazine* 17 (3): 37–54.

9 Piatetsky, G. (2014). KDnuggets 15th annual analytics, data mining, data science software poll: rapidminer continues to lead. http://www.kdnuggets. com/2014/06/kdnuggets-annual-software-poll-rapidminer-continues-lead.html (retrieved 1 August 2014).

10 Papakyriakou, D. and Barbounakis, I.S. (2022). Data mining methods: a review. *International Journal of Computer Applications* 183 (48): 5–19.

11 Arasteh, H., Vahidinasab, V., Sepasian, M.S., and Ghaderi, A. (2019). Integrated generation, transmission and energy efficiency planning. *34th International Power System Conference (PSC)*, Tehran, Iran (9–11 December 2019), pp. 571–578. IEEE. https://doi.org/10.1109/PSC49016.2019.9081514.

12 Liu, X., Zhao, T., Deng, H. et al. (2023). Microgrid energy management with energy storage systems: a review. *CSEE Journal of Power and Energy Systems* 9 (2): 483–504.

13 Chen, C. (2023). Framing energy-efficiency programs: a survey experiment. *Energy Policy* 183: 113776.

14 Hamidreza Arasteh, Mohammad Sadegh Sepasian, Vahid Vahidinasab, "An aggregated model for coordinated planning and reconfiguration of electric distribution networks", *Energy*, Volume 94, pp. 786-798, 1 January 2016.

15 Mohammadreza Shekari, Hamidreza Arasteh, Alireza Sheikhi Fini, Vahid Vahidinasab, "Demand response requirements from the cultural, social, and behavioral perspectives", *Applied Sciences*, Volume 11, Issue 23, pp. 1-23, 3 December 2021, 11456, https://doi.org/10.3390/app112311456.

16 Arasteh, H.R., Moghaddam, M.P., Sheikh-El-Eslami, M.K., and Abdollahi, A. (2013). Integrating commercial demand response resources with unit commitment. *International Journal of Electrical Power and Energy Systems (IJEPES)* 51: 153–161.

17 Hashemi, S.M., Arasteh, H., Shafiekhani, M. et al. (2023). Multi-objective operation of microgrids based on electrical and thermal flexibility metrics using the NNC and IGDT methods. *International Journal of Electrical Power and Energy Systems (IJEPES)* 144: 1, 108617–16. https://doi.org/10.1016/j.ijepes.2022.108617.

18 Hamidreza Arasteh, Niki Moslemi, S. Mohsen Hashemi, "Chapter 7: Demand response measurement and verification approaches: analyses and guidelines," In *Industrial Demand Response: Methods, Best practices, Case Studies, and Applications* (ed. Hassan Haes Alhelou, Antonio Moreno-Muñoz, Pierluigi Siano), pp. 133-161, 30 June 2022, The Institution of Engineering and Technology (IET), https://doi.org/10.1049/PBPO215E_ch7, ISBN: 9781839535611, e-ISBN: 9781839535628.

19 Khan, H.A.U., Ünel, B., and Dvorkin, Y. (2023). Electricity tariff design via lens of energy justice. *Omega* 117: 102822.

20 Kia, M., Shafiekhani, M., Arasteh, H. et al. (2020). Short-term operation of microgrids with thermal and electrical loads under different uncertainties using information gap decision theory. *Energy* 208: 1. https://doi.org/10.1016/j.energy.2020.118418.

21 Rahman, T., Othman, M.L., Noor, S.B.M. et al. (2024). Methods and attributes for customer-centric dynamic electricity tariff design: a review. *Renewable and Sustainable Energy Reviews* 192: 114228.

22 Tlenshiyeva, A., Tostado-Véliz, M., Hasanien, H.M. et al. (2024). A data-driven methodology to design user-friendly tariffs in energy communities. *Electric Power Systems Research* 228: 110108.

23 Ruan, G., Dawei Qiu, S., Sivaranjani, A.S.A., and Awad, G.S. (2024). Data-driven energy management of virtual power plants: a review. *Advances in Applied Energy* 14: 100170.

24 Yassim, H.M., Abdullah, M.N., Gana, C.K., and Ahmed, A. (2024). A review of hierarchical energy management system in networked microgrids for optimal inter-microgrid power exchange. *Electric Power Systems Research* 231: 110329.

25 Razieh Darshi, Saeed Shamaghdari, Aliakbar Jalali, Hamidreza Arasteh, "Decentralized energy management system for smart microgrids using reinforcement learning", *IET Generation, Transmission & Distribution*, Volume 17, pp. 2142-2155, 23 February 2023. https://doi.org/10.1049/gtd2.12796.

26 Zheng, Z., Shafique, M., Luo, X., and Wang, S. (2024). A systematic review towards integrative energy management of smart grids and urban energy systems. *Renewable and Sustainable Energy Reviews* 189 (Part B): 114023.

27 Darshi, R., Shamaghdari, S., Jalali, A., and Arasteh, H. (2023). Decentralized reinforcement learning approach for microgrid energy management in stochastic environment. *International Transactions on Electrical Energy Systems* 2023: 1, 1190103–15.

5

Data-Driven Tariff Design for Equitable Energy Distribution

Salah Bahramara[1], Hamidreza Arasteh[2,3], Asrin Seyedzahedi[4], and Khabat Ghamari[5]

[1] *Department of Electrical Engineering, Sa. C., Islamic Azad University, Sanandaj, Iran*
[2] *Power Systems Operation and Planning Research Department, Niroo Research Institute, Tehran, Iran*
[3] *Center for Research on Microgrids (CROM), Huanjiang Laboratory, Zhuji, Shaoxing, Zhejiang, China*
[4] *Department of Science, Campus of Bijar, University of Kurdistan, Bijar, Iran*
[5] *Iran Meteorological Organization, Tehran, Iran*

5.1 Introduction

The main role of the power systems is to meet the electrical energy demand of the consumers with high quality and reliability. From the viewpoint of the electrical energy price, the consumers prefer to purchase electrical energy at the minimum price, and the producers prefer to sell their generated electrical energy at the maximum price. In the market environment, the price can be determined regarding the bids/offers of these players considering the technical constraints of the networks. The market mechanism is implemented in the transmission networks in the most developed and even developing counties. The major problem is related to the consumers with low power capacity located in the distribution networks. It should be noted that in most countries, the consumers with medium power capacities, e.g., 1 or 0.5 MW, can purchase their required energy from the market. For other consumers, there are three main solutions. The first one is to design a local energy market similar to the wholesale market where the consumers can purchase their required energy from this market. The different conceptual frameworks are proposed for this market while it is not implemented in the real world broadly. The performance of the local energy market is simulated through proposing a software [1] in which the behavior of the market players and the operation of the distribution networks are modeled. A survey and analysis is done [2] to evaluate the performance of the blockchain-based local energy markets.

The second solution is supplying the demand of the consumers through different retailers. In this case, the consumers have different options to manage their billing energy cost. This solution is being implemented in the most developed countries. A review is done [3] on the different aspects of the consumers' participation in the retail electrical energy market. The last solution is selling energy to the consumers by the distribution company, which is being implemented in developing countries such as Iran. It should be noted that, in the first and the second solutions, the electrical energy price is very close to the real values. This is while, in the third solution, the electrical energy price is determined by the government, and some discounts may be considered in these values. The aim of this chapter is to design the equitable electrical energy price tariff for the consumers of the electrical distribution networks in the third solution.

Data-Driven Energy Management and Tariff Optimization in Power Systems: Shaping the Future of Electricity Distribution through Analytics, First Edition.
Edited by Hamidreza Arasteh, Pierluigi Siano, Niki Moslemi, and Josep M. Guerrero.

5.1.1 Literature Review and Contributions

The most popular electricity tariffs used in the world are fixed, time-of-use (TOU), and real-time tariffs. These tariffs are used considering different prices for the residential, commercial, and small industrial consumers in different counties. The real-time tariffs are mostly used for the industrial loads, and they cannot be used broadly for the residential consumers. The effect of the different electricity tariffs on the energy management of consumers is investigated in [4]. In this review paper, 22 electricity tariffs used in different counties are discussed, where these tariffs are divided into the energy-based and power-based, time-dependent and time-independent ones, and it is mentioned that implementing any of these tariffs requires smart meters. The 29 dynamic distribution tariffs are analyzed [5], considering the technical, economic, social, and regulatory indices. These tariffs are mostly based on real-time and TOU strategies, and they are considered for both the distributed energy resources and electrical consumers. The effect of the TOU tariff on the energy management of the residential consumers in Indonesia is addressed [6].

In the several studies, it is mentioned that there are two main factors that affect the energy consumption of the residential consumers, i.e., the residential consumers' behavior and the weather conditions. The effect of the weather conditions on the electrical energy consumption of the residential consumers is investigated [7] through analyzing the data of 3827 consumers' smart meters in Ireland. The effect of the thermal comfort of the residential consumers on their energy consumption is analyzed through a comprehensive review [8]. In this study, both the conventional comfort model and modern approaches such as machine learning are discussed. The authors [9] analyzed mathematical models for forecasting the electrical energy consumption of the residential consumers considering the ambient temperature. In this study, a monthly based data related to the energy consumption and the temperature is used in Ukraine. The effect of climate change on both heating and cooling energy demand of residential consumers is investigated through reviewing several studies [10, 11]. A machine learning-based approach is employed [12] to model the effect of the residential consumers on the cooling electrical demand. The effect of climate change in the future on the cooling and heating energy consumption of the residential consumers in Ontario is investigated [13]. The results of this research showed that the need of the residential consumers for heating energy will decrease in the future, while the cooling energy demand is increasing regarding the climate change scenarios.

There are few studies that address the design of the electricity tariff considering the equitable indices in the distribution networks. A comprehensive review of the previous studies is done [14] on the equitable electricity tariff at the distribution network level considering the renewable energy sources. Although the concept of equitable electricity tariffs is investigated [14], this issue is mostly used for renewable energy resources instead of residential consumers.

Reviewing the previous studies shows that coupling the present TOU tariffs with the weather conditions can lead to designing equitable electrical tariffs for the residential consumers. In fact, when the government sells energy to the consumers with some discounts, the main difference among these consumers is their locations, which leads to their different cooling requirements of electrical energy demand regarding the weather conditions. For this purpose, in this chapter, a new framework is proposed to design the equitable electricity tariffs for the residential consumers. In this framework, at first, the real data related to the temperature and humidity is collected, regarding which the heat index (HI) is calculated for all considered regions in a country. Then, the relation between the cooling energy demand and the HI is obtained, and consequently the cooling required energy demand for all regions is obtained. Regarding the obtained results, the electrical distribution companies can divide the residential consumers into several zones, and they can also calculate the required cooling electrical demand for each zone. Since this framework is mainly based on analyzing the big data related to the weather conditions, this framework is named as a data-driven approach for designing the equitable electricity tariff. The effectiveness of the proposed approach in this chapter is investigated through applying this method to the residential consumers of Iran. Therefore, the main contributions of this chapter are as follows:

- Proposing a data-driven approach to model the effect of the weather conditions on the cooling electrical energy demand of the residential consumers.
- Dividing the residential consumers of a country into some zones regarding the HI and the electrical required cooling demand.

5.1.2 Chapter Organization

In the next section, the proposed approach in this chapter and its related formulations are described. The case study is described in Section 5.3. The simulation results are described with some discussions in Section 5.4 and the conclusions and future works are presented in the final section.

5.2 Proposed Approach and Formulations

The proposed approach in this chapter to design the equitable electrical energy tariff is shown in Figure 5.1. To design the equitable energy tariff in a country, the required energy

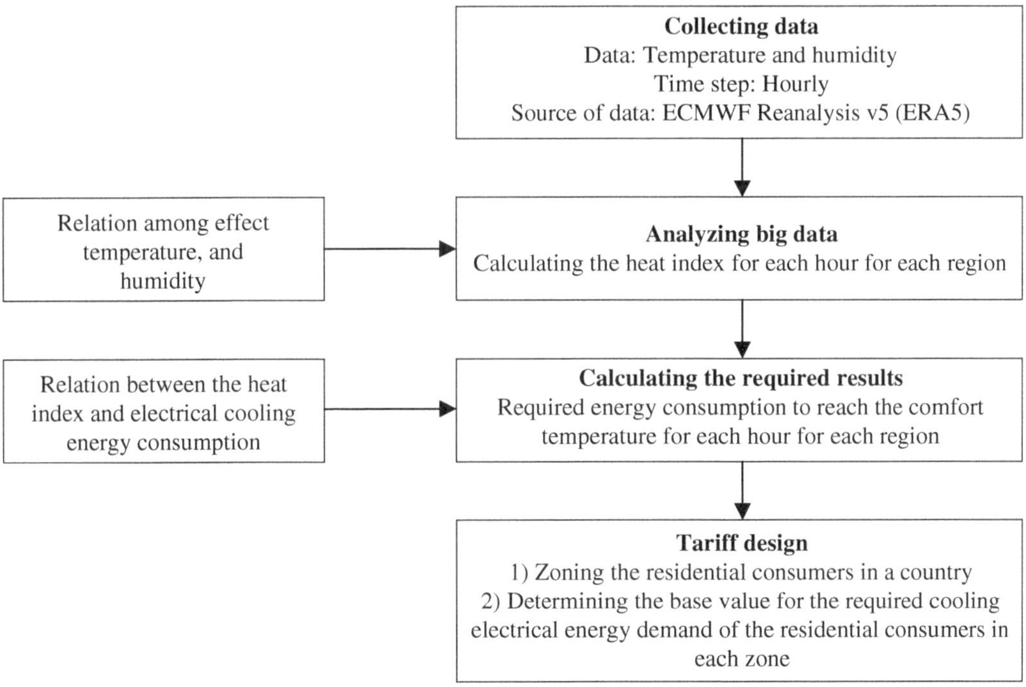

Figure 5.1 Proposed approach in this chapter to design equitable electricity tariffs regarding the data-driven approach.

consumption of the residential consumers in each region should be calculated. This parameter is considered the base value to separate the consumers from each other and to assign the pre-defined tariff for them. For this purpose, two steps need to be implemented. First, the relation between weather conditions of different regions in that country and the comfort temperature of the residential consumers must be analyzed. For this purpose, there are two main approaches, including the HI and apparent temperature. Then, the cooling required energy demand of each region is calculated regarding the obtained results from these approaches.

In this chapter, the HI is used, which this index shows the actual feeling of humans from temperature in which the effect of the humidity on the temperature is considered. This index is appropriate for temperatures between 26 and 53 °C and humidity between 40% and 100%. This index is calculated as given in Eq. (5.1) [15],

$$\mathrm{HI} = \alpha_1 + \alpha_2 T + \alpha_3 H_R + \alpha_4 H_R T + \alpha_5 T^2 + \alpha_6 H_R^2 + \alpha_7 H_R T^2 + \alpha_8 T H_R^2 + \alpha_9 T^2 H_R^2 \tag{5.1}$$

where

α_1–α_9 arc cocfficients whose values are presented in [15]
T is the temperature (°C)
H_R shows the relative humidity (%)

To calculate the HI, two parameters consisting of 2 m of dry temperature and relative humidity are required. The relative humidity can be calculated regarding the 2 m dry temperature and dew point temperature. This data can be extracted from ERA5, which is the fifth-generation ECMWF atmospheric reanalysis of the global climate. Then, the HI can be calculated for each region and each time step using Eq. (5.1).

In the next step, the required cooling electrical energy consumption of residential consumers in each region needs to be calculated. The real results are obtained by installing the equal cooling devices in each region and obtaining the relation between the HI and the electrical energy consumption of each region. Since this approach is very hard and sometimes impossible, in this chapter, the real data reported [16] is used. In this reference, the relation among temperature, relative humidity, and cooling electrical energy consumption of residential consumers in the United States (US) is calculated. Regarding these results, the HI is calculated, and the relation between HI and electrical energy consumption is obtained as Figure 5.2. As shown in this figure, the coefficient of determination, or R-squared, is 0.982, which shows the obtained Eq. (5.2) can properly model the relation of the HI and electrical cooling demand.

$$E = 0.0034\mathrm{HI}^3 - 0.9089\mathrm{HI}^2 + 107.57\mathrm{HI} - 2068 \tag{5.2}$$

In this equation, E shows the yearly electrical cooling energy demand. Regarding Eq. (5.2), the required energy consumption of each region to reach the comfort temperature is calculated.

These results will be sent to the tariff design part. In this step, the residential consumers in a country can be divided into several zones regarding the HI and the required cooling electrical energy demand. Also, the base required cooling electrical energy demand of residential consumers in each zone is determined.

5.3 Describing the Case Study

In Iran, the electrical energy price for the consumers of the distribution networks, including residential, agricultural, commercial, and industrial, is determined by the government. Different TOU

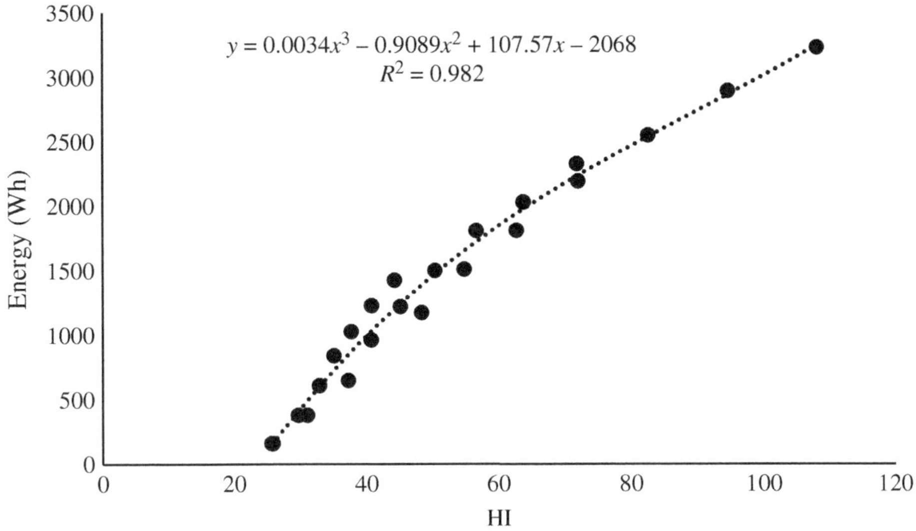

Figure 5.2 The relation between yearly cooling energy consumption and the HI.

tariffs are determined for each type of consumer. For example, some tariffs are defined for the industrial consumers with different values for the electrical energy price and demand. Since the focus of this chapter is the residential consumers, the electricity tariffs defined for these consumers are investigated in this section. Some types of electrical energy tariffs are determined for the residential consumers in different locations of Iran. In each type, a base value is considered for the electrical energy consumption of the consumers in one month, named as "consumption pattern of residential consumers" (CPRC). The type of energy tariff for each consumer in each region is considered regarding the CPRC. When the energy consumption of the consumers in one month (kWh/month) is more than the CPRC, the electrical energy price increases in some steps. Also, to encourage the consumers to shift their energy consumption from the peak hours to the off-peak ones, the incentive/penalty coefficients are considered. Although the consumers in the electrical distribution networks of Iran are divided into some types regarding the weather conditions, it seems that the appropriate values regarding the weather conditions need to be defined for the CPRC. In fact, regarding the climate change in these years, there may be some near regions that are now considered in equal energy tariff type, while with detailed calculation of the effect of the weather conditions, these regions may have different values of CPRC.

The variety of weather conditions in Iran leads to different energy consumption by the consumers in different climates. Therefore, to implement justice among the residential consumers in Iran, the flexible energy tariff in different months needs to be defined by the government. The weather parameters, such as temperature and humidity, have a major effect on the electrical energy consumption of the residential consumers. The consumers in the climate with high temperature and humidity need to increase their energy consumption to reach their comfort temperature. Therefore, the high CPRC may be considered for these consumers. As mentioned before, in this chapter, a new data-driven approach is proposed to design the equitable electrical energy price tariff for the residential energy consumers. This approach is applied in Iran. Since in this framework a relation between the electrical cooling energy consumption, temperature, and humidity is obtained, the CPRC for each region of Iran can be obtained regarding this framework. The CPRC is defined as the required electrical energy consumption of residential consumers in each region to reach the temperature of their homes to comfort temperature.

5.4 Simulation Results

As mentioned before, to investigate the effectiveness of the proposed approach in this chapter, Iran as the test country is considered. The data related to the temperature and humidity is collected for 10 years, from 2014 to 2023, with hourly resolution for Iran. The HI is calculated for all regions of Iran in different months for the years 2014–2023. The results of the HI for January and July, as the warmest month, at 15:30 (local time of Iran) are presented in Figure 5.3. As shown in this figure, for January, the HI for all regions of Iran is below the comfort temperature (25 °C considered in this simulation), and therefore no electrical cooling energy demand is required for all regions. Therefore, it can be concluded that the amount of the CPRC for all regions can be equal in this month. This is while, in July, most of the regions have the HI greater than the comfort temperature so that for some regions in the Southeast, the HI reaches 54 °C. Therefore, these regions need more values for the CPRC this month in comparison with the other regions. The results of the average daily required cooling electrical energy demand for some months for the years 2014–2023 for different regions of Iran are shown in Figure 5.4. As shown in this figure, the cooling electrical energy demand for all regions of Iran is very low in January, as is expected regarding the obtained results for the HI. However, the results are different for July so that in some regions the required daily energy is lower than 6000 Wh, while is some regions, the required energy reaches the values greater than 30,000 Wh. Therefore, in this month, different values of the CPRC must be defined for the regions of Iran to consider the equitable electrical tariffs for the residential consumers.

To better analyze the results and to compare the results of the proposed approach in this chapter with the real case implemented in Iran, the results of the HI and the required cooling electrical energy demand for some cities in Iran are presented in Tables 5.1 and 5.2, respectively. In the present Iran's defined price tariff, Ahvaz, Boshehr, and Bandar Abbas are considered in one region. Although the HI index may seem equal for these cities, the HI for different months and, consequently, the required cooling electrical energy demand in different months are different for these cities, as the results shown in Tables 5.1 and 5.2. The required cooling electrical energy demand for Ahvaz, Boshehr, and Bandar Abbas for August is 694.8, 949.5, and 779.1 kW, respectively. Therefore, it is not equitable to consider equal CPRC for these cities.

Also, in the present Iran's defined price tariff, Gorgan and Zahedan are considered in one region. The required cooling energy demand for these cities for August is 52.6 and 180.3 kW, and it is not equitable to consider equal CPRC for these cities. However, the results presented in these tables show that for some cities, such as Tehran and Sanandaj, which are now considered in a region, the difference in the average daily required energy in different months is very low, and therefore, the equal CPRC can be considered for them.

5.5 Conclusions and Future Works

In this chapter, a data-driven method is developed to design the equitable electrical energy tariffs for residential consumers in all regions of a country. The basic idea of this method is modeling the effect of the weather parameters, temperature and humidity, on the required cooling electrical energy demand of the residential consumers. For this purpose, the HI is calculated regarding which the required demand of the consumers is obtained. There are two main outputs from this approach. First, divide the residential consumers of a country into some zones so that the required cooling energy demand of the consumers in each zone is near to each other, and second, determine the base value for the required cooling energy demand of the consumers in each zone.

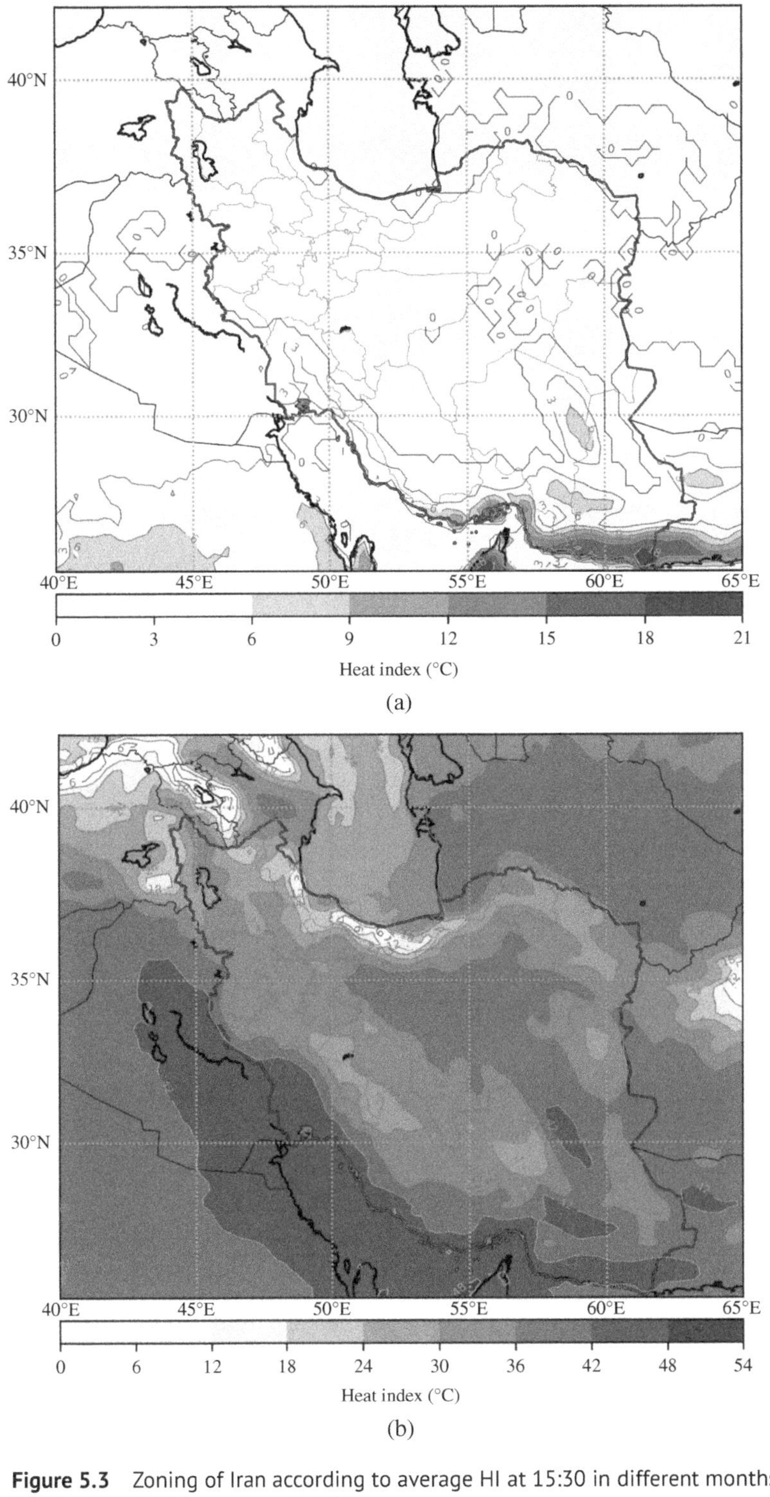

Figure 5.3 Zoning of Iran according to average HI at 15:30 in different months (local time) for the years 2014–2023. (a) January and (b) July.

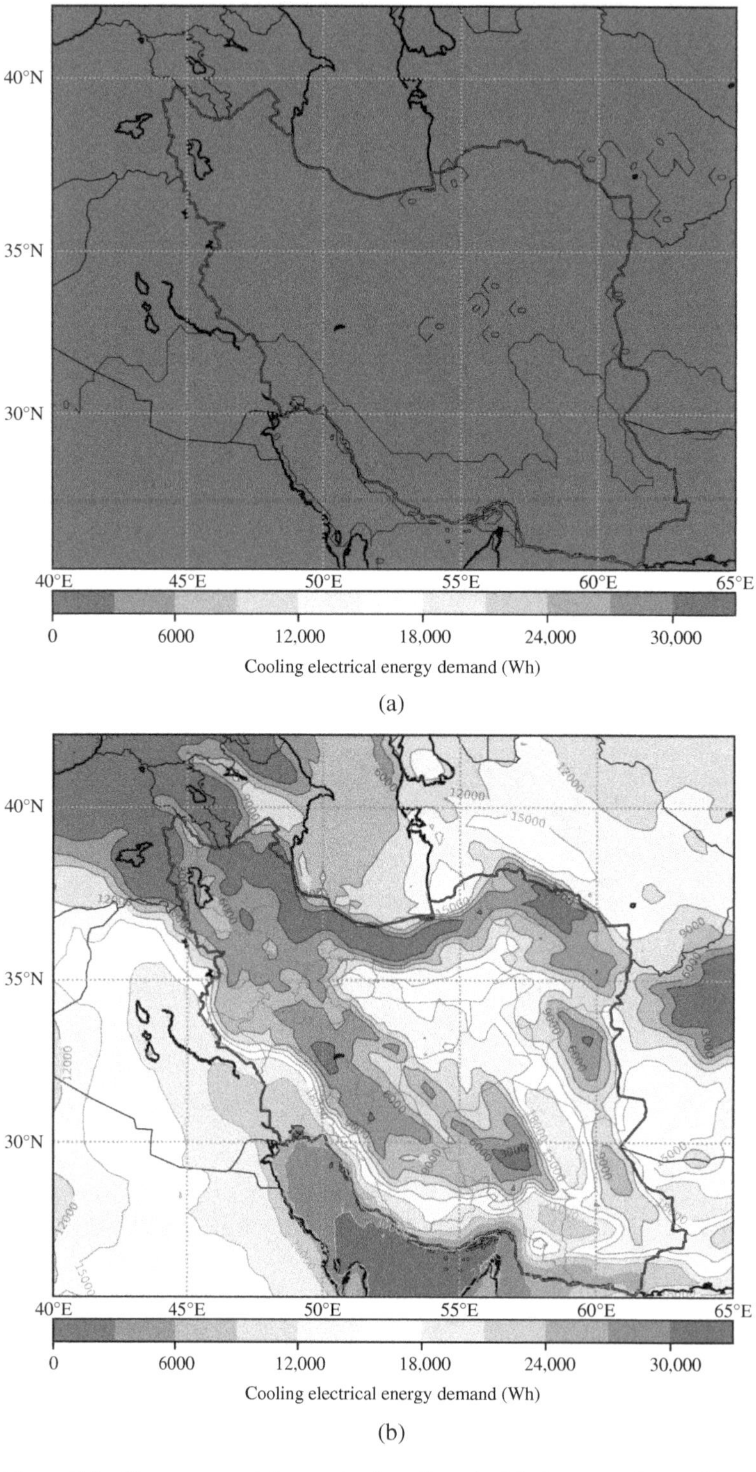

Figure 5.4 The average daily required cooling electrical energy demand for different regions of Iran for the years 2014–2023. (a) January and (b) July.

Table 5.1 Average daily HI (sum of 24 hours) in different months for some cities in Iran.

	January	February	March	April	May	June	July	August	September	October	November	December
Tehran	0	0	0	3.21	98.43	396.83	539.2	440.07	235.26	11.28	0	0
Sanandaj	0	0	0	2.29	66.25	300.13	438.37	417.03	237.2	23.99	0	0
Ahvaz	2.67	33.35	155.58	429.9	729.46	862.54	931.77	939.13	817.18	536.51	137.93	11.65
Booshehr	0.08	1.39	25.8	292.52	738.72	891.2	1047.1	1101.4	990.03	771.98	229.92	4.13
Bandar Abbas	1.25	31.73	138.25	511.29	773.56	924.3	1013.2	986.32	904.18	734.70	295.82	55.73
Kazron	0.74	7.88	60.5	261.6	509.78	696.5	800.87	768.90	588.04	363.81	67.65	8.52
Gilan-e Gharb	0	0	0.49	25.05	188.64	422.3	586.48	568.76	380.34	108.52	0.25	0
Gorgan	0	0	0	1.72	26.917	116.38	183.08	171.36	61.98	4.3	0	0
Zahedan	0	1	29.39	161.48	332.84	543.51	606.59	458.98	309.78	127.33	7.45	1.06

Table 5.2 The average daily required electrical cooling energy in different months for some cities in Iran (Wh).

	January	February	March	April	May	June	July	August	September	October	November	December
Tehran	0	0	0	17	750	4312	6613	4848	2050	68	0	0
Sanandaj	0	0	0	12	477	3335	6212	5627	2428	155	0	0
Ahvaz	17	241	1476	5912	13103	183,84	22068	22,414	16,677	8570	1434	87
Booshehr	1	9	206	2964	12233	20,195	28076	30,629	25,217	14,190	2420	34
Bandar Abbas	8	233	1284	6463	13794	21940	265,22	25,132	20,852	12,471	2967	411
Kazron	4	52	494	2784	7466	12284	15380	14,668	10,303	4694	538	50
Gilan-e Gharb	0	0	3	193	1928	5526	8983	8765	4703	918	1	0
Gorgan	0	0	0	11	239	1136	2010	1696	549	30	0	0
Zahedan	0	5	248	1467	3717	7403	8437	5816	3547	1069	45	5

This approach is applied to Iran as the test case. The results show the minor difference among required cooling electrical demand of the consumers in different regions of Iran in some months, such as January. However, the major difference is obtained for other months, such as July, in which the required cooling electrical energy demand for some regions is five times that of some other ones. Also, to investigate the equitable index in the present defined tariffs in Iran, the results are presented for some cities that are located in a zone. The results show that for cities with high HI, such as Ahvaz, Boshehr, and Bandar Abbas, the present CPRC considered for them is not equitable. This is while, for the cities with low HI, such as Sanandaj and Tehran, the considered CPRC is appropriate.

The proposed approach in this chapter can be extended as follows:

- In this chapter, the heat index is used to calculate the effect of the humidity on the temperature. In future works, the other methods, such as apparent temperature, can be used.
- The effect of the other parameters, such as the economic conditions of the people in each city, on determining the base value of CPRC for each region can be considered.
- To design the equitable electrical energy tariff, the effect of the energy demand of the residential consumers in the peak period on the penalty coefficient of the peak prices can be modeled. This is while, in the present tariffs, the fixed coefficients are considered to penalize the energy demand of the consumers in peak periods.
- The relation between the HI and cooling electrical energy demand can be obtained through the real data of Iran.

References

1 Essayeh, C. and Morstyn, T. (2024). OPLEM: open platform for local energy markets. *Applied Energy* 373: 123848.

2 Okwuibe, G.C., Brenner, T., Tzscheutschler, P., and Hamacher, T. (2023). Survey and analysis of local energy markets based on distributed ledger technologies. *IEEE Access* 11: 23771–23791.

3 Hampton, H., Foley, A., Del Rio, D.F. et al. (2022). Customer engagement strategies in retail electricity markets: a comprehensive and comparative review. *Energy Research & Social Science* 90: 102611.

4 Zaki, D.A. and Hamdy, M. (2022). A review of electricity tariffs and enabling solutions for optimal energy management. *Energies* 15: 8527.

5 Christensen, K., Ma, Z., and Jørgensen, B.N. (2021). Technical, economic, social and regulatory feasibility evaluation of dynamic distribution tariff designs. *Energies* 14: 2860.

6 Yuniarto, H.A., Normasari, N.M.E., Silalahi, S.F. et al. (2024). Customer behaviour towards energy usage with time of use tariff: a systematic literature review. *Journal of Revenue and Pricing Management* 23: 44–61.

7 Kang, J. and Reiner, D.M. (2022). What is the effect of weather on household electricity consumption? Empirical evidence from Ireland. *Energy Economics* 111: 106023.

8 Arowoiya, V.A., Onososen, A.O., Moehler, R.C., and Fang, Y. (2024). Influence of thermal comfort on energy consumption for building occupants: the current state of the art. *Buildings* 14: 1310.

9 Stanytsina, V., Zaporozhets, A., and Artemchuk, V. (2024). Demand forecasting mathematical models for residential electricity consumption considering ambient temperature. *Nexus of Sustainability: Understanding of FEWSE Systems I* 127–145.

10 Li, Y., Wang, W., Wang, Y. et al. (2021). A review of studies involving the effects of climate change on the energy consumption for building heating and cooling. *International Journal of Environmental Research and Public Health* 18: 40.

11 Campagna, L.M. and Fiorito, F. (2022). On the impact of climate change on building energy consumptions: a meta-analysis. *Energies* 15: 354.

12 Amasyali, K. and El-Gohary, N. (2021). Machine learning for occupant-behavior-sensitive cooling energy consumption prediction in office buildings. *Renewable and Sustainable Energy Reviews* 142: 110714.

13 Berardi, U. and Jafarpur, P. (2020). Assessing the impact of climate change on building heating and cooling energy demand in Canada. *Renewable and Sustainable Energy Reviews* 121: 109681.

14 Ansarin, M., Ghiassi-Farrokhfal, Y., Ketter, W., and Collins, J. (2022). A review of equity in electricity tariffs in the renewable energy era. *Renewable and Sustainable Energy Reviews* 161: 112333.

15 National Weather Service (2022). Weather prediction center. https://www.wpc.ncep.noaa.gov/html/heatindex_equation.shtml (accessed August 2024).

16 Thunder Said Energy. Air conditioning: energy consumption? https://thundersaidenergy.com/downloads/air-conditioning-energy-consumption/ (accessed August 2024).

6

Applying Artificial Intelligence to Improve the Penetration of Renewable Energy in Power Systems

Abbas Marini and Seyed Mohsen Hashemi

Power System Operation and Planning Research Group, Niroo Research Institute (NRI), Tehran, Iran

6.1 Introduction

Global demand for energy is rapidly increasing, and energy systems must be upgraded to improve efficiency, flexibility, and sustainability. The smart grid integrates digital information and communication technologies (ICT) with traditional electrical grid systems. It is characterized as a network of interconnected, self-sufficient systems that support the efficient combination of conventional and renewable energy sources (RES). This integration helps lower costs while delivering safe, reliable, high-quality, and sustainable electricity to end users. Smart grids, supported by digital measurement and communication systems, incorporating smart meters, intelligent devices, renewable energy sources, and energy-efficient resources, allow for a two-way exchange of both electrical energy and data. The bidirectional communication between energy producers and consumers, along with the flow of power, improves the security, reliability, and efficiency of the system.

Smart grids are characterized by advanced measurement, control, and communication infrastructure, which are interconnected and integrated. This results in the production of large volumes of high-dimensional data in diverse formats. Such large volumes of data require efficient collection, storage, and management. The enormous amount of data available through measurement systems provides significant potential for managing the system and conducting operational and planning processes. However, traditional methods and technologies used in power grids face many limitations in processing and utilizing this data effectively. Therefore, implementing artificial intelligence (AI) techniques, particularly machine learning (ML), in various applications within the smart grid has become essential. These methods provide an effective approach to analyzing data, extracting insights, and making informed decisions to maintain the grid's desired performance. AI and ML have significant potential to revolutionize the energy sector, acting as the intelligence core of the smart grid. They enable continuous data collection and integration while making prompt decisions to improve the grid's sustainability and reliability [1–3].

Conversely, the electric power industry is increasingly transitioning to RES due to their cost efficiency and their crucial contribution to lowering carbon emissions. RES include solar, wind, biomass, geothermal, hydro, and ocean energy, among others. Solar and wind energy are generated by means of technologies such as concentrated solar power (CSP), photovoltaic (PV) panels, and wind turbines. These technologies are being utilized in many parts of the world.

Solar panels generate electricity through solar radiation, while wind turbines generate energy using wind energy. The main features of these energy sources include imperfect controllability

Data-Driven Energy Management and Tariff Optimization in Power Systems: Shaping the Future of Electricity Distribution through Analytics, First Edition.
Edited by Hamidreza Arasteh, Pierluigi Siano, Niki Moslemi, and Josep M. Guerrero.

and predictability, as well as the variability and intermittency of output power. This is because the power generated from RES is entirely dependent on environmental factors such as solar radiation, temperature, humidity, and wind speed. For instance, PV panels generate more energy when there is a clear sky and produce minimal energy under cloudy skies or at night. Conversely, wind turbines generate minimal energy when wind speed is lower than the cut-in speed and above the cut-out speed. Thus, the significant fluctuations in electricity production from PV and WT power plants create several challenges, including voltage instabilities and issues with power storage and distribution. The integration of RES complicates the operation of the electric grid and creates challenges in preserving the balance between energy production and consumption. Thus, precise prediction of energy production from RES and electrical load demand is a crucial need in the current era of power grids. Reliable prediction of energy production on one side and load/demand on the other side enables electricity companies to effectively manage demand-based supply and generate excess power from conventional resources when RESs are incapable of meeting consumer demand.

Accurately predicting electricity generation from RES presents a significant challenge, as it is heavily influenced by weather conditions (e.g., humidity, temperature, solar radiation, etc.). Additionally, electricity demand from consumers is unpredictable, varying due to factors like environmental changes and individual preferences. Unlike conventional simulation methods typically used in power grid management, forecasting can be enhanced through physical models and AI techniques, such as ML and deep learning (DL). Over the last ten years, ML and DL methods have been widely utilized across various domains of computational intelligence and predictive analytics. In fact, the main advantage of these new approaches in managing RES integration is the use of the massive volumes of various data collected within the smart grid to address the challenges posed by RES. This approach contrasts with probabilistic simulations, which require precise computational models and complete information about operational or planning issues [4–6].

It is important to recognize that each forecasting method comes with its own set of strengths and limitations. For instance, physical weather forecasting models are useful for predicting atmospheric dynamics but often demand substantial computational power, as they require vast amounts of data for calibration. On the other hand, ML/DL-based forecasting methods typically offer more precise predictions compared to physical models, thanks to their advanced data mining techniques and ability to extract relevant features from large datasets. This advantage positions these approaches as a promising future for supporting RES integration into smart grids. This potential has been demonstrated in Figure 6.1, where ML and DL approaches, by means of data collected from smart grid measurements, assist in addressing the challenges of the integration of RES in power systems.

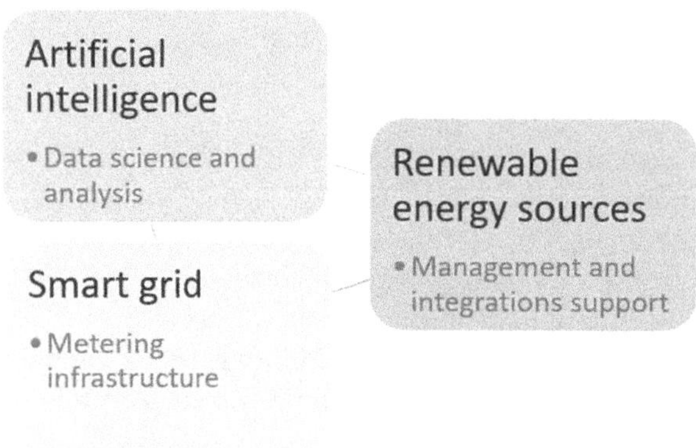

Figure 6.1 AI application for RES integration support.

This chapter examines the role of ML/DL techniques in supporting the integration of RES into the power grid. To this end, an introduction to various ML/DL methods and techniques will first be presented in Section 6.2. Following this, the general framework for using these techniques in the forecasting and management of RES will be introduced in Section 6.3, and detailed discussions will be covered in Sections 6.4 and 6.5.

6.2 Machine Learning Techniques

ML is a set of methods and computer algorithms that use the analysis of preceding data to forecast upcoming performance. Over time, ML has become one of the main techniques in AI, widely applied in various industries such as energy, meteorology, and management. In general, it can be said that the main tool for many ML methods is an artificial neural network (ANN). However, other ML methods such as support vector machines (SVM) and Gaussian process regression (GPR) have also been employed. Studies indicate that ML methods have been employed to forecast energy consumption, forecast prices, and estimate energy production. ANN and clustering, as alternative energy analysis tools, have proven effective in calculating energy performance. GPR has also been used to determine heating, ventilation, and air conditioning (HVAC) loads. A summary of various ML models that have been developed is presented in Figure 6.2.

Furthermore, ML algorithms are capable of recognizing objects in images, converting speech to text, matching items with user preferences, and selecting relevant search results. These tasks are powered by DL, a subset of ML that relies on ANNs. The key difference between ML and DL is that DL can identify intricate patterns, such as those found in images. Unlike traditional ML, DL involves multiple hidden layers arranged in a deep, nested structure. DL mimics human learning by training computers through examples. In general, convolutional neural networks (CNN) and recurrent neural networks (RNN) are the primary models used in DL. However, other DL architectures, including deep belief networks (DBN), autoencoders (AEs), and long short-term memory (LSTM) networks, have also been applied. DL techniques have been employed in various studies

Figure 6.2 ML models for RES applications.

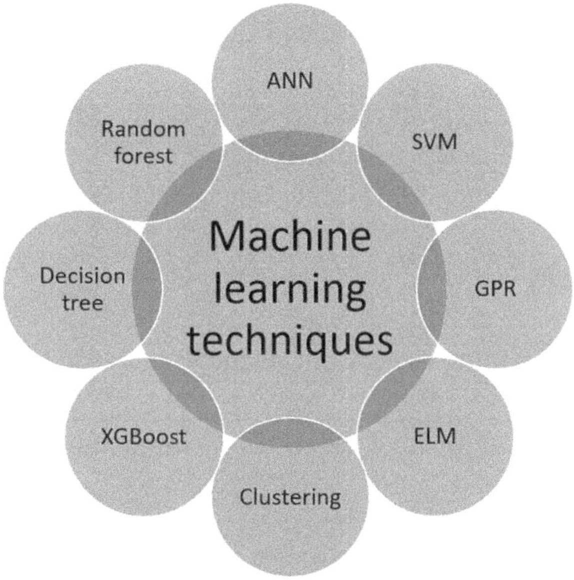

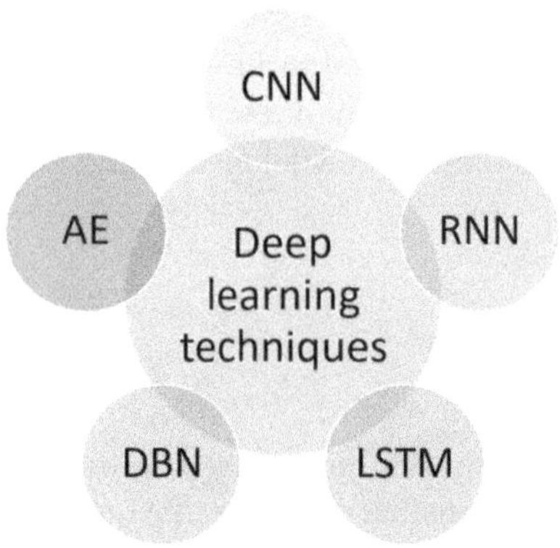

Figure 6.3 DL models for RES applications.

to estimate energy consumption and predict the capacity of RES, including solar, wind, biomass, and hydropower. Figure 6.3 summarizes DL models in RES research.

The following provides a brief introduction to some of the ML/DL methodologies.

6.2.1 Artificial Neural Network and Deep Neural Network

ANNs are computational models inspired by the structure and function of the human brain, specifically the nervous system. ANN is based on a network of neurons, which are the fundamental units where communication occurs. The structure of an ANN includes input, hidden, and output layers, along with an activation function that governs data processing within the hidden layer. The ANN structure has been shown in Figure 6.4. Neurons receive inputs, process them using internal activation functions, and produce outputs. The parameters that control these activation functions, such as weights, are adjusted through a process known as learning. Important factors in learning for ANNs include the learning rate, the number of hidden layers, and the maximum number of iterations. Activation functions like Sigmoid, Rectified Linear Unit (ReLU), and Softmax are commonly used to perform calculations. Common ANN architectures include multilayer perceptrons (MLP), feedforward neural networks, and radial basis function networks (RBFN).

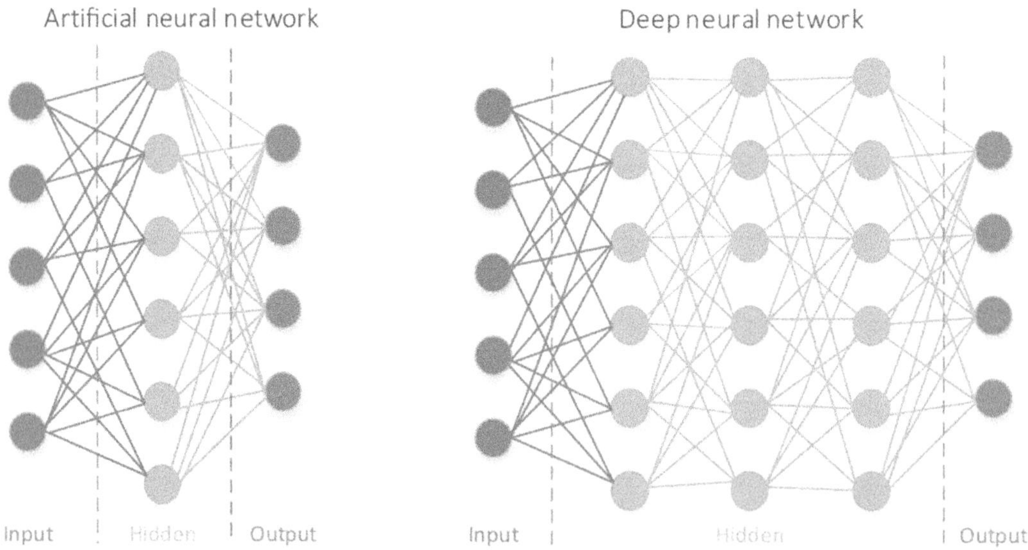

Figure 6.4 Artificial neural network and deep neural network.

Other DL models, such as CNN, LSTM, and RNN, are also derived from neural networks. Generally, ANNs have the ability to model large, nonlinear, and intricate systems. These models are also robust to errors, resistant to noise, and effective at reducing data dimensionality. Information is distributed throughout the network, ensuring that the loss of any single data point does not significantly impact the overall performance. Furthermore, ANNs have parallel processing capabilities. However, a notable disadvantage is that ANNs require large datasets for effective training, and determining the optimal number of hidden neurons can be a complex task. Additionally, DL models can be computationally intensive, often dependent on high-processing power hardware, and susceptible to overfitting.

A deep neural network (DNN), as shown in Figure 6.4, includes not just input and output layers, but also several hidden layers. A neural network with two or more hidden layers is referred to as a DNN. To generate an output, the DNN processes input data through a series of mathematical operations. The network is trained using a training dataset, typically by calculating probabilities for each possible output. DNNs share many of the same advantages and disadvantages as ANNs, but because DNNs contain more layers, they generally require larger datasets to produce more accurate results compared to ANNs.

6.2.2 Convolutional Neural Network

Convolutional neural networks (CNN) are among the most widely used methods in energy management, pattern recognition, and image processing. CNN is an enhanced version of the MLP. While MLP is a fully connected (FC) layer network where each neuron is connected to neurons in other layers, this full connectivity often leads to overfitting. To address this issue, CNN employs various regularization techniques. CNN provides high accuracy, particularly in managing image data, but it requires large datasets, resulting in significant computational costs and the essential need for powerful graphics processing units (GPUs).

CNN operates as a neural network with input, output, and some hidden layers. However, unlike traditional ANNs, CNNs use a series of specialized layers, including convolutional, pooling, FC, flatten, dropout, and normalization layers. The hidden layers in CNN use an activation function to process inputs and outputs. The ReLU is the activation function most commonly used in CNNs, and it incorporates a backpropagation method to optimize the weights. The convolution layer in CNN is for identifying patterns and features in the input data. Filters are used on the input, generating activation maps. Following the convolution layer, the pooling layer reduces the data's dimensions, simplifying processing while preserving the essential information. This dimensionality reduction decreases computational complexity and highlights key features, which aids in the efficient training of the model. The data is then transmitted to the FC layer. In this layer, similar to the MLP, each neuron in one layer is connected to every other neuron in the subsequent layer. This full connectivity increases the number of parameters, which can lead to overfitting. To mitigate this issue, a dropout layer is incorporated. Following the dropout layer, the data is passed to the flatten layer, which reshapes the data into a column vector. Finally, feedforward neural networks and backpropagation techniques are applied during each training phase to adjust the model's weights and improve performance.

6.2.3 Recurrent Neural Network

To process sequential data, a specialized type of neural network known as a recurrent neural network (RNN) is employed. CNNs typically train each sample independently; however, this method of independent training is inadequate, especially for data like sound, text, images, and time-series data. In RNNs, data is processed sequentially, with feedback connections in the hidden layer units that allow the network to maintain memory of previous inputs, enabling it to capture temporal dependencies. As a result, RNNs process data over time and learn sequentially. Additionally, RNNs utilize a hidden layer that acts as memory to retain sequential information over time. RNNs also use the same parameters for each layer.

Key advantages of RNNs include remembering complete information over time, handling sequential data effectively, and providing adequate accuracy when forecasting based on time-series data. Nevertheless, long-range learning with RNNs is challenging due to exploding or gradient vanishing problems. RNNs are developed for processing sequential data and can establish temporal relationships between the current state and past information. RNNs make decisions based on $t-1$ and t at time step t. These features of RNNs make them effective in predicting load and energy production from solar/wind energy sources. Additionally, RNNs are trained over time using backpropagation.

6.2.4 Long Short-Term Memory

LSTM networks are among the most effective neural network architectures for forecasting and modeling time-series data. Unlike standard neural networks, which learn input-output relationships from a static perspective, LSTMs are designed to capture temporal dependencies. In traditional neural networks, when time-series data are treated independently as inputs and outputs, valuable sequential information is lost. In contrast, LSTMs establish connections between each input-output pair, allowing the model to retain and utilize the temporal information across different time steps for more accurate predictions. RNNs are well-suited for tasks involving short-term dependencies, as they are designed to process sequential data with immediate context. However, for tasks that require capturing long-term dependencies over extended periods, LSTMs are preferred due to their ability to retain information over longer sequences. This capability is attributed to the specialized gates in LSTM networks, which help mitigate issues like vanishing gradients that RNNs face when dealing with long-term memory.

LSTM networks consist of three gates: the forget gate, the input gate, and the output gate. The forget gate determines which information from the previous state should be retained, using a sigmoid activation function. The input gate decides which information should be added to the current state as input. The output gate determines which information should be output from the current state. These gates work together to effectively manage the flow of information, allowing LSTM networks to capture long-term dependencies in sequential data.

6.3 General View of ML/DL Methods for RES Integration

Energy management plays a crucial role in the effective integration of RES and Energy Storage Systems (ESS) into power systems. It involves monitoring, planning, and controlling the operation of energy generation and consumption units. Through effective energy management, customers can lower their energy costs, while utility companies can minimize the need for peak power generation. The optimal use of RES can be attained by implementing strategies such as shifting energy loads and charging ESS with solar power during the day, rather than relying on grid electricity. Additionally, efficient energy management can help extend the lifespan of ESS by optimizing their usage.

Having an accurate forecast of the energy that renewable resources can produce is essential for effective energy management because of the fluctuating nature of energy production from RES. Researchers have established numerous prediction methods to predict energy demand as well as energy generation from renewable resources based on their characteristics, for example, wind speed, solar radiation, temperature, etc. Due to the variable nature of weather, the output energy generated from solar and wind resources is always intermittent. As a result, predicting the power output of these resources is challenging and requires progressive approaches. The methods used in this field can be broadly categorized into four main types:

- Physical
- Statistical
- ML/DL
- Hybrid

Physical methods, also known as numerical weather prediction (NWP) methods, are mathematical frameworks that simulate atmospheric processes based on physical and mechanical laws. These models depend on computer simulations, requiring significant computational power, which makes them more suitable for long-term forecasting. In contrast, statistical models aim to identify mathematical correlations between inputs and outputs, typically assuming linear relationships. While these models have been extensively used in various studies, their performance often falls short, as they are not effective at capturing nonlinear relationships.

ML/DL techniques have gained considerable interest because of their capability to reveal nonlinear relationships and deliver enhanced performance. It has been noted that deep learning, in particular, holds a promising future in this field due to its generalizability and unsupervised feature learning. ML/DL techniques can provide deeper insights into consumer behavior, allowing electricity suppliers to make accurate predictions about their customers' power demand. This also facilitates accurate billing for electricity consumption. On the other hand, integrating ML/DL into smart grids helps consumers access data on their energy consumption and the price they pay for it. This allows them to manage their energy usage and, if necessary, respond to requests from the center to decrease load demand in peak hours. This process, in turn, improves the operational efficiency of smart grids.

Forecasting wind energy, solar energy, and load using ML/DL consists of three key stages:

- Preprocessing
- Model training
- Prediction

Initially, the preprocessing stage involves cleaning and normalizing the input data, which is then divided into training, validation, and test datasets. The next step involves training the model to develop a robust and accurate forecasting model. Lastly, the trained model is used to perform forecasting, followed by necessary validation. This process is shown in Figure 6.5.

In a more general sense, the procedure and general settings for performing forecasting using ML/DL techniques are shown in Figure 6.6. In this figure, the three main sections shown in Figure 6.5 are broken down into their respective subcategories. In the data preprocessing stage, it is necessary to consider the forecasting horizon and prepare the data accordingly for training. The forecasting horizon consists of four categories: very short-term, short-term, medium-term, and long-term. Typically, predictions made by ML/DL methods are performed within a time

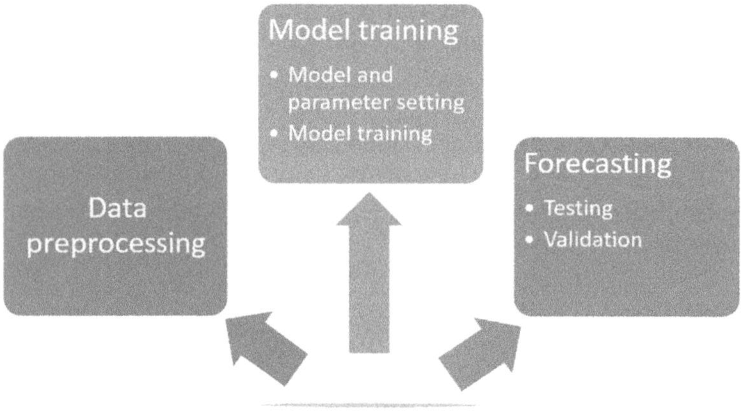

Figure 6.5 DL models for RES applications.

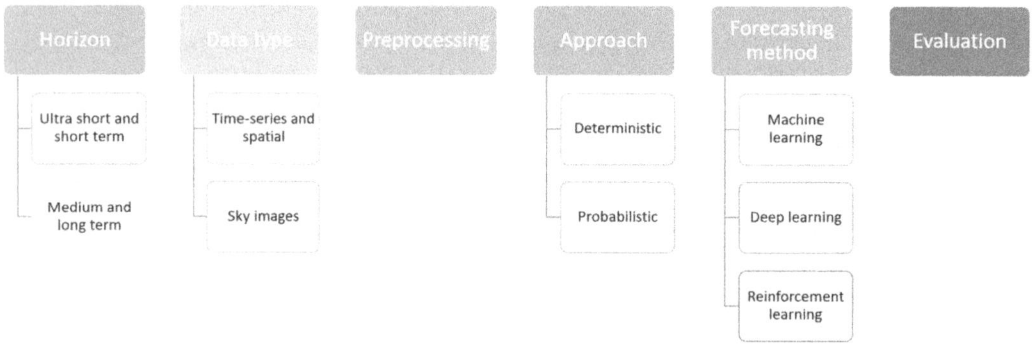

Figure 6.6 General procedures of ML/DL techniques for prediction of RES performance.

frame ranging from a few minutes to several days, which are considered part of the network operation problems. The data employed for forecasting can include spatial images, time series, or sky-related data. For instance, it might involve historical wind speed or wind power data to forecast wind energy or solar radiation data for predicting solar energy. This data can be utilized either independently or in conjunction with other meteorological information. The methods for preprocessing the data will be explained in detail below.

In the model training section, a deterministic or probabilistic approach is selected to handle the problem, and the appropriate ML/DL methodology has been selected for the forecasting process. Deterministic and probabilistic approaches are elaborated in Section 6.3.2. Various ML/DL models were introduced earlier in Section 6.2. It is worth noting that the DL models commonly used are CNN, RNN, SAE, and DBN. These DL techniques can also be integrated with data decomposition methods, feature extraction techniques, or data correction approaches to create hybrid models.

6.3.1 Data Preprocessing

In order to prepare the data for the learning process of the algorithm, several stages are required to synchronize and normalize the data. These stages are normalization, wrong/missing data, data resolution, inactive time data, data augmentation, data correlation, and data clustering, as shown in Figure 6.7.

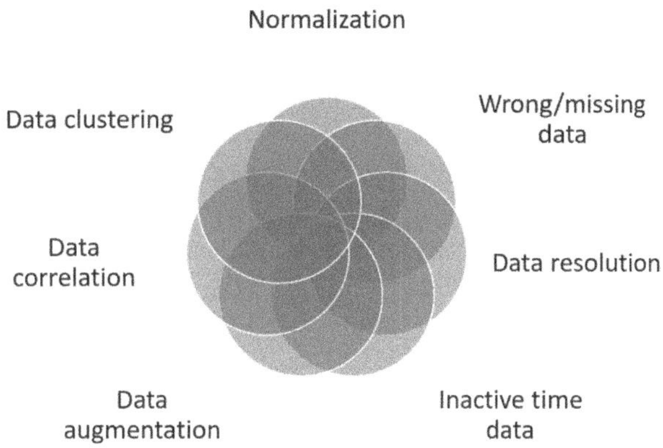

Figure 6.7 Data preprocessing stages.

6.3.1.1 Normalization

Normalization is a crucial stage while dealing with varying scales of numerical input, particularly for prediction inputs. Skipping this step, particularly in gradient descent-based algorithms, can impede the learning process, slow down convergence to the optimal solution, and ultimately degrade performance outcomes. A common normalization technique used in studies is Min-Max scaling, which maps the data to a range between 0 and 1. Examples of this method include normalizing the output data using the rated capacity of each PV plant or using zero-mean normalization for variables that can change, like wind direction, or variables such as temperature that can be both positive and negative. The logical post-processing step, after obtaining the predicted results from the model, is to convert them back to the original scale for better interpretability. This process is known as denormalization or anti-normalization.

6.3.1.2 Wrong/Missing Values and Outliers

To perform the training process, wrong, missing, and outlier data in the dataset must be identified. In cases where such data is encountered, values before and after it may be used to estimate the missing data. Several methods, such as the linear interpolation method and the mean value of previous years (which can also be taken as a median within any desired window), are used in this regard. Outliers can be managed using various criteria. One such criterion is comparing them with the standard deviation of the data, and any data points that fall outside the three times the standard deviation range are removed from the dataset.

6.3.1.3 Data Resolution

When high-resolution datasets (e.g., collected every second or minute) are used for algorithms designed for longer forecasting horizons (such as 1 or 24 hours), it may be necessary to adjust the resolution of the data for specific applications. One common approach is averaging, which is often cited as a way to smooth the data and reduce fluctuations. However, there are concerns that this method could lead to the loss of valuable information. To address this, some studies, such as [7], have employed two techniques – averaging and stacked sparse autoencoders – to reduce data resolution from three seconds to one minute while preserving key information. This combination of resolution adjustment and feature extraction helps maintain important details while lowering the data resolution. Averaging has been applied in various studies to adjust the resolution of data, such as from 15 minutes to 1 hour, 2 to 10 minutes, or from 1 hour to 1 day. For example, in Ref. [8], linear interpolation was used to change the resolution of NWP data from 1 hour to 10 minutes. Aside from time-series data, image resolution also plays a critical role in prediction quality, especially in weather forecasting.

6.3.1.4 Inactive Time Data

For RES generation, it is possible that at certain times of the day, the power output of PV or wind turbines may drop to zero. Zero power generation could occur for maintenance purposes or due to malfunctions in the RES. The common approach in this case is to remove these records to improve prediction accuracy. However, zero-generation times may differ from region to region and from season to season. For example, variations in sunrise times in different regions can present challenges in this regard. In any case, the zero-generation data should either be removed or converted to negative values before and after the training process.

6.3.1.5 Data Augmentation

Data augmentation is a prevailing tool in data management when the focus is on a specific feature of the input and there is little data available for this specific feature. For example, in Ref. [9], the inputs only included wind direction, but the authors computed the sine and cosine of the wind direction and included them as additional inputs.

6.3.1.6 Correlation

For data analysis, particularly in statistical and machine learning contexts, one of the basic prerequisites is that input data should not be correlated with each other. This dependence can lead to incorrect results and misinterpretations during analysis. Therefore, identifying and managing correlations between variables is of special importance. Various methods, such as mutual information for spatial and temporal correlations and Pearson correlation coefficients for showing relationships between input and output variables, can be used to determine correlation between data.

6.3.1.7 Data Clustering

Data clustering in renewable energy data analysis is of particular importance due to the significant differences between types of input data. Seasons, different times of the day, and various seasonal conditions with their unique characteristics, such as sunlight and cloud cover, fall into different categories. Additionally, if a spatial dimension needs to be added to wind or solar data, and if location information from wind or solar farm data is used in forecasting models, a spatiotemporal network or graph is created for representation.

6.3.2 Deterministic/Probabilistic Forecasting Methods

As shown in Figure 6.6, the general approaches that may be used to forecast the output power of RES are deterministic and probabilistic approaches. In the deterministic approach, the prediction results are expressed precisely, with no probability assigned to them. On the other hand, in the probabilistic approach, the prediction results are provided along with a probability value. Many studies in the field of applying different ML/DL approaches to RES have utilized deterministic approaches, with DL being more commonly used. This section will examine the use of these approaches in supporting the integration of RES into the grid.

6.3.2.1 Deterministic Methods

Deterministic models have been shown in Figure 6.8, which will be discussed in the following sections.

CNN Models CNN-based prediction models leverage convolutional layers to extract various features from the data, with regression typically carried out in the final fully connected layer. These models are commonly applied to forecast the power output of RES. CNN is often the first choice

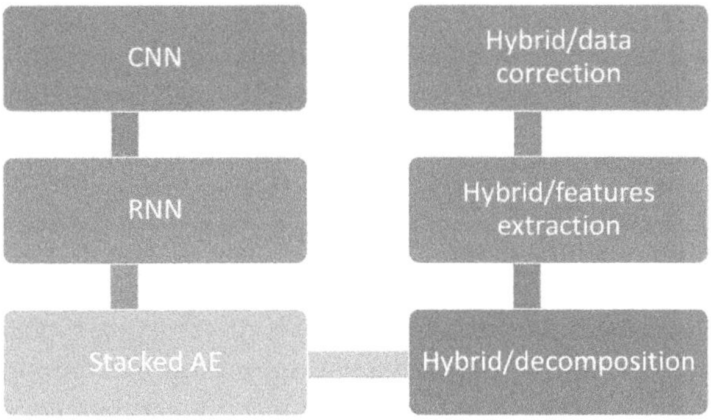

Figure 6.8 Deterministic ML/DL models.

for processing image-like inputs, such as images captured from the sky, which provide valuable information about cloud cover, thereby improving prediction accuracy. Different CNN architectures and input image resolutions for predicting PV output generally achieve high accuracy under clear skies, but the accuracy tends to decrease under partly cloudy or fully cloudy conditions. As previously stated, CNN is typically employed for data that can be represented in two or three dimensions, such as images or videos, although it can also be applied to one-dimensional data. By designing prediction models that incorporate both spatial and temporal features of wind turbines in a given area, the output of wind power can be forecasted for specific regions. In Ref. [10], the influence of such features on wind speed and the accuracy of wind power predictions was examined. Data gathered from wind turbines was mapped onto a network space, where the time series represented a multi-channel image that illustrated the spatiotemporal characteristics of wind in a particular region and period. The researchers noted that their proposed model outperformed existing methods in terms of accuracy, and they found that predicting wind speed is generally easier than predicting wind power, as the latter depends on specific turbine characteristics.

RNN Models Time-series data refers to sequential data collected over a precise time frame, often ranging from a few seconds to an hour, especially in prediction applications. This type of data is typically handled using RNN and their advanced variants, such as LSTM and gated recurrent units (GRU). In stacked LSTM networks, each LSTM layer's output serves as the input for the next LSTM layer, creating a deep structure that allows for a more intricate depiction of sequential data across time. Stacked LSTM approaches have been applied to wind speed forecasting. For instance, in Ref. [11], spatiotemporal information was represented as a graph, where nodes corresponded to data-generating units, and edges indicated the interactions between these nodes. This method helps incorporate data from adjacent stations to enhance predictions for a target station.

One effective approach in this context is the attention mechanism, which enables DL methods to emphasize the most pertinent features within a sequence of inputs. For example, in Ref. [12], a sequence-to-sequence model for multi-step prediction of wind power was introduced, utilizing a multi-input multi-output model. This approach generates numerous outputs at different time steps by employing a vector of values as input. The model's architecture includes two sets of GRU blocks functioning as an encoder and decoder. The model's performance highlighted its superior prediction accuracy.

Stacked AE Autoencoders are particularly effective for dimensionality reduction and for generating a compact representation of the original data. Denoising autoencoders, a specialized variant, are designed to handle noisy data by intentionally introducing noise into the input during training. This technique helps the model become more robust and improves its ability to learn important features from noisy inputs, making it more capable of filtering out irrelevant information compared to a standard autoencoder. Stacked denoising autoencoders (SDAE) were introduced in Ref. [13] for wind speed forecast through a transfer learning (TL) approach. The input layer and the two hidden layers are common across multiple wind farms, with each farm having its own distinct output layer to account for different data distributions. This structure allows the model to train using data from a farm with abundant information and then transfer the learned knowledge to a new, data-poor farm.

Hybrid/Data Decomposition Time-series data are typically composed of four key elements: level, trend, seasonality, and noise. For precise prediction, it is crucial to decompose the data to isolate these elements or at least distinguish the level and noise elements. In hybrid forecasting models, decomposition techniques are applied as a preprocessing step to break down the time-series data into smaller, more manageable subsets. Each of these subsets is then trained independently, and the predictions from different models are combined to produce the final forecast. This process is illustrated in Figure 6.9. For instance, in Ref. [14], a hybrid approach for wind speed forecasting was introduced, which integrates wavelet transform, DBN, and quantile regression. This method uses a particular type of discrete wavelet transform with four filters that split wind speed data into an approximation component and several detail components. The DBN model processes the approximation with four layers, while the detail components are handled by a 10-layer DBN.

Hybrid/Feature Extraction One of the advantages of DL models is their capability to extract nonlinear relationships from data. For example, CNN models are widely used to extract deep features in RES forecasting models, particularly when the data contains spatial information. In this approach, CNN is first employed for feature extraction of the data, and then these features are fed into other machine learning methods for final prediction. An example of this forecasting can be found in Ref. [15], where a doubled CNN with filters of different sizes is employed to extract short-term features from the input data. The output features obtained at this stage are subsequently input into an LSTM model to generate solar power predictions. Moreover, feature selection methods can be applied after using decomposition techniques. For instance, in Ref. [16], the MI method was applied to decomposed time series before the reconstruction stage to rank the candidate inputs based on their informational value for load forecasting. Feature selection can also serve as a preprocessing step before the predictive model in machine learning methodologies. An example of this method is implemented in Ref. [17], where four classification approaches are used to cluster similar wind turbine data and filter hierarchical features. The researchers claim that spectral clustering, along with an enhanced LSTM model structure, provides better accuracy and faster convergence compared to traditional LSTM models.

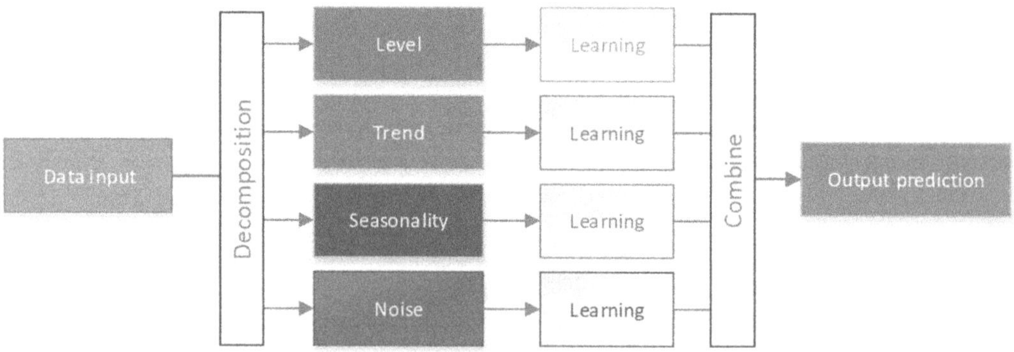

Figure 6.9 The structure of hybrid data decomposition.

Hybrid/Data Correction In some hybrid prediction techniques, the error from the initial prediction, along with the results from the first stage, is used as input for a final model that generates the final forecast. A case of this approach is illustrated in Ref. [18], where the error time series is decomposed, and an LSTM network is trained to predict the error. Additionally, certain studies have incorporated weather-related data to improve or adjust the initial predictions. For example, in Ref. [19], a combined model for PV power forecasting was proposed, consisting of two sub-models. The first sub-model is an LSTM network, and the second is a time-correlation model. These models are combined within a daily relative pattern forecasting framework.

6.3.2.2 Probabilistic Forecasting Methods

In probabilistic prediction, the objective is to determine prediction intervals within which actual values are likely to fall or to assign probabilities to the forecasted outcomes. Understanding the uncertainty associated with RES forecasts is crucial for the effective planning and management of electrical grids. As a result, probabilistic forecasting models have been developed. These models use both parametric and nonparametric methods. For instance, in Ref. [20], the output of an LSTM model for forecasting wind turbine power was analyzed for uncertainty using a Gaussian mixture model. This approach allows for probabilistic forecasts expressed as confidence intervals. The study found that the Gaussian Mixture model provided more precise results compared to two other uncertainty models – mixture density neural networks and relevance vector machines.

6.3.3 Evaluation Measures

As shown in Figure 6.6, after the forecasting process, it is necessary to evaluate the quality of the ML/DL algorithm's output and compare the predicted and actual values. Three common evaluation metrics for deterministic forecasting outputs are root mean squared error (RMSE), mean absolute error (MAE), and mean absolute percentage error (MAPE). The coefficient of determination (R^2) and standard deviation of error (SDE) are also frequently used. These metrics are introduced below.

Mean Squared Error (MSE) is the most generally employed metric to compare different ML methods, and it is defined as Eq. (6.1) where y_i and $\hat{y}_i$ represent the ith actual and predicted values, respectively, and N_s refers to the number of training samples used in the model training process. The RMSE metric is calculated through the square root of MSE. RMSE is a common metric used to evaluate the accuracy of predictions made by a regression model. It measures the average magnitude of errors between predicted values and actual observations and is calculated as the square root of the mean squared differences. RMSE is particularly sensitive to large errors, and in fields where such discrepancies are critical, like financial forecasting or weather modeling, it is valuable. A lower RMSE indicates better alignment between predicted and actual values, and an ideal RMSE equal to zero represents perfect predictions.

$$\text{MSE} = \frac{1}{N_s}\sum_{i=1}^{N_s}(y_i - \hat{y}_i)^2 \tag{6.1}$$

The MAE metric is calculated using Eq. (6.2). MAE measures the average of the absolute differences between predicted values and actual results. Unlike RMSE, MAE treats all errors equally without squaring them, and therefore, it is less sensitive to large errors. This feature allows MAE to provide a direct interpretation of the prediction accuracy in the same units as the data. MAE is particularly useful in scenarios where maintaining a fixed error margin is more important than minimizing large errors, such as in supply chain management or inventory forecasting.

$$\text{MAE} = \frac{1}{N_s}\sum_{i=1}^{N_s}|y_i - \hat{y}_i| \tag{6.2}$$

The MAPE metric is defined as Eq. (6.3). MAPE represents the prediction accuracy as a fraction and provides a relative measure of error, making its interpretation easier across different scales. This metric computes the mean of the absolute percentage differences between the predicted and actual values, providing a useful way to evaluate and compare the performance of various models or datasets. It is particularly helpful in assessing the accuracy of forecasts by quantifying how far off the predictions are from the true values on average. MAPE is commonly used in business and economic forecasting, where understanding errors as percentages can assist in decision-making processes. However, MAPE can lead to inflated percentage errors when actual values are close to zero.

$$\text{MAPE} = \frac{100\%}{N_s} \sum_{i=1}^{N_s} \left| \frac{\hat{y}_i - y_i}{y_i} \right| \tag{6.3}$$

The R^2 metric is defined as Eq. (6.4). R^2 is a statistical measure that represents the share of variance in the dependent variable that can be clarified by the independent variables in a regression model. This metric ranges from 0 to 1; higher values of this metric indicate a better fit of the model to the data, suggesting that the forecasts are closer to the actual values on average. This implies a more accurate model performance. R^2 is useful for evaluating model performance and comparing different models; however, this metric does not account for the number of predictors, which can lead to misleading interpretations if not considered alongside the adjusted version of R^2.

$$R^2 = 1 - \sum_{i=1}^{N_s} \left(\frac{y_i - \hat{y}_i}{y_i - \overline{y}_i} \right)^2 \tag{6.4}$$

The goal of probabilistic prediction is to determine prediction intervals that are likely to contain the actual values, with the goal of keeping these intervals as thin as possible. The coverage rate of these intervals corresponds to the confidence level, commonly referred to as the prediction interval nominal confidence (PINC). To evaluate the performance of probabilistic forecasting models, two commonly used metrics are the average coverage error (ACE) and the interval score (IS), as detailed below [21].

The ACE metric quantifies the discrepancy between the PINC and the actual coverage rate of an interval. It is calculated as shown in Eq. (6.5), where i and u represent the lower and upper bounds of the ith prediction interval, respectively. An ACE value close to zero suggests that the prediction interval reliably covers the actual values, indicating good accuracy in the probabilistic forecasts. The ACE metric is used to evaluate the precision of these predictions, quantifying the average difference between the predicted coverage probability and the actual proportion of observations that fall within the specified intervals. Specifically, this metric evaluates how well the prediction intervals cover the actual results. For example, if a model predicts a 90% coverage interval, ACE checks how much of the actual values fall within that range. A lower ACE indicates better calibration of the prediction, and this metric is particularly useful in areas such as weather forecasting and financial risk assessment, where understanding the reliability of predictions is critical for decision-making.

$$\text{ACE} = \frac{1}{N_s} \sum_{i=1}^{N_s} \begin{cases} 1 & y_i \in [\hat{l}_i, \hat{u}_i] \\ 0 & y_i \notin [\hat{l}_i, \hat{u}_i] \end{cases} \times 100\% - \text{PINC} \tag{6.5}$$

Additionally, the width of each interval is an important pointer in probabilistic prediction. However, if the actual value falls outside the prediction interval, the width of the interval cannot be effectively assessed using the width indicator alone, as it fails to account for situations where the interval does not encompass the true value. Therefore, the IS metric can be defined as Eq. (6.6). IS refers to the width of the prediction intervals generated by a probabilistic forecasting model.

This metric measures how narrow or wide these intervals are while still maintaining an adequate level of coverage for the actual outcomes. Sharper intervals are desirable because they provide more precise predictions and reflect higher confidence in the predicted values. However, excessively sharp intervals may lead to a lack of coverage, where actual results fall outside the predicted range too often. The sharpness of intervals is crucial in applications such as energy demand forecasting and supply chain management, where decision-makers benefit from both accuracy and reliability in their predictions.

$$\text{IS} = \frac{1}{N_s}\sum_{i=1}^{N_s}\begin{cases} -2\alpha(\widehat{u}_i - \widehat{l}_i) - 4(\widehat{l}_i - y_i) & y_i < \widehat{l}_i \\ -2\alpha(\widehat{u}_i - \widehat{l}_i) & y_i \in [\widehat{l}_i, \widehat{u}_i] \\ -2\alpha(\widehat{u}_i - \widehat{l}_i) - 4(y_i - \widehat{u}_i) & y_i > \widehat{u}_i \end{cases} \tag{6.6}$$

The value of the IS is always non-negative, with lower values indicating better prediction accuracy. When IS is close to zero, it suggests that the predicted intervals are highly accurate in capturing the actual values. Alpha represents a confidence level parameter ($\alpha < 0.5$), which is defined as follows:

$$\alpha = 1 - \text{PINC} \tag{6.7}$$

6.4 ML/DL Application for Integration of RES

ML/DL have garnered significant research interest due to their ability to address complex forecasting tasks, particularly in time-series analysis, regression, and clustering. In the context of energy applications, these data-driven approaches are increasingly used to forecast future trends and outcomes. Traditionally, power grid management relies on mathematical optimization techniques for tasks such as capacity planning, unit commitment scheduling, energy network planning, operational strategy, market trading, and power regulation. These tasks are typically framed as optimization problems with an objective function that must be minimized or maximized, subject to various constraints. Solvers such as GUROBI or CPLEX are then employed to find the optimal decision variables within the feasible solution space.

Optimization methods are well-established tools that engineers and decision-makers use to solve complex problems. Compared to simulation approaches, optimization models offer a direct route to an optimal solution and can accommodate uncertainties in stochastic scenarios. However, optimization techniques have significant limitations, including high computational time, expensive hardware, and cost. The efficiency of optimization algorithms is influenced by factors like hardware specifications (e.g., RAM, processor speed, number of cores) and the problem's inherent complexity, such as the number of iterations required for convergence. Certain problems, such as integer programming (IP), are NP-hard and pose significant challenges even for advanced solvers. The complexity also depends on whether the problem is convex or non-convex, linear or non-linear, and the number of decision variables and constraints. Consequently, real-time decision-making, which demands quick responses, may be hindered by the computational burdens of optimization.

An alternative to optimization techniques is the use of ML/DL methods, which are increasingly applied to tackle challenges in traditional power grid planning, especially in the integration of renewable energy sources (RES). ML/DL models, which leverage historical data, can quickly generate forecasts for RES performance and other key metrics. Over the last decade, ML/DL approaches have found applications across various forecasting fields, providing accurate predictions and strong feature extraction capabilities. ML-based forecasting techniques typically perform well but rely on relatively simple models, often referred to as "shallow" models, such as decision trees, regression models, or neural networks with one or no hidden layers. Training such models requires substantial expertise and experience.

In contrast, DL-based forecasting models excel because they do not rely on manual feature engineering and can handle complex datasets with higher sample complexity and better generalization performance compared to ML approaches. Despite their advantages, ML/DL methods come with several limitations. These include (1) the "black-box" nature of these models, where training often involves trial-and-error adjustments to achieve desired results, which means global optimization is not guaranteed; (2) their applicability being limited to forecasting and prediction tasks, rather than decision-making or optimal planning; and (3) potential discrepancies between the predicted results and real-time observations, as accuracy is typically defined by the minimal difference between predicted and experimental data, which may not hold in practice.

In this section, general applications of ML/DL for supporting the integration of RES into power grids are introduced. It is pointed out that, considering the challenges in integrating RES into grids, the benefits of using ML/DL methods in addressing these challenges and facilitating the planning and operation of the grid in the presence of these resources are highlighted. These applications are shown in Figure 6.10 and include renewable resources data prediction, RES power generation prediction, electric demand forecasting, and stability analysis.

Next, various applications of ML/DL in supporting the integration of RES into the grid are explained.

6.4.1 Renewable Resources Data Prediction/Planning

In energy studies, one of the most important components is the collection of data required for various applications of renewable energy. The production of renewable energy, particularly from wind and solar power, is highly dependent on environmental weather conditions. As a result, predicting weather data that can affect renewable energy plants forms a significant part of these studies. Accurate forecasting of weather data, including global radiation, sunlight intensity, wind speed, and direction, presents a major challenge due to the complexities involved in atmospheric systems. ML and DL methods, with their advanced data analysis capabilities, are increasingly being used in weather forecasting. These techniques can process massive volumes of meteorological data and recognize complex patterns that were less visible in traditional methods.

Renewable energy plants, unlike traditional power plants, require significantly larger areas for operation. For instance, the Shams-1 CSP-PV hybrid plant in Abu Dhabi, the largest solar power

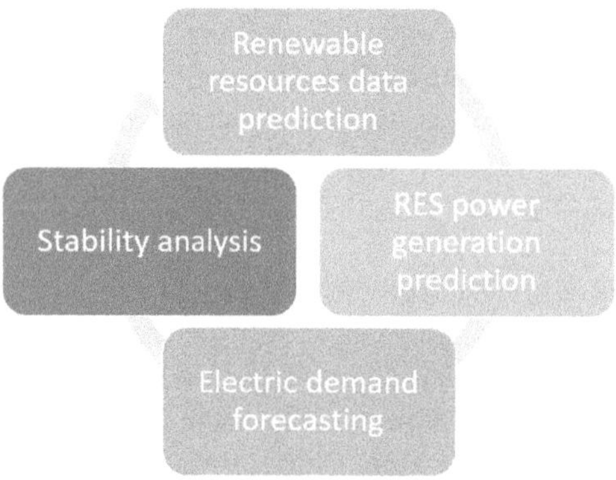

Figure 6.10 General applications of ML/DL for supporting the integration of RES.

facility globally, spans 77 km^2, while a conventional plant with comparable capacity only occupies a few square kilometers. As a result, it is essential to assess the required land area for renewable energy plants based on the energy demand. Furthermore, renewable energy plants such as photovoltaic (PV) solar and concentrated solar power (CSP) have specific site selection and orientation requirements. Solar panels, for example, need to be positioned optimally to capture solar radiation efficiently for maximum energy production. In this context, machine learning techniques play a pivotal role in aiding decision-making processes during these planning and design stages.

Traditional methods for determining the location of PV plants generally rely on weather data such as radiation, temperature, humidity, transparency index, and wind speed. However, these methods are not applicable to photovoltaic (PV) systems located in remote areas where the necessary data is not readily available. In such cases, ML/DL techniques are needed for estimation purposes. In practice, when a plant is not yet installed in a location, ML/DL methods can forecast the weather conditions of the intended installation site and predict the energy output of renewable energy systems, thereby assisting in planning scenarios.

One of the key methods in this context is the use of deep learning models capable of identifying nonlinear relationships among meteorological variables. These models, by analyzing historical data and real-time inputs, can provide predictions for extreme weather events such as storms and heavy rainfall. For example, companies like DeepMind have developed models based on radar data that offer highly accurate predictions, outperforming traditional models. In addition to deep learning models, hybrid models are also employed, which combine multiple algorithms to improve prediction accuracy. This approach, especially in conditions with diverse input data, can provide better results. Furthermore, using weather APIs to collect real-time and historical data is crucial for improving the accuracy of AI-based predictions.

Challenges in using ML/DL for weather forecasting include the quality and completeness of the data. ML/DL models require large volumes of high-quality data to function effectively. Additionally, the computational complexity of these models requires strong infrastructure and suitable computational resources. Therefore, meteorological organizations must make the necessary investments in this area. It is also worth noting that combining ML/DL with human expertise can significantly accelerate the weather forecasting process. This collaboration not only improves the accuracy of predictions but also facilitates the identification of new and rare patterns. Given the increasing climate changes and the need for more accurate weather predictions, the use of AI in this field is becoming an inevitable necessity.

When evaluating potential wind or solar farm sites, especially when historical energy production data is unavailable, energy output from RES can be estimated by forecasting factors such as wind speed or solar radiation. These forecasts are then used in combination with specific formulas for each type of RES power plant to calculate the expected output. This indirect approach offers greater flexibility, as it can be adapted to different conditions. For instance, because the electricity generated by wind turbines depends on factors like hub height and blade pitch angle, forecasting wind speed can provide a more straightforward method for estimating power output, as wind speed forecasting is easier than wind power prediction, which is influenced by turbine characteristics. Moreover, indirect methods can incorporate additional meteorological variables, such as wind speed and solar radiation, to further improve prediction models. This is particularly beneficial when data is sparse or inconsistent, as it helps mitigate the effect of irregular data on model accuracy.

Variables with a high correlation to PV power production include air temperature, solar radiation, and dew point, while cloud type and relative humidity are rather negatively correlated. In the case of solar energy, the blade pitch angle has a greater impact on power prediction compared to

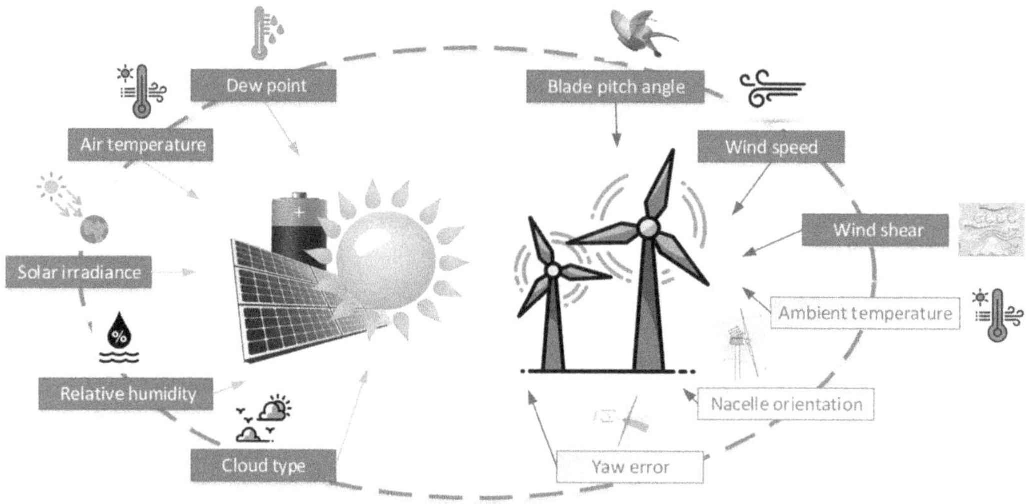

Figure 6.11 Various parameters affecting RES energy output

wind shear and wind speed. On the other hand, environmental temperature, wind direction, and deviation error have minimal influence on wind power forecasts and can be excluded from the inputs. These influences have been illustrated in Figure 6.11. Several studies highlight that the key factors for predicting Global Horizontal Irradiance (GHI) typically include temperature, clear sky index, relative humidity, and time of day, whereas other variables such as pressure, wind speed, and wind direction tend to have a lesser impact. However, the correlation between output power and weather conditions is also influenced by the specific location and its climate. As a result, it is essential to employ a feature selection technique to assess the relevance of each variable before training forecasting models.

Some studies that have been published on the prediction of meteorological parameters have been introduced. In Ref. [22], an ANN was employed to estimate global, direct, and diffuse solar radiation, as well as wind speed and direction across the Arabian Peninsula. The input variables for solar radiation included parameters such as time of day, day, month, solar azimuth, solar elevation, and cloud identification quality. For wind speed prediction, the input features consisted of time of day, day, month, air temperature, relative humidity, atmospheric pressure, and precipitable water. The model demonstrated strong predictive performance, achieving correlation coefficients (R) greater than 0.96 and an average percentage absolute error (MAPE) below 3%. The study also concluded that the ANN Levenberg-Marquardt (LM) function delivered superior prediction accuracy compared to the ANN scaled conjugate gradient (SCG) function. In Ref. [23], short-term forecasts for solar radiation and global wind speed in the USA were made using a multi-head convolutional neural network (MH-CNN). The study compared the performance of MH-CNN with traditional intelligent models and found that MH-CNN outperformed conventional machine learning models. Specifically, it achieved reductions in RMSE by 44.94% for wind data and 7.68% for solar data. Various ML/DL models, such as ANN, CNN, RNN, polynomial regression, SVM, and random forests, were employed to predict global and diffuse solar radiation components in Nigeria in Ref. [24]. The models used input data including year, month, day, hour, air temperature, wind speed, and solar height. All models, except for SVM, were effective in predicting global and diffuse radiation. Among them, the RNN model performed the best, with correlation coefficient (R), RMSE, and MAE values of 0.954, 82.22, and 36.52 W/m^2, respectively. In Ref. [25], an ANN model was developed to predict the luminous efficiencies of straight, diffuse, and worldwide radiation. The findings indicated that the ANN could effectively replace conventional empirical modeling methods for forecasting luminous efficacies. Moreover, the model attained an RMSE of less than 2% for each of the forecast luminous efficiencies. These studies demonstrate the increasing use and success of ML/DL in forecasting meteorological parameters critical to the performance of renewable energy systems.

6.4.2 RES Power Generation Prediction/Operation

The second category of applications of ML/DL methods involves using them to predict the power and energy produced by RES (renewable energy sources) plants in the operational time horizons of the grid. In terms of general principles, this application shares many similarities with the first category, with the difference being that in these studies, the output power of RES is predicted as the dependent output variable based on various input data. Predicting electricity production from renewable sources, due to the natural fluctuations of these sources, is a main challenge in the electricity industry. Since the production of this type of energy is dependent on weather conditions, the use of ML/DL techniques can help improve the accuracy of predictions. These technologies analyze historical data and environmental conditions to provide more accurate predictions of energy production and assist network operators in better resource management. Among these technologies, deep learning models are one of the most effective tools for predicting renewable energy production. These models, using complex neural networks, are capable of identifying nonlinear and complex patterns. For example, in predicting solar energy production, these models can simultaneously analyze the effects of variables such as sunlight intensity, temperature, and humidity. This capability makes the predictions not only more accurate but also more reliable.

Collecting accurate and comprehensive data from weather factors and historical energy production is a crucial step in the process of predicting RES energy. The use of advanced devices and Internet of Things (IoT) devices can help gather real-time information. This data acts as input for AI models and provides a basis for more precise analyses. Additionally, big data processing using ML/DL techniques can help identify new patterns and improve the accuracy of predictions. Furthermore, the use of hybrid methods is another effective approach in this area. Combining different ML algorithms, such as regression, random forests, and neural networks, can help increase the accuracy of predictions. These methods, by leveraging the advantages of each algorithm, can provide better results than using a single model. This approach not only improves accuracy but also enables the identification of the strengths and weaknesses of each model.

In this field, AI is capable of predicting future fluctuations by analyzing historical power production data and simulating different weather scenarios. This information helps network operators to plan energy supply accurately and activate storage resources or backup plants when needed. By integrating ML/DL technologies with energy management systems, the development of smarter electrical grids is also facilitated. These grids, using real-time information, will be able to make quicker decisions regarding energy dispatch. For example, during times of increased demand or reduced renewable energy production, the system can automatically activate alternative resources to ensure the sustainability of energy supply. Below, some studies in the field of RES production prediction are introduced.

Each forecast model brings its own advantages and disadvantages when forecasting solar radiation and PV power production. Therefore, determining which model is the best is difficult. However, it has been claimed that for a single model, the LSTM method performs better than the RNN, as LSTM has inherent memory for solving gradient vanishing issues caused by RNN.

Also, hybrid models perform better than independent models in predicting solar radiation. However, in view of computational efficiency and training time, the GRU model outperforms LSTM. When considering both training duration and prediction accuracy, the GRU model delivers promising results for forecasting PV power and solar radiation.

6.4.3 Electric Load and Demand Forecasting

Accurate prediction of electrical energy demand is one of the key factors in the optimal management and operation of electrical networks. This prediction helps electric companies plan for energy production and distribution. If the demand is not accurately predicted, it may lead to shortages or overloads, which can result in unwanted blackouts or waste of resources. Therefore, accuracy in demand forecasting helps maintain grid stability and ensures the energy required by consumers is provided. From an economic perspective, accurate energy demand prediction can reduce costs and improve the efficiency of power supply systems. With accurate information about future needs, electricity companies can optimize their investments and avoid additional costs associated with overproduction or energy shortages. This also helps lower the energy cost for consumers, ultimately leading to increased customer satisfaction and economic well-being.

Despite the high importance of demand forecasting, there are challenges in this area. Sudden changes in consumer behavior, weather conditions, and economic developments can significantly affect consumption patterns. To address these challenges, using advanced analytical models and data mining techniques can help improve the accuracy of predictions. Additionally, close cooperation with consumers and the collection of more accurate data can provide the necessary information for better decision-making. Due to changes in consumer behavior, weather patterns, and market dynamics, obtaining an accurate forecast of future electrical energy consumption has become complex. In this field, various ML/DL techniques, as advanced tools, provide new capabilities for analyzing and forecasting this demand. These techniques, by processing historical data and identifying patterns, are capable of providing accurate forecasts of future energy consumption. In particular, the use of nonlinear models can enhance prediction accuracy and assist electric companies in more efficiently managing their resources.

One important aspect of energy demand forecasting is incorporating external factors, such as weather conditions, public holidays, and industrial activities, into the forecasting process of electrical energy consumption. These factors can have a significant impact on consumption patterns. By combining data related to these factors with historical energy consumption data, models can identify more complex relationships and provide more accurate forecasts. Moreover, online learning enables models to be continuously updated with new data. This method is particularly useful in situations where sudden changes in demand or the integration of renewable energy sources occur. This type of learning ensures that models are always up-to-date and capable of providing accurate predictions in dynamic environments.

There are various methods for forecasting energy demand using ML/DL. The use of deep belief networks (DBN) has gained significant attention due to their ability to process large and complex data. Additionally, ensemble learning methods, which combine several prediction models, can improve the accuracy of results. These methods help reduce errors and increase prediction accuracy by merging the outputs of different models. It is important to highlight the significance of transparency in the prediction results to build trust among different stakeholders. By using techniques that allow for result interpretation, one can ensure that decisions are based on reliable data and analyses. This approach not only increases prediction accuracy but also boosts consumer and investor confidence in smart energy systems. Below, some studies in this field are introduced.

6.4.4 Stability Analysis

In power systems, grid stability is dependent on maintaining a balance between energy demand and supply. When demand exceeds supply, it causes grid instability, which can result in power quality issues or even blackouts in specific areas. On the other hand, when supply exceeds demand, it leads to wasted energy and unnecessary high costs. Generating the appropriate amount of energy at the right moment is essential for ensuring optimal grid operation. To preserve this balance, much of the research has focused on energy supply and demand forecasting, as it helps predict the required energy and influences grid stability. RES, which are inherently variable, pose additional challenges for grid stability. Fluctuations in RES output, driven by environmental factors like wind speed, solar radiation, and cloud cover, can cause grid disruptions. Furthermore, renewable energy systems are impacted by daily and seasonal cycles; for instance, solar energy is only available during daylight hours. As such, it is essential to produce energy when resources are abundant and store it for later use.

The challenges outlined above have driven the adoption of ML and DL techniques to enhance the management of energy production and consumption while supporting the integration of RES. In power grids where RES contribute a substantial share of the energy supply, it is vital to forecast both short-term and medium-term energy demand. This allows for more informed energy policy decisions. Similarly, forecasting the energy output from renewable power plants is crucial, as their production is influenced by various uncontrollable environmental factors. Consequently, ensuring grid stability in the presence of RES becomes a significant challenge. In this context, AI techniques offer valuable solutions, enabling more accurate predictions and improving the integration of RES into the grid.

The goal of stability analysis is to determine whether a power grid can maintain its stability during disturbances and to identify the type of instability that may arise. Currently, two common methods for stability analysis are time domain simulation (TDS) and the transient energy function (TEF). While TDS provides a detailed simulation, it is computationally inefficient and requires complex models with highly accurate parameters to replicate real power grid behaviors. On the other hand, the TEF method is less computationally demanding, but extracting the energy function for large-scale power networks is a complex task. Additionally, the second-order classical models typically used in TEF, which overlook the intricate transient processes, may compromise the accuracy of the stability analysis.

In summary, there is a need for a new stability analysis approach that aligns with the advancements in smart grids. The rapid progress in ML/DL technologies offers a promising solution. Stability analysis can be framed as a classification problem addressed by supervised learning algorithms. With the integration of phasor measurement units (PMUs) and wide area measurement systems (WAMS) in smart grids, high-resolution real-time electrical data collected with a synchronized time framework can be leveraged as a database for AI-based techniques. These data-driven approaches can efficiently extract stability-related features from the input data using their advanced learning capabilities. Additionally, AI methods eliminate the need for complex modeling and extensive domain knowledge in power networks. This section explores the role of ML/DL techniques in analyzing and ensuring network stability.

Artificial intelligence is being developed and used in many aspects of electrical grids. In Ref. [26], the capabilities provided by various ML/DL techniques in the topic of stability for grids have been reviewed. These capabilities are shown in Figure 6.12. In this section, security assessment and stability assessment are introduced as the most important topics in stability analysis.

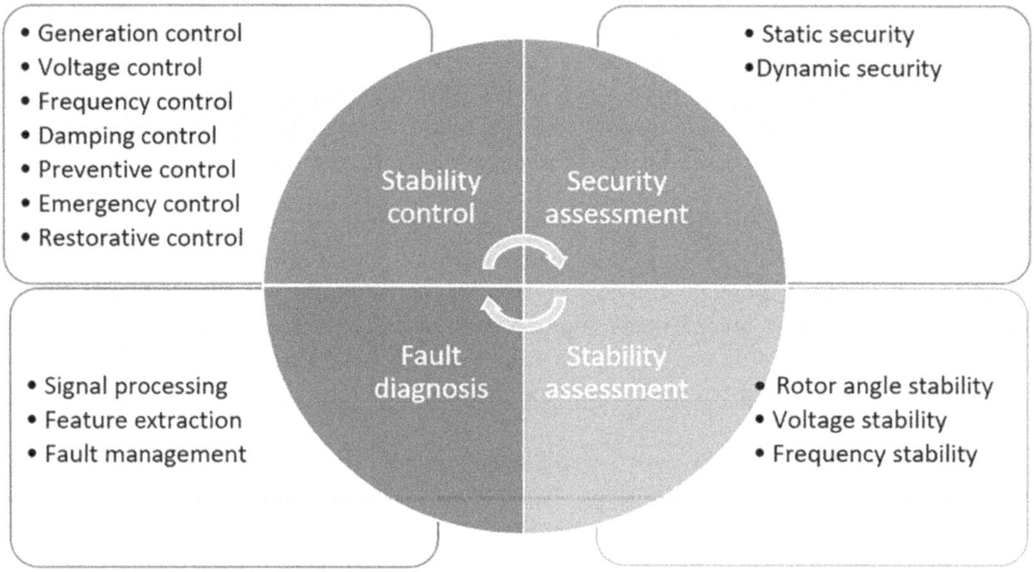

Figure 6.12 Application of ML/DL techniques for stability analysis and control.

6.4.4.1 Security Assessment

Security in power systems refers to the level of risk associated with the system's capability to endure impending contingencies without disrupting service to consumers. It is a critical aspect of smart grid operations. Security analysis typically includes two key components: static security assessment (SSA) and dynamic security assessment (DSA), both of which are explained in the following sections.

SSA evaluates a power system's capacity to handle potential contingencies without breaching system constraints such as equipment limits, bus voltages, or line power flows. SSA focuses on steady-state conditions, ignoring transient dynamics, and typically involves calculating power flows for N-1 contingency scenarios. This approach provides detailed insights into voltage profiles and power distribution within the grid. AI presents an effective solution for addressing SSA challenges. By training classifiers to map operational features to safe or unsafe states, AI can help identify specific conditions that ensure system stability. Through the analysis of recorded network measurements, critical features influencing stability are extracted. In ML/DL-based methods, most computations are carried out offline, enabling the trained model to quickly assess system security in real-time operations.

DSA examines the ability of a power system to remain secure under both transient and steady-state conditions. It ensures that the system meets critical security criteria, such as avoiding thermal overloads, maintaining stable voltage and frequency levels, and addressing different types of stability, including transient stability, small-signal stability, voltage stability, and frequency stability. DSA can be classified into pre-contingency and post-contingency assessments. Pre-contingency DSA focuses on analyzing the current operating conditions to determine if they are secure against a potential contingency that has not yet occurred. If the assessment identifies potential risks, preventive control actions are implemented to mitigate them. In contrast, post-contingency DSA evaluates the system's security after an event has taken place. By using dynamic measurements such as rotor angles and voltage profiles, this approach involves applying emergency control measures to prevent instability from propagating throughout the interconnected grid. Numerous studies have explored the application of machine learning and deep learning methods for dynamic security assessment. These methods leverage real-time data to improve the system's ability to predict and address instability effectively.

6.4.4.2 Stability Assessment

Stability assessment refers to monitoring and early detection of events that could lead to system instability, playing a crucial role in the safe and stable operation of electrical grids. Stability issues can be classified into three types: rotor angle stability, voltage stability, and frequency stability. Due to the advantages of speed, accuracy, and adaptability, AI is applicable in all of these areas.

Rotor angle stability refers to the ability of synchronous generators to maintain synchronization with the grid after a functional disturbance. The use of ML/DL approaches in rotor angle stability can provide an effective tool for grid operators. ML/DL can extract the mapping relationship between power grid variables and the network's stability state using its specific features. Once this model is trained offline, it can be used in online applications to perform rotor angle stability analyses by inputting real-time data. Various ML/DL methods, such as SVM, DBN, LSTM, and CNN, have been used in studies for this purpose.

Voltage stability refers to the ability of the grid to maintain steady-state bus voltages after a disturbance occurs. This is particularly significant in current grids, which are increasingly facing higher consumption, making the grids operate near stability margins. By using ML/DL techniques, nonlinear relationships between system variables and the corresponding voltage stability state can be identified. As a result, after offline training of the model, it can be used in online applications to evaluate the grid's voltage stability based on real-time system measurements and conditions.

Frequency stability refers to the ability of the system to maintain the grid frequency after a large imbalance between energy consumption and production. In this field, ML/DL methods also provide several advantages due to their speed and accuracy. The operational principle of this analysis is similar to the previous cases, relying on offline training of the model to discover complex relationships between frequency stability and system operation variables and then performing real-time predictions of the grid's stability based on the system's real-time conditions.

6.5 Integrated Machine Learning and Optimization Approach

ML and DL approaches have been highlighted as alternatives to traditional optimization techniques for uncovering hidden patterns in data and predicting future operational states. However, integrating optimization methods with ML/DL can significantly enhance their performance and accuracy. By leveraging this combination, AI-based models gain greater capability in handling complex problems. The accuracy and reliability of ML models are generally assessed based on minimizing prediction errors. The error value of a model is influenced by hyperparameters such as the number of neurons, the number of hidden layers, the training algorithm used, and batch size, among others. To achieve globally optimal results, some studies have redefined the development of ML algorithms as an optimization problem. This involves setting the chosen evaluation metric as the objective function, expressing the ML algorithm as constraints, and treating the hyperparameters as decision variables. Through this formulation, the process ensures a more robust and effective approach to improving ML model performance. A schematic of how the hyperparameter optimization problem of ML is modeled through optimization is shown in Figure 6.13.

As a result, by optimally determining the hyperparameters of ML/DL methods, their performance can be enhanced. This approach has been used in various studies. In this area, in Ref. [27], a

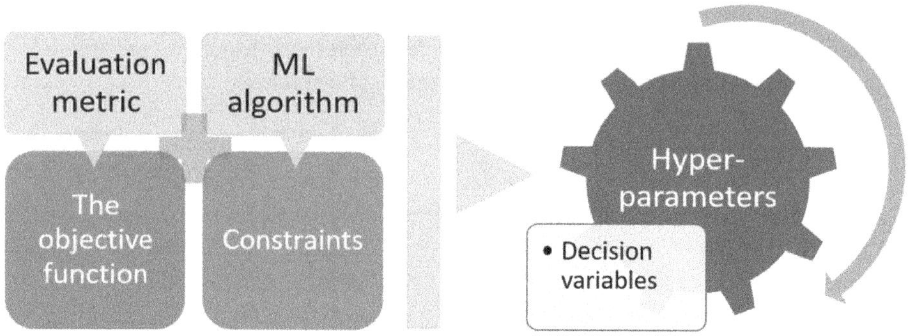

Figure 6.13 Definition of an optimization problem for ML hyperparameters.

GA was combined with an ANN model to enhance the accuracy of short-term PV power prediction. The ANN model has been formulated as a mathematical optimization problem, with GA serving as the optimizer to identify the optimal mixture of ANN hyperparameters. Similarly, in Ref. [28], a GA-ANN hybrid model was applied to short-term load forecasting using data from Portugal, New York, and Rio de Janeiro, achieving an average prediction error of less than 2%.

Deep neural networks (DNNs) are among the most widely used methods for wind energy forecasting. When paired with optimization techniques to fine-tune numerous network parameters, the overall system accuracy can be significantly enhanced. Consequently, there has been notable growth in research focusing on these combined methods, aimed at improving the forecasting process through parameter optimization to achieve greater precision. In scenarios where data availability is limited or weather stations are absent near the wind farms being studied, these hybrid models have proven effective. They provide alternatives to traditional statistical approaches that rely on historical data and address challenges posed by climate variability in real-world wind farms. Studies have consistently demonstrated that such combined approaches deliver superior accuracy in wind energy forecasting.

In addition to using optimization for tuning ML/DL hyperparameters, some combined methods have proposed using ML/DL and optimization techniques to estimate model uncertainties. The intermittent nature of renewable resource generation and the fluctuation of unforeseen demand during real-time operations need to be specifically considered in the operation and planning of IES (Integrated Energy Systems). In traditional uncertainty analysis approaches, this issue is usually addressed by using measured statistical methods, and then the output is used as input parameters for optimization problems in various scenarios. However, some researchers have considered the use of ML/DL methods to address uncertainties, mostly in the planning domain. In Ref. [29], an extensive uncertainty analysis of decarbonization pathways for Canada was conducted using ML and DL techniques. This study successfully captured the complexities and nonlinear behaviors of the power system, which are difficult to model with traditional methods. By applying statistical techniques and K-means clustering to identify correlations, the research assessed the impact of uncertainties in key parameters such as capital costs, demand, and carbon taxes, while optimizing the power generation mix.

On the other hand, considering the power of ML/DL for forecasting, as well as the suitability of optimization methods for attaining globally optimal decisions in energy systems, combining these two approaches can provide an appropriate mechanism for optimal IES planning and operation. Optimization techniques have long been established as effective tools for decision-making in IES. For instance, in Ref. [30], a combined autoregressive model and Cholesky decomposition were used to make predictions, which were then followed by optimal decision-making based on consumer preferences. It is also noted that ML techniques, particularly DL, offer a viable alternative to traditional statistical methods for time-series forecasting.

6.6 Conclusion

The intermittent nature of RES presents considerable challenges for their seamless integration into power grids. This chapter examines the role of ML and DL tools in facilitating the integration of RES into the grid. Accurate and reliable forecasting of solar and wind energy plays a crucial role in enhancing service quality and enabling efficient energy management. ML/DL-based prediction methods are increasingly recognized as effective and efficient approaches for predicting energy generation, leveraging historical data to provide insights into future energy production patterns. These advanced techniques help mitigate the uncertainties associated with renewable energy generation, supporting grid stability and optimizing the operation of power systems. However, despite the widespread use of these methods, there are several challenges in their application. Some of these challenges include:

- The optimal performance of various ML and DL approaches is heavily dependent on the availability of large and high-quality datasets. The effectiveness of these models in learning and configuration is determined by the quantity and quality of historical data, as ML/DL models typically have numerous adjustable parameters that require substantial and well-structured data for training. However, this data challenge remains significant in the context of integrating RES into power grids and managing EMS. High-quality labeled datasets for energy and load forecasting are still scarce. This scarcity is largely due to utilities and service providers' reluctance to share real-time and historical data, driven by concerns related to security and privacy. Additionally, the data collection process can be problematic due to discrepancies such as duplication, incorrect labeling, and temporary data loss, which can further complicate the accurate forecasting and management of RES in the grid.
- ML/DL-based methods are entirely dependent on historical data and make predictions based on this data. However, the heavy reliance on large datasets requires powerful servers. In addition, when using ML/DL-focused approaches, high processing requirements represent another challenge. Furthermore, more processing time is needed for training with additional data compared to training with clean data. As a result, computational technologies such as in-memory processing, database processing, and parallel processing need to be utilized.
- Probabilistic prediction of energy consumption and electricity generation from RES is crucial for optimizing the operation of upcoming smart microgrids. Forecasting energy production from sources like PV panels and wind turbines typically relies on local, on-site information. However, energy farms are geographically distributed, forming a network structure within the broader distribution system. This distribution introduces additional complexity, as the energy output from different farms can be influenced by varying local weather conditions and environmental factors. Therefore, effective forecasting must account for spatial variability and integrate data from multiple locations to improve the accuracy of predictions and ensure the efficient operation of the entire grid. Load forecasting must also perform adequately in different locations, considering residential, commercial, and industrial loads. Spatiotemporal forecasting approaches for future smart microgrids are more accurate and efficient than single-location techniques. Therefore, there is a need to develop new DL models that handle the spatial and temporal dynamics of solar and wind energy along with load demand.

- ANNs are more efficient and effective than physical and statistical methods for short-term wind speed and power forecasting. However, for long-term forecasting, the reliance on historical data grows, which can reduce the accuracy of ANNs. This limitation calls for further refinement of ANN-based techniques to enhance their accuracy in long-term forecasting.
- Heterogeneous users with varying skill levels need to utilize different ML methods that can be applied to these energy sources in a manner that is accessible and clear for both beginners and experts. Additionally, ML models should be capable of supporting both large and small heterogeneous datasets and remain consistently efficient for both small and large datasets.

Currently, DL methods are typically used for forecasting the next day's or real-time energy production of solar and wind plants. However, there are also other renewable energy sources, such as hydropower, geothermal energy, ocean energy, and biomass. While ML and DL algorithms can also be applied to RES, there are few applications for forecasting energy production, which represents an unexplored and valuable research area.

References

1 Alabi, T.M., Aghimien, E.I., Agbajor, F.D. et al. (2022). A review on the integrated optimization techniques and machine learning approaches for modeling, prediction, and decision making on integrated energy systems. *Renewable Energy* 194: 822–849.

2 Alkhayat, G. and Mehmood, R. (2021). A review and taxonomy of wind and solar energy forecasting methods based on deep learning. *Energy and AI* 4: 100060.

3 Aslam, S., Herodotou, H., Mohsin, S.M. et al. (2021). A survey on deep learning methods for power load and renewable energy forecasting in smart microgrids. *Renewable and Sustainable Energy Reviews* 144: 110992.

4 Abualigah, L., Zitar, R.B., Almotairi, K.H. et al. (2022). Wind, solar, and photovoltaic renewable energy systems with and without energy storage optimization: a survey of advanced machine learning and deep learning techniques. *Energies* 15 (2): 578.

5 Ibrar, M., Hassan, M.A., Shaukat, K. et al. (2022). A machine learning-based model for stability prediction of decentralized power grid linked with renewable energy resources. *Wireless Communications and Mobile Computing* 2022 (1): 2697303.

6 Mostafa, N., Ramadan, H.S.M., and Elfarouk, O. (2022). Renewable energy management in smart grids by using big data analytics and machine learning. *Machine Learning with Applications* 9: 100363.

7 Liu, H. and Chen, C. (2019). Multi-objective data-ensemble wind speed forecasting model with stacked sparse autoencoder and adaptive decomposition-based error correction. *Applied Energy* 254: 113686.

8 Zhang, Y., Li, Y., and Zhang, G. (2020). Short-term wind power forecasting approach based on Seq2Seq model using NWP data. *Energy* 213: 118371.

9 Wang, K., Qi, X., and Liu, H. (2019). A comparison of day-ahead photovoltaic power forecasting models based on deep learning neural network. *Applied Energy* 251: 113315.

10 Yu, R., Liu, Z., Li, X. et al. (2019). Scene learning: deep convolutional networks for wind power prediction by embedding turbines into grid space. *Applied Energy* 238: 249–257.

11 Ghaderi, A., Sanandaji, B.M., and Ghaderi, F. (2017). Deep forecast: deep learning-based spatio-temporal forecasting. *arXiv preprint arXiv:1707.08110.*

12 Niu, Z., Yu, Z., Tang, W. et al. (2020). Wind power forecasting using attention-based gated recurrent unit network. *Energy* 196: 117081.

13 Hu, Q., Zhang, R., and Zhou, Y. (2016). Transfer learning for short-term wind speed prediction with deep neural networks. *Renewable Energy* 85: 83–95.

14 Wang, H., Wang, G., Li, G. et al. (2016). Deep belief network based deterministic and probabilistic wind speed forecasting approach. *Applied Energy* 182: 80–93.

15 Lee, W., Kim, K., Park, J. et al. (2018). Forecasting solar power using long-short term memory and convolutional neural networks. *IEEE Access* 6: 73068–73080.

16 Memarzadeh, G. and Keynia, F. (2020). A new short-term wind speed forecasting method based on fine-tuned LSTM neural network and optimal input sets. *Energy Conversion and Management* 213: 112824.

17 Yu, R., Gao, J., Yu, M. et al. (2019). LSTM-EFG for wind power forecasting based on sequential correlation features. *Future Generation Computer Systems* 93: 33–42.

18 Ma, Z., Chen, H., Wang, J. et al. (2020). Application of hybrid model based on double decomposition, error correction and deep learning in short-term wind speed prediction. *Energy Conversion and Management* 205: 112345.

19 Wang, F., Xuan, Z., Zhen, Z. et al. (2020). A day-ahead PV power forecasting method based on LSTM-RNN model and time correlation modification under partial daily pattern prediction framework. *Energy Conversion and Management* 212: 112766.

20 Zhang, J., Yan, J., Infield, D. et al. (2019). Short-term forecasting and uncertainty analysis of wind turbine power based on long short-term memory network and Gaussian mixture model. *Applied Energy* 241: 229–244.

21 Zang, H., Cheng, L., Ding, T. et al. (2020). Day-ahead photovoltaic power forecasting approach based on deep convolutional neural networks and meta learning. *International Journal of Electrical Power & Energy Systems* 118: 105790.

22 Shboul, B., AL-Arfi, I., Michailos, S. et al. (2021). A new ANN model for hourly solar radiation and wind speed prediction: a case study over the north & south of the Arabian Peninsula. *Sustainable Energy Technologies and Assessments* 46: 101248.

23 Alhussein, M., Haider, S.I., and Aurangzeb, K. (2019). Microgrid-level energy management approach based on short-term forecasting of wind speed and solar irradiance. *Energies* 12 (8): 1487.

24 Bamisile, O., Oluwasanmi, A., Ejiyi, C. et al. (2022). Comparison of machine learning and deep learning algorithms for hourly global/diffuse solar radiation predictions. *International Journal of Energy Research* 46 (8): 10052–10073.

25 López, G. and Gueymard, C.A. (2007). Clear-sky solar luminous efficacy determination using artificial neural networks. *Solar Energy* 81 (7): 929–939.

26 Shi, Z., Yao, W., Li, Z. et al. (2020). Artificial intelligence techniques for stability analysis and control in smart grids: methodologies, applications, challenges and future directions. *Applied Energy* 278: 115733.

27 Izidio, D.M., de Mattos Neto, P.S., Barbosa, L. et al. (2021). Evolutionary hybrid system for energy consumption forecasting for smart meters. *Energies* 14 (7): 1794.

28 Viegas, J.L., Vieira, S.M., Melício, R. et al. (2016). GA-ANN short-term electricity load forecasting. In: *Technological Innovation for Cyber-Physical Systems: 7th IFIP WG 5.5/SOCOLNET Advanced Doctoral Conference on Computing, Electrical and Industrial Systems, DoCEIS 2016, Costa de Caparica, Portugal, April 11–13, 2016, Proceedings* (ed. L.M. Camarinha-Matos, A.J. Falcão, N. Vafaei, and S. Najdi), 485–493. Springer.

29 Jahangiri, Z., Miri, M., Yi, K.M., and McPherson, M. (2024). Machine learning-based uncertainty analysis in power system planning: insights and pathways for decarbonization. *Energy Reports* 12: 942–954.

30 Zhong, J., Li, Y., Cao, Y. et al. (2020). Stochastic optimization of integrated energy system considering network dynamic characteristics and psychological preference. *Journal of Cleaner Production* 275: 122992.

7

Machine Learning-Based Solutions for Renewable Energy Integration

Applications, Optimization, and Grid Stability

Ali Paeizi[1,2], *Mohammad Mehdi Amiri*[1,2], *Sasan Azad*[1,2], *and Mohammad Taghi Ameli*[1,2]

[1] *Department of Electrical Engineering, Shahid Beheshti University, Tehran, Iran*
[2] *Electrical Networks Research Institute, Shahid Beheshti University, Tehran, Iran*

7.1 Introduction

Fossil fuels have always been the main source of energy, though, due to their limited supply and negative environmental consequences, scientists are now compelled to investigate renewable energy alternatives that support an ecosystem with cleaner energy. During the recent years, wind, solar, and hydro energies have heightened anticipations for the future of sustainable energies. However, with all the advancements, a significant proportion of energy continues to derive from fossil fuels. The pressing requirement for a complete shift to clean and renewable energy systems cannot be fulfilled solely by renewable sources, notwithstanding their potential to address current demands and facilitate the adoption of cleaner alternatives. Moreover, a significant distinction between fossil fuels and renewable energy sources (RESs) is their availability. Traditional fuel sources have a finite supply; however, RESs are inexhaustible in the long term. There are huge capacities for using ML algorithms in RESs due to their capability to manage the complexity and variability of renewable resources, including solar and wind power, which are heavily influenced by weather conditions. Forecast accuracy, system reliability, and operating efficiency would be increased using ML algorithms in energy generation, storage, and distribution of power grids. These techniques endeavor to unveil concealed patterns inside datasets and make robust methods; nonetheless, it is essential to recognize that a profound comprehension of this domain results in more precise outcomes. Consequently, these models exhibit significant potential for making more precise projections, especially when applied to novel data [1].

The interaction of artificial intelligence (AI) and deep learning (DL) in these fields highlights their crucial importance in influencing the present and future landscape of renewable energy. AI and DL have not only penetrated but also revolutionized numerous sectors, with renewable energy being no exception. In the realm of photovoltaic (PV) power plants, significant advancements have been demonstrated in system normal and fault condition detection. Machine learning has a huge potential to be utilized in the domain of RESs. Machine learning algorithms can detect and identify faults in renewable energy systems, such as a malfunctioning solar panel or a defective wind turbine blade. This can decrease maintenance and downtime expenses. Machine learning may be employed

Data-Driven Energy Management and Tariff Optimization in Power Systems: Shaping the Future of Electricity Distribution through Analytics, First Edition.
Edited by Hamidreza Arasteh, Pierluigi Siano, Niki Moslemi, and Josep M. Guerrero.

to anticipate maintenance requirements in renewable energy systems, including wind turbines and solar panels. Machine learning systems may detect possible issues by evaluating data related to weather conditions. These models can forecast the output of future energy by historically analyzing weather, power, and other relevant energy production data. This assists energy grid operators in regulating energy supply and demand. Furthermore, employing machine learning-based techniques allows for the estimation of energy consumption and loads. ML models may assess the stability of the power grid with the integration of RESs, identify possible concerns, and maintain grid stability to prevent outages, maintaining reliability. This can facilitate the prediction and detection of abnormalities in the energy system and regulate changes in energy supply [1].

The main contributions provided by this chapter include the following:

1) This study presents a complete categorization of ML algorithms and their importance and applications in RESs integration.
2) Recent developments in RESs optimization and stability forecasting, including ML methods, are reviewed.
3) Recommendations for future research, challenges, and gaps in RESs studies are indicated.

Figure 7.1 illustrates the graphical abstract/outlook of this chapter, encompassing machine learning applications in RESs, optimization, grid stability, and challenges and opportunities. In this regard, the remainder of this chapter is as follows: In Section 7.2, ML importance in RESs is explained along with a categorization of different practical algorithms with their applications in the domain. In Section 7.3, the role of ML in optimizing RESs generation is introduced with the application of different ML techniques in optimization methods in RESs. In Section 7.4, the grid stability methods and their applications in ensuring grid stability through ML-based forecasting is presented. The future challenges direction in ML-based approaches in RESs studies are presented in Section 7.5, and finally, the conclusion of the chapter is presented in Section 7.6.

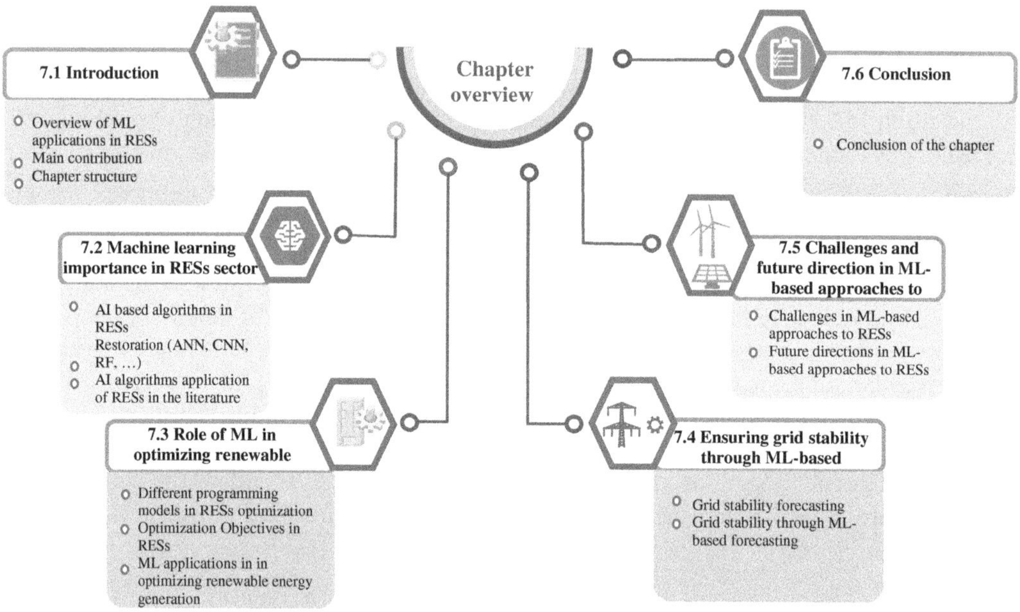

Figure 7.1 Overview of the chapter.

7.2 Machine Learning Importance in RESs Sector

This section reviews the prevalent applications of machine learning techniques across several renewable energy sectors that influence sustainability. The machine learning algorithms employed in renewable energies mostly concentrate on wind and solar energy. The subsequent discussion focuses on machine learning applications and their influence on enhancing sustainability across diverse energy sectors [2].

7.2.1 AI-Based Algorithms in RESs

One of the key phases in the developing procedure of ML models and an appropriate preprocessing approach guarantees that the data utilized for the deployment of ML is consistent and acceptable for further analysis. This may encompass numerous essential phases, including [3]:

- **Data Collection**: Combining data from different sources such as smart meters, sensors, and weather stations and using historical demand data provide a complete dataset that boosts the model's capacity to learn effectively.
- **Data Cleaning**: Different approaches, including imputation using mean and median, together with smoothing, filtering, and outlier identification, could be used to ensure the integrity of data and avoid the negative impact of incomplete data on the model efficacy.
- **Data Transformation**: The accuracy prediction of the model could be enhanced using new feature generation from raw data or integrating existing features. Decreasing the number of features while preserving critical information and computational complexity will be decreased, hence enhancing model interpretability.
- **Data Segmentation**: Dividing the dataset into two subsets of training and testing and evaluating the model efficacy with a prevalent strategy of allocating 70%–80% of data for training and 20%–30% for testing will help to increase the model accuracy.
- **Data Augmentation**: In the case of unbalanced or insufficient data, methods such as synthetic minority over-sampling can be used to enhance the dataset.

In the following, the most common AI-based algorithms in the domain of RESs are provided.

- **Supervised Learning**: This approach involves training the model on a labeled dataset, where the input–output pairings are established, with the primary objective of mapping inputs to outputs. Prevalent algorithms contain [1]:
  - **Linear Regression (LR)**: The primary assumption in LR is having a linear correlation between the inputs and the output of the labeled dataset, which can be used for predicting a continuous output variable based on input factors.
  - **Support Vector Machine (SVM)**: SVM identifies the optimal hyperplane that distinguishes the classes inside the feature space and can be utilized for classification and regression applications. They are effective for small to medium-sized datasets with few characteristics [4].
  - **Artificial Neural Networks (ANN)**: With an advanced pattern recognition capability, this technique is employed for intricate tasks, enabling the learning of nonlinear correlations of inputs and outputs.
  - **Decision Trees (DT)**: DT is an interpretable algorithm for both classification and regression tasks, where it is based on data splitting into branches with respect to decision rules.
  - **Random Forests (RF)**: RF is an ensemble learning technique that can be used in classification and regression tasks and constructs the aggregation of several decision tree outputs to generate predictions.

- **Gradient Boosting Machines (GBMs)**: GBMs are an ensemble learning technique with superior prediction accuracy that constructs numerous decision trees sequentially, with each successive tree fixing the errors of the prior one.
- **XGBoost**: An improved version of GBM recognized for speed and performance, which may be utilized for both classification and regression tasks [5].
- **K-Nearest Neighbors (KNN)**: The KNN is a straightforward but efficient classification algorithm that identifies the KNN of a new data point and tries to forecast the class based on the categorization of those neighbors. They are effective in a dataset with small to medium size and limited characteristics.
- Naive Bayes is a straightforward and effective method for classification problems that utilizes Bayes' theorem, relying on independence assumptions among the features.
- **Ensemble Learning (EL)**: The EL is a supervised approach that merges the capabilities of many models. With distanced forms of bagging, boosting, and stacking, the objective is to independently train many models on identical data and subsequently amalgamate their forecasting to derive the final outcome [6]. Bagging is an ML ensemble meta-algorithm aimed at enhancing the stability and accuracy of the model. The technique also diminishes variance and assists in mitigating overfitting. Boosting is a straightforward method that significantly enhances overall performance by integrating several weak classifiers to create a robust committee, employing an iterative process that focuses on poor classifications. Stacking entails training many models and using their predictions as input for a superior model that learns to amalgamate them [7].
- **Support Vector Regression**: It is employed in regression analysis in the domain of ML. It operates by identifying the optimal hyperplane that can distinguish the data points inside a high-dimensional space.
- **Logistic Regression (LR)**: The LR is a classification technique that uses one or more independent variables to forecast a binary-dependent variable [8].
- Generalized Regression Neural Network (GRNN) can widely be employed in regression problems that need continuous variable prediction.
- **Adaptive Boosting (AdaBoost)**: AdaBoost is a fundamental boosting method commonly utilized as an ensemble learning strategy. A base learner is taught in each iteration of the algorithm, and according to the performance of previous iterations, the weights of the training sample are adjusted [9].

- **Unsupervised Learning**: This method addresses unlabeled data when output variables lack categorization. The primary function of this model is to recognize patterns or clusters within data. It is appropriate for feature extraction and exploratory data analysis. Prevalent algorithms of technique comprise:
  - **Clustering Algorithms**: Methods such as K-means that categorize data points into clusters based on their resemblance to one another. K-means has several applications across diverse domains, including renewable energy forecasts [10].
  - **Dimensionality Reduction Techniques**: Techniques like principal component analysis (PCA) are employed to diminish the amount of characteristics in data while maintaining its fundamental structure. This facilitates the visualization of high-dimensional data while simultaneously reducing processing complexity, particularly for large datasets. Dimensionality reduction techniques employed in renewable energy forecasting include auto-encoders, PCA, and t-SNE [10]. The t-SNE is a nonlinear dimensionality reduction approach that effectively visualizes high-dimensional data in a lower-dimensional environment. It employs a probabilistic method to associate analogous data points with proximate locations in a space with low dimensions. Auto-encoders are a category of neural networks capable of encoding and decoding high-dimensional data into a lower-dimensional representation.

The encoder network is taught to compress the input data into a lower-dimensional form, while the decoder network is trained to recover the original data from this compressed representation [10].

- o A Markov process is a stochastic process characterized by the Markov property, indicating that future states are independent of past states when conditioned on the current state.
- o **Auto-encoder**: Auto-encoder models have been employed in renewable energy forecasting to extract characteristics from input data, including historical energy output data, meteorological data, and other related parameters. These attributes are then employed to train ML models for predicting the output energy.
- o **Deep Belief Networks (DBN)**: DBNs are deep neural architectures composed of many layers of Restricted Boltzmann Machines.

- **Adaptive Neuro-Fuzzy Inference System (ANFIS)**: ANFIS, as a hybrid model that integrates ANN with fuzzy logic, was developed in the early 1990s. It employs a fuzzy inference technique to estimate nonlinear functions, rendering it a robust generic estimator. ANFIS comprises five layers: the data input layer and fuzzification, rule, normalization, and output layers [11].
- **Reinforcement Learning (RL)**: This approach entails training a model to make a series of decisions by incentivizing desirable actions and penalizing undesirable ones. Through this method, the model acquires the ability to optimize cumulative rewards during the time and identify useful information, which generally includes Q-learning and deep reinforcement learning [12].
  - o **Q-Learning**: This is an RL model-free algorithm that learns the value of actions in a certain state through reward maximization.
  - o **Deep Reinforcement Learning**: Integrates RL with deep neural networks (DNN) to navigate extensive and intricate state spaces in domains such as gaming and robotics.

  Deep learning is a branch of machine learning that uses layered neural networks to recognize intricate patterns inside extensive datasets. In contrast to conventional ANN in deep machine learning (ML), DNN possess additional hidden layers, endowing deep learning (DL) with a superior capacity to model intricate connections. Prevalent architectures comprise:
  - o **Convolutional Neural Networks (CNNs)**: Mostly employed for video and image recognition assignments. In this manner, convolutional layers may be automatically identified from unprocessed data.
  - o **Recurrent Neural Networks (RNNs)**: The RNNs are recommended for sequential data, including natural language processing and time series. RNNs utilize the memory of prior inputs to improve subsequent forecasting.
  - o **Long Short-Term Memory (LSTM)**: LSTM is effective in time series forecasting applications, and each neuron has a memory cell that discards the data or remembers prior data.
- **Extreme Learning Machines (ELM)**: The ELM was introduced as an alternative to traditional gradient-based neural networks. ELM consists of a single layer that has hidden neurons. In the layer, the links across the input and the hidden layers are randomly established and endure fixed. The output layer of an ELM generally executes LR or classification, with the links among the hidden and the output layers established by analytical methods including matrix inversion.
- **Transfer Learning (TL)**: TL is a machine learning methodology that employs a pretrained model as a benchmark for a new task, rather than constructing a new model from the ground up. The pretrained model is often trained on an extensive dataset and has acquired valuable characteristics applicable to other similar tasks. A prevalent method of transfer learning in RESs prediction is utilizing pretrained models from other analogous forecasting activities.

The summary of a typical flow diagram of the ML model in the domain of RESs, including data preprocessing, different algorithm types, and evaluation metrics, is depicted in Figure 7.2. Moreover, the evaluation metric formulas (Eqs. 7.1–7.10) are shown in Table 7.1. The details of

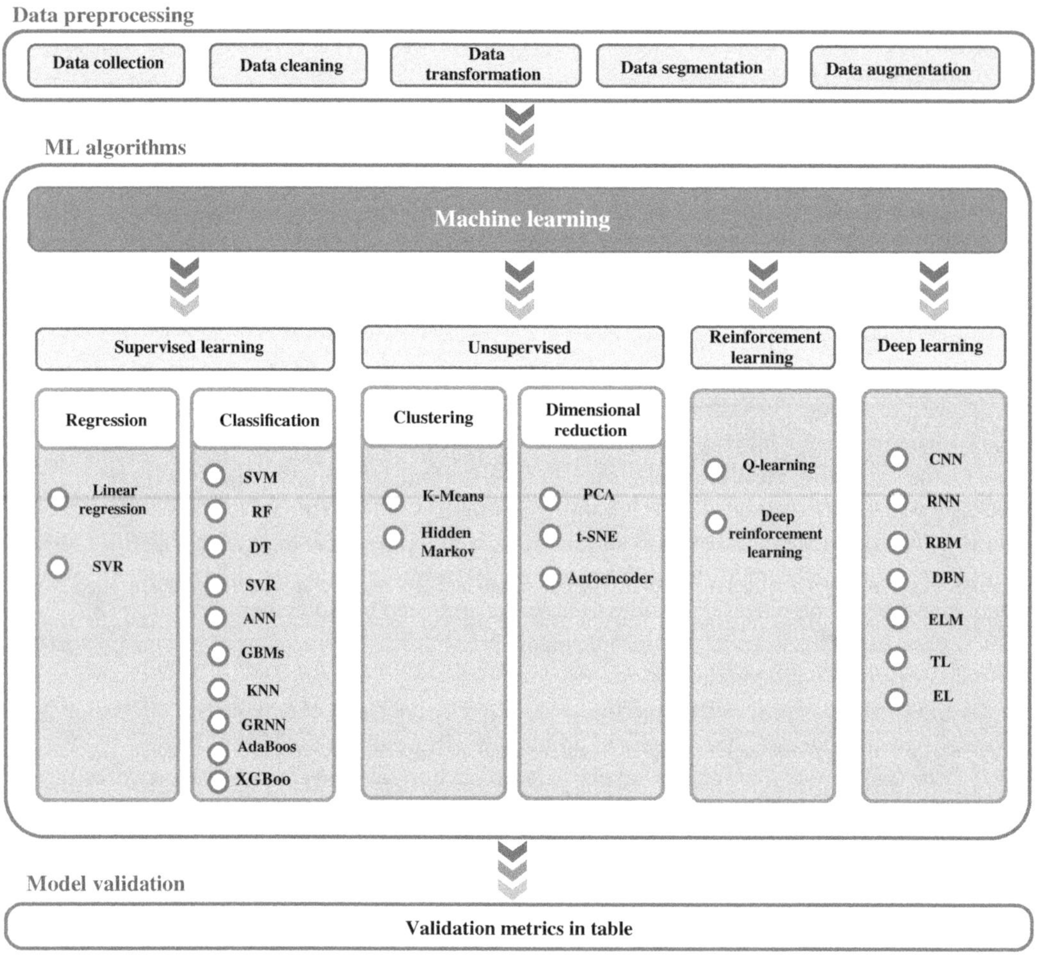

Figure 7.2 A typical flow diagram of the ML model in the domain of RESs.

more common ML algorithms in RESs with schematic architecture, related equations, and their advantages and limitations are provided in Table 7.2.

7.2.2 ML Algorithms Application in RESs

The following is a summary of some of the most significant fields in which machine learning may make a contribution to the field of RESs:

- **Energy Production, Load, and Demand Forecasting**: ML methods can be employed to predict RESs energy. These methods can forecast the energy output of the future considering historical data related to weather conditions and power generation output, enabling energy grid operators to more effectively manage load and generation. In order to better integrate RESs into the power grid, machine learning (ML) can be used to predict energy demand. This is particularly crucial as RESs like solar and wind are variables, and precise forecasts of energy demand can facilitate sustaining a stable electricity supply. Load prediction helps to enhance resource optimization and minimize downtime.

Table 7.1 The evaluation metric formulas.

Metric	Definition		Description
Accuracy	$\text{Accuracy} = \frac{\text{TP} + \text{TN}}{\text{TP} + \text{TN} + \text{FP} + \text{FN}}$	(7.1)	• Total successful prediction ratio (optimal value = 1) • TP: true positive; TN: true negative; FP: false positive; FN: false negative
Precision	$\text{Precision} = \frac{\text{TP}}{\text{TP} + \text{FP}}$	(7.2)	• True positive to total positive prediction ratio (optimal value = 1) • Appropriate when the cost of FP is high
Recall	$\text{Recall} = \frac{\text{TP}}{\text{TP} + \text{FN}}$	(7.3)	• True positive to total actual positive ratio (optimal value = 1) • Appropriate when the cost of FN is high
F1 score	$\text{F1 score} = 2 \times \frac{\text{Precision} \times \text{Recall}}{\text{Precision} + \text{Recall}}$	(7.4)	• The harmonic mean of precision and recall (optimal value = 1) • Appropriate for an imbalanced dataset
MSE	$\text{MSE} = \frac{1}{n}\sum_{t=1}^{n}(y_i - y_t)^2$	(7.5)	• Average squared difference of predicted and value • More sensitive to outliers (optimal value = 0)
RMSE	$\text{RMSE} = \sqrt{\frac{1}{n}\sum_{t=1}^{n}(y_i - y_t)^2}$	(7.6)	• The square root of MSE (optimal value = 0)
MAE	$\text{MAE} = \frac{1}{n}\sum_{t=1}^{n}\lvert y_i - y_t\rvert$	(7.7)	• Average magnitude of error (optimal value = 0) • Less sensitive to outliers than MSE
R-Squared	$R^2 = 1 - \frac{\sum_{i=1}^{n}(y_t - y_i)^2}{\sum_{t=1}^{n}(y_i - \hat{y})^2}$	(7.8)	• Variance in the predictable dependent variable from the independent variable portion (optimal value = 1)
RSE	$\text{RSE} = \frac{\sum_{i=1}^{n}(y_t - y_i)^2}{\sum_{t=1}^{n}(y_t - \hat{y})^2}$	(7.9)	• Relative squared difference of predicted and actual values (optimal value = 0)
MAPE	$\text{MAPE} = \frac{1}{n}\sum_{t=1}^{n}\frac{\lvert y_i - y_t\rvert}{y_i}$	(7.10)	• Average percentage difference of predicted and actual values (optimal value = 0)

Table 7.2 The detailed overview of more common ML algorithms in RES.

Algorithm schematic architecture	Related equation	Pros and cons
(a) LR 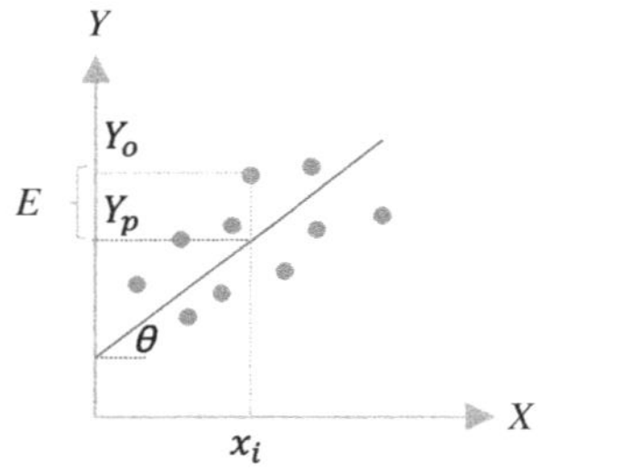	$\hat{y} = \theta_1 + \theta_2 X$ (7.11) $\widehat{P}(t, \vec{\alpha}) = \alpha_0 + \alpha_1 I(t-1) + \alpha_2 T(t-1)$ (7.12) $\min_{\vec{a}} \left(\sum_t \mathcal{L}(\widehat{P}(t, \vec{\alpha}) - P(t)) \right)$ (7.13) Equation (7.11): Linear relation between X and Y E: error; θ_1: slope; θ_2: X coefficient; Y_o: observed value; Y_p: predicted value; x_i: input value; $P(t)$: the estimation of the PV power the estimation of the PV power; $T(t-1)$: the temperature at the previous time step. The coefficients of the linear regression $\alpha^{\rightarrow}$ are found by Eq. (7.13)	• Easy implementation • Rapid training • Linear relationship limitation
(b) SVM 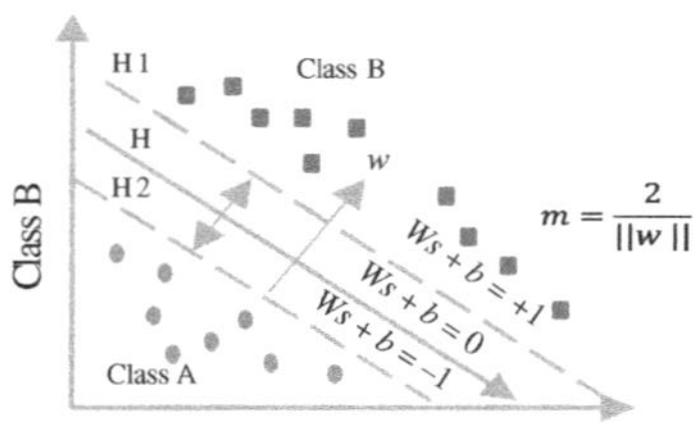	The optimization problem of the optimal hyperplane [13]: $\text{Minimize } \frac{1}{2}\lVert w \rVert^{-2} + C\sum_{i=1}^{M} \xi_i$ (7.14) S.t: $O_i(w^T s + b) \geq 1 - \xi_i$ and $\xi_i \geq 0$ for i $\lVert w \rVert^{-2}$: geometrical distance; ξ_i: slack variable; C: regularization parameter; w: the weight vector for the hyperplane orientation; b: the bias for the hyperplane position; O_i : class label	• Efficient handling of high-dimensional datasets • Training could be slow
(c) DT 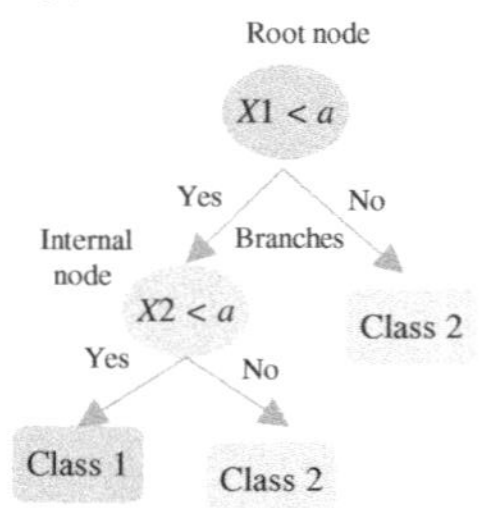	$\text{Gini}(S) = 1 - \sum_{i=1}^{n} p_i^2$ (7.15) The Gini index is a widely utilized metric for determining splits in the classification and regression trees [14]. Where S: sample; n: number of unique classes; p: proportion of samples	• Easy to interpret • Overfitting may happen

(d) K-means 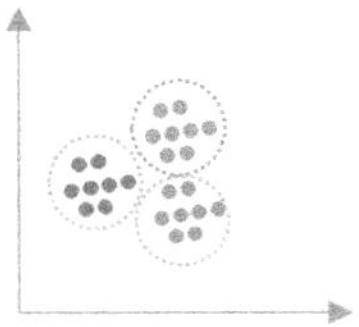	It quantifies the linear distance between both spots in a multidimensional space. The Euclidean distance between two points (x_1, y_1) and (x_2, y_2) is determined as $$d = \sqrt{((x_2 - x_1)^2 + (y_2 - y_1)^2)} \quad (7.16)$$	• Simple and quick for data categorization • Cluster numbers are needed
(e) RL 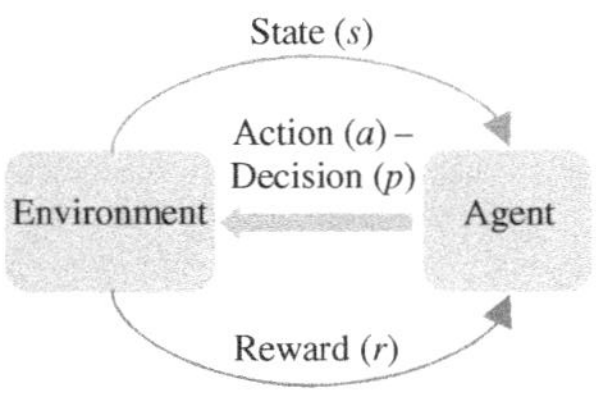	The objective function of RL can be written as $$\max_a \sum_{t=1}^{\infty} T(s_t, a(s_t), s_{t+1})[R_a(s_t, a(s_t)) + \sigma \cdot V(s_{t+1})] \quad (7.17)$$ $R_a(s_t, a(s_t))$: immediate reward after taking section $a(s_t)$ in state s_t; $V(s_{t+1})$: long-term average return of the system in the state s_{t+1}; σ: discount factor State: current condition; reward: feedback signal; environment: system; action: agent choice	• Effective in complex and dynamic spaces • Requires a huge volume of data
(f) TL 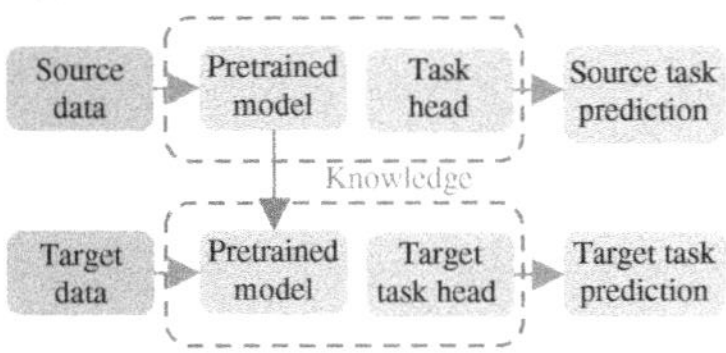	Fine-tuning includes establishing the target model using the parameters of a pretrained model. Equation (7.18) shows the relation of the parameter of the target model (θ_T) and loss function L_T. The objective is to minimize loss in the target parameter $$\theta_T = \arg\min_{\theta} L_T(\theta) \quad (7.18)$$	• Less data is needed using pretrained model • Limitation in similar tasks

(Continued)

Table 7.2 (Continued)

Algorithm schematic architecture	Related equation	Pros and cons
(g) ANN 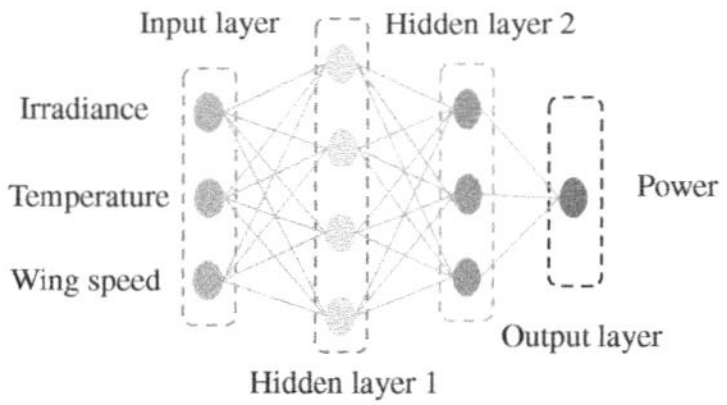	If x_i is the input of neuron j, the output is : $$u_j = \phi\left(\sum_{i=1}^{N} w_i x_i + b_j\right) \quad (7.19)$$ ϕ: activation function; b_j: bias neuron j; w_i: weight factor of input and neuron	• Handle nonlinear and complex relationships • Requires a huge volume of data
(h) CNN 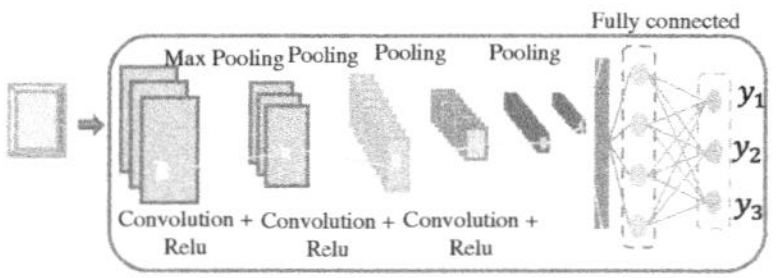	Given an input picture denoted as I and a filter represented as F, the resultant output expression is $$O = I * F \quad (7.20)$$ ReLU is one of the most common activation functions with a lower computational load, as shown in Eq. (7.21) $$f(x)_{ReLU} = \max(0, x) \quad (7.21)$$	• Automatic feature extraction • Computationally expensive
(i) RNN 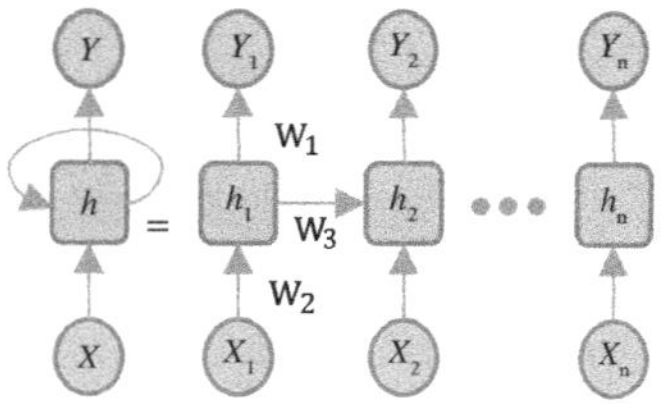	The output of an RNN model is shown in Eq. (7.22) $$\hat{y}_t = W_1 \tanh(W_2 h_{t-1} + W_3 x_t) \quad (7.22)$$ $\hat{y}_t$: the predicted output; W_1 : the weight matrix connecting the hidden state; W_3: the weight matrix for the recurrent connection; W_2: the weight matrix connecting the input to the hidden state; h_{t-1}: the hidden state; x_t: the input	• Sequential data handling • Slow training and overfitting may happen

(j) Naïve Bayes 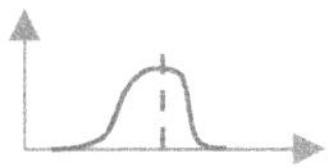	Probability to generate the PV power (P) with irradiance (I) and temperature (T): $$p(P \mid I, T) = \frac{1}{p(I, T)} p(P) p(I, T \mid P) \quad (7.23)$$ $p(I, T)$: the distribution of meteorological conditions	• Computationally efficient • High correlated features limitation
(k) GRNN 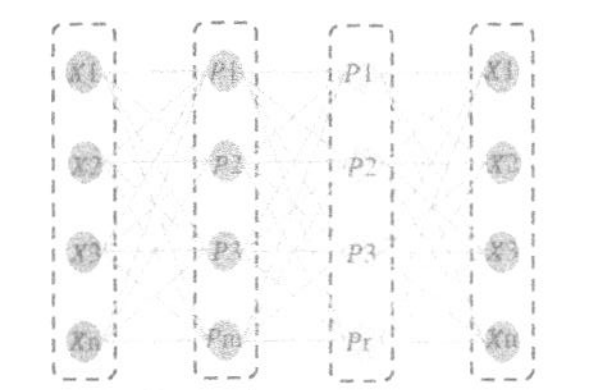	The predicted output for the input x is $$\hat{y}_i = \frac{\sum_{i=1}^{n} y_i \cdot \exp[-D(x, x_i)]}{\sum_{i=1}^{n} \exp[-D(x, x_i)]} \quad (7.24)$$ $$D(x, x_i) = \sum_{k=1}^{m} \left(\frac{x_i - x_{ik}}{\sigma_k} \right)^2 \quad (7.25)$$ y_i: the predicted output for the input x; $\exp[-D(x, x_i)]$: kernel function	• Fast training • May not be a satisfactory scale for large datasets
(l) Auto-encoder 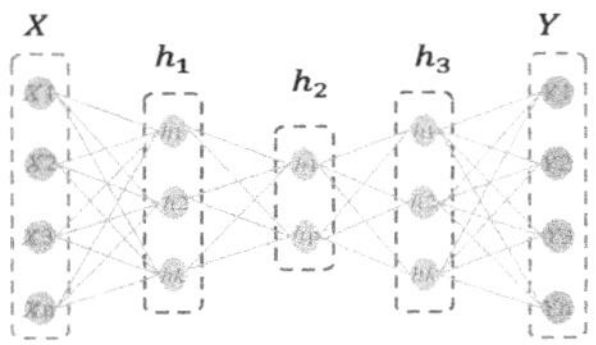	Auto-encoder loss function is written in Eq. (7.26): $$L(\theta, \phi) = \frac{1}{n} \sum_{i=1}^{n} (x^i - f_\theta(g_\phi(x^i)))^2 \quad (7.26)$$ $f_\theta(g_\phi(x^i))$: reconstruction image; g_ϕ: encoder; f_θ: decoder	• Dimension and noise reduction • Requires a huge volume of data

(Continued)

Table 7.2 (Continued)

Algorithm schematic architecture	Related equation	Pros and cons
(m) ELM 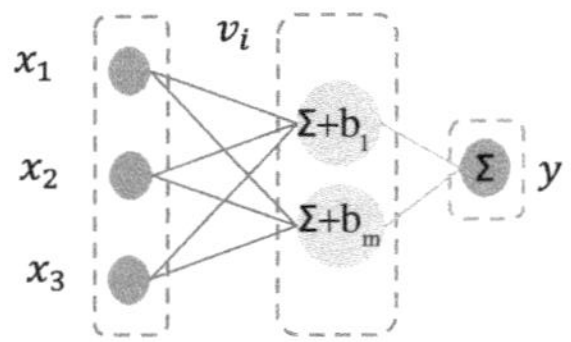	The kth hidden layer output is given in Eq. (7.27) and the output in Eq. (7.28) [15]. $$H_{ik} = \phi\left(\sum_{i=1}^{n} x_{ik}v_{ik} + b_k\right) \quad (7.27)$$ $$y_k = \sum_{j=1}^{n} H_{jk}w_j \quad (7.28)$$ where ϕ: the activation function; H: input weigh matrix; b: bias of hidden layer neurons	• Good handling of large datasets and rapid training • Sensitive to random input weights
(n) LSTM 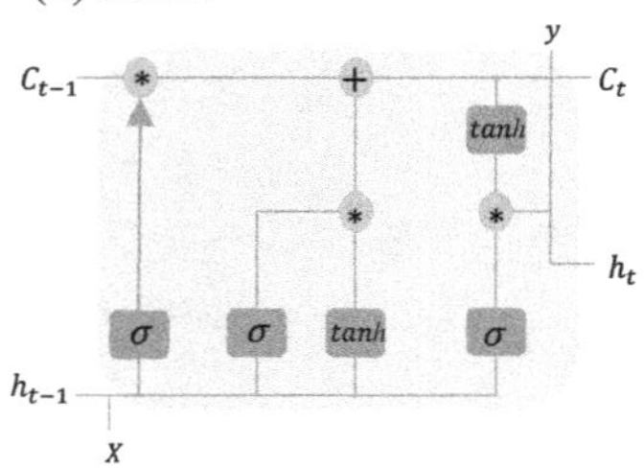	The forget gate layer filters superfluous data from the cell state that is written in Eq. (7.29). Moreover, Eq. (7.30) shows the output of LSTM $$f_f = \sigma[w_f(h_{t-1}, x_t) + b_f] \quad (7.29)$$ $$h_i = O * \tanh(c_i) \quad (7.30)$$ where σ: the sigmoid activation function; w_f: weight matrix; h_{t-1}: hidden state from the previous time step; b: bias; c_i: cell state; O: output gate	• Long sequence capturing • Expensive computationally

- **Resource and Energy Efficiency Optimization**: ML algorithms can enhance the utilization of RESs, such as solar and wind, by forecasting periods and locations of peak energy output. This can assist utilities in optimizing their energy supply. Moreover, ML could be leveraged to enhance the efficiency and effectiveness of energy storage systems, including batteries. By analyzing data on energy generation, demand, and meteorological conditions, ML can be used to improve RESs for maximum efficiency and cost-effectiveness, generating the maximum energy at the lowest possible cost [16].
- **Grid Stability Analysis**: By analyzing the power grid's stability while adding RESs, machine learning models could help integrate RESs into the grid more effectively. ML models may detect potential problems and maintain grid stability by preventing outages and guaranteeing a consistent and reliable electricity supply, even with the integration of additional RESs.
- **Predictive Maintenance and Fault Diagnosis**: ML can be employed to forecast maintenance requirements in RESs, including wind turbines and solar panels. ML algorithms can detect possible issues by analyzing data on variables such as energy generation, weather conditions, and the performance of components, facilitating proactive maintenance, as well as identifying faults in wind turbine blades or malfunctioning solar panels [17].
- **Smart Grid (SG) Administration**: ML could play a key role in regulating the balance between supply and demand, overseeing energy storage, and maintaining system stability in smart grids. To enhance real-time energy allocation, ML algorithms utilize data from smart meters, weather predictions, and energy usage patterns.
- **The Implementation of Demand Response (DR)**: ML algorithms enhance advanced demand response methods by forecasting consumption patterns and autonomously managing loads. Machine learning algorithms evaluate data from smart meters and Internet of Things devices to predict the date and location of peak energy consumption.
- **Security and Resilience of Smart Power Grids**: The growing dependence on digital technology and interconnected systems makes cybersecurity a key issue for smart grids. Through real-time cyber threat identification and mitigation, ML enhances the security and resilience of smart grids, which may examine patterns in network traffic to detect abnormalities that could signify cyberattacks.
- **Energy Trading and Market Operations**: A prospective area of research is the merging of ML in energy trading and market operations that improve trading strategies by examining market movements, pricing, and consumption patterns. Through decentralized energy trading, AI-driven systems make it easier to exchange excess energy generated from renewable sources.

The rotor blade power generation is shown in Eq. (7.31) [18], where p_T is the turbine power output; ρ is the air density; R is the radius of the rotor and v denotes wind speed; $C_p(\lambda, \beta)$ is the power coefficient; λ represents the tip-speed ratio; and β is the pitch angle.

$$p_T = \rho\pi R^2 v^3 C_p(\lambda, \beta) \tag{7.31}$$

The direct radiation of solar rays is assessed using Eq. (7.32) [18], where G_{sc} is the solar constant; P is the transparency factor, and M is the mass of the air. The solar altitude angle is α_s.

$$I_{br} = G_{sc}p^M, \quad M = \frac{1}{\sin \alpha_s} \tag{7.32}$$

The overview of ML algorithms application in RESs is illustrated in Figure 7.3. The authors [19] utilized an ANN to extract regression rules from meteorological data, and the forecasts from every model were merged using XGBoost, facilitating the measurement of individual model error and data noise uncertainty, hence improving prediction accuracy. The study [20] proposes

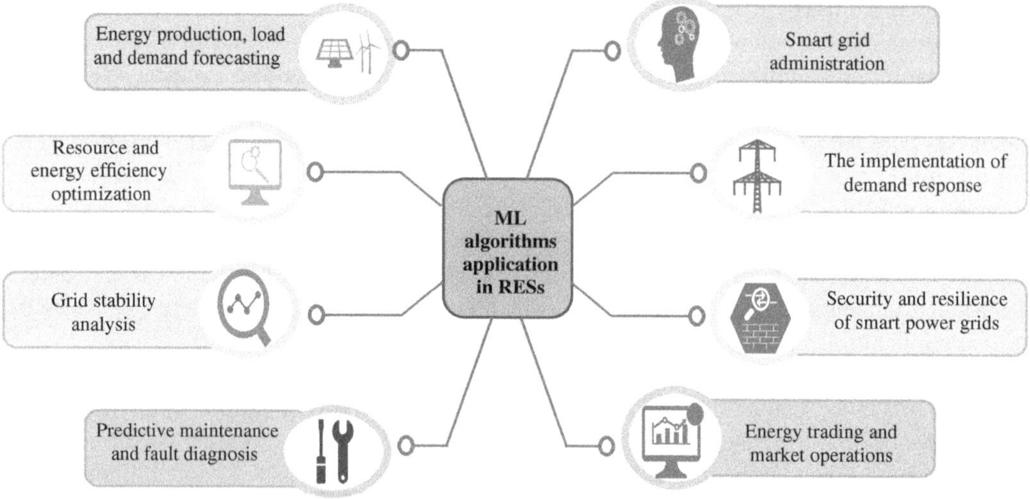

Figure 7.3 The overview of ML algorithms application in RESs.

a data-driven approach for wind power forecasting that encompasses preprocessing, feature engineering, anomaly detection, and tuning of hyperparameters utilizing gated recurrent deep learning methods. The authors [21] presented a DBN model utilizing clustered numerical weather prediction data to improve wind power prediction, achieving over 44% greater accuracy compared to Morlet wavelet neural networks and back-propagation neural networks in tests conducted at the Sotavento wind farm.

The authors [22] created a predictive system for wind speed and temperature employing a mix of ANN, a genetic algorithm integrated with ANN, and ANFIS. The use of ANFIS with a trapezoidal membership function produced the most favorable outcomes, achieving an ideal MSE of 3.8364°C for temperature and 7.2989 m/s for wind speed. Reference [23] proposed an ensemble model for predicting hourly global horizontal irradiance that combines DNN and extreme gradient-boosting forests. The study [24] presents a grey wolf optimization-based GRNN for forecasting energy production in solar power systems, influenced by unpredictable parameters including cloud cover and weather conditions. The authors [25] established an ELM model to forecast wind power utilizing kernel mean p-power error loss, addressing the constraints of conventional back-propagation neural networks. The strategy utilizes PCA to minimize duplicate data components, resulting in reduced prediction errors while maintaining accuracy. This study introduces an innovative continuous learning framework for online static security assessment (SSA) that facilitates the regular updating of the assessment model. The system comprises a Mondrian forest (MF)-based training model and a database creation method that analyzes operating condition characteristics. The employed security index is a novel performance metric influenced by weather conditions, with ambient temperature impacting line ratings [26]. In a study [27], power system resilience is defined and examined, encompassing risk and vulnerability assessment, resilience metrics, and operational and planning strategies. Additionally, the modeling of vulnerability and resilience evaluation indices, along with cost-benefit analysis, is also addressed. Furthermore, an expanded hypothesis on resilience enhancement, incorporating confounding variables within the distribution system domain, is offered in Ref. [28]. The study [29] delineates an approach for forecasting load and pricing to balance power demand and supply. A hybrid methodology is developed utilizing a multistage forecast engine and dual-tree complex wavelet transform. The signals undergo wavelet processing followed by an advanced feature selection filtering procedure. The ML applications in RESs in the literature are summarized in Table 7.3.

Table 7.3 The summarization of ML applications in RESs in the literature.

Reference	Goal/strategy	Algorithm and method	Results and description
[29]	Solar irradiance prediction	CNN	• CNN method for solar prediction based on meteorological data • GA/PSO for hyperparameter optimization • Decreases the annual average mean absolute error
[30]	Wind speed forecasting	ANN, kernel ELM	• Three different wind speed data stations are used • Decreases the MAPE values
[31]	Wind speed forecasting	Deep learning	• Using preprocessing and a deep learning-based nonlinear combination forecasting module • Enhances the wind speed forecasting performance
[32]	Solar radiation forecasting	CNN	• Predicting complex time series with a high degree of irregularity • Three one-dimensional convolutional layers and three pooling layers
[19]	Improving PV generation forecasting	Stacked ensemble algorithm using ANN and LSTM	• High stability and consistency on weather variations
[33]	Solar power forecasting	CNN and LSTM	• CNN for weather conditions classification, and LSTM for power generation patterns with respect to the weather conditions • The mean absolute percentage error of 4.58 on a sunny day
[34]	Solar energy generation prediction	LSTM	• It models the PV time series data, and the best recurrent model achieved a normalized MSE value of 0.014
[35]	Hydroelectric energy generation for water supply	Integrated permanent magnet generator	• Using a permanent magnet generator in the micropower generation to minimize the total operating hours
[25]	Wind power forecasting	ELM with kernel mean *p*-power error loss and PCA	• Can achieve lower prediction error • Reduces the computational problems caused by huge volumes of data
[23]	Solar irradiance forecasting	Ensemble model	• Using extreme gradient boosting forest and DNN
[36]	Predict energy generation and demand patterns	LSTM, SVM, RF	• Random forest achieved the lowest MAE • CO_2 emission reductions are quantified
[37]	Solar power forecasting	Light gradient boosting machine and KNN	• The LGBM model shows better accuracy (RMSE: 5.77 vs. 6.93) • Training times for the LGBM model are longer (120 s vs. 90 s) and increased memory consumption

7.3 Role of ML in Optimizing Renewable Energy Generation

The optimization process throughout the RESs development planning and utilization phase includes tasks such as assessing potential, establishing development objectives, and refining configuration. The modeling of RES optimization decisions is often based on economic and technical objectives and limitations. Studying the different programming and ML models, algorithms, and objective functions in optimization problems could enhance the effectiveness of using ML in RESs.

7.3.1 Different Programming Models in RES Optimization

Technical issues include evaluations of stability and safety. Moreover, there could be social and environmental concerns that would be reflected in the objectives. In the following, different models of programming in RESs optimization problems are provided [38].

- **Linear Programming (LP)**: This paradigm is predominantly utilized in all aspects of RESs. LP models exhibit constrained linear relationships and expressions, and a stringent dependence on data accuracy may pose a restriction.
- **Nonlinear Programming (NLP)**: This encompasses iterative techniques and diverse heuristic algorithms. It may encompass optimal power flow solvers. The constraint is that NLP frequently identifies only local optima, and significant scaling challenges may emerge owing to linearization.
- **Mixed Integer Linear Programming**: This model yields integer outcomes and assists in decision-making, such as evaluating whether to undertake a task (e.g., the facility location problem in renewable energy systems). Limitations encompass the stringent accuracy requirements of algorithms and the challenges associated with resolving large-scale models.
- **Stochastic Programming (SP)**: This model is often utilized in optimization problems characterized by risks and uncertainties. Nonetheless, its constraints encompass the curse of dimensionality and substantial spatial demands.
- **Dynamic Programming (DP):** This model is appropriate for uncertain decision-making in RESs and offers adaptable models for decision processes. However, analyzing runtime is challenging and may yield inaccurate solutions with little likelihood.
- **Bi-level Programming (BP)**: This paradigm emphasizes the interplay among various decision-makers and is applicable to diverse energy sectors or subsystems.
- **Multi-Objective Decision-Making**: This model evaluates economic, technological, environmental, and social dimensions. It is appropriate for energy management and dispute resolution in decision-making. The disadvantage is that these models may be challenging to manage when addressing contradictory objectives, making it difficult to achieve optimal Pareto fronts.
- **Hybrid Models/Techniques in Artificial Intelligence**: These models integrate predictive or simulation-based methodologies with evaluative frameworks, enhancing their applicability and validity. They enhance the theory of optimization mechanisms and provide support for data and outcome evaluation. Hybrid models entail the intricacies of system and information flow and may encounter difficulties in reconciling diverse models.

7.3.2 Optimization Objectives in RESs

Optimization strategies frequently concentrate on the load-shaping curve to align with improved supply availability. This may entail redistributing demand from peak periods to off-peak periods, decreasing total energy usage, or mitigating demand surges. Dynamic pricing schemes, including critical peak pricing and time-of-use rates, are employed to encourage users to adjust their electricity consumption to off-peak hours. This aids in load balancing on the grid and diminishes the necessity for peaking power facilities. The objective function is determined based on the task that has been tried to be optimized. Common objective functions in RESs optimization include [3]:

- **Minimize Overall Costs**: This includes optimization techniques designed to decrease the total energy consumption costs.
- **Minimizing Energy Losses**: This alternative could enhance efficiency and diminish the necessity for expanding the generation capacity.
- **Enhancing Efficiency of Energy**: This goal includes increasing the efficiency of energy use, reducing energy waste, and improving the functionality of machinery and procedures.
- **Minimizing Voltage Deviation**: The power factor close to 1 p.u. could result in enhanced voltage improvement.
- **Minimizing Greenhouse Gas Emissions**: Reducing the use of fossil fuels and, thus, greenhouse gas emissions is one of the main goals of optimizing electricity use to protect the environment.
- **Minimization of Load Within Peak Hours**: Using demand-side management and load-shedding techniques, diminished the peak demand and can mitigate costs of necessity for additional infrastructure expenditures.
- **Ensuring System Stability**: To prevent possible instabilities and outages, this entails keeping the power grid's voltage and frequency in equilibrium.

Figure 7.4 shows in detail a flow diagram of optimization problems in RESs, including decision variables, objective function constraints, and programming methods.

Some indices and important features in RESs (with Eqs. 7.33–7.41) are provided in Table 7.4 that could be helpful in constructing effective optimization problems [3].

7.3.3 ML Applications in Optimizing Renewable Energy Generation

Optimization algorithms derived from animal social behavior are categorized as swarm intelligence (e.g., PSO, Artificial Bee Colony, and Firefly Algorithm); those based on physical laws are classified as "physics-based" algorithms (e.g., SA and GSA); algorithms inspired by human social behavior fall under "human-based" algorithms (e.g., TLBO, Harmony Search, and Tabu Search); and evolution-based algorithms (e.g., GA and DE) are grounded in Darwinian evolutionary principles [39]. Moreover, there are recent meta-heuristic algorithms that can be used in the domain of RE, such as [40]. Hybrid methods utilizing meta-heuristic algorithms and Ml-based algorithms have recently absorbed attention due to their potential and capabilities. Ref. [41] determines the optimal mix of energy sources, taking into account cost, generation capacity, and environmental effects. Particle swarm optimization (PSO) is evaluated against six more optimization algorithms: Grey Wolf Optimizer, Harmony Search, Genetic Algorithm, Mayfly Algorithm, Flower Pollination Algorithm, and CUKO Search. Reference [42] integrated a Multi-Layer Perceptron Neural Network with the Electromagnetic Field Optimization algorithm. The proposed model was evaluated against the Shuffled Frog Leaping Algorithm and Shuffled Complex Evolution. The results of RMSE, MAE, and R^2 indicated that EFO demonstrated a rapid convergence rate. A hybrid PSO-ANN strategy was developed to enhance the accuracy of photovoltaic power prediction and to optimize the tracking of maximum global energy [43]. The study [44] seeks to model, build, and optimize integrated

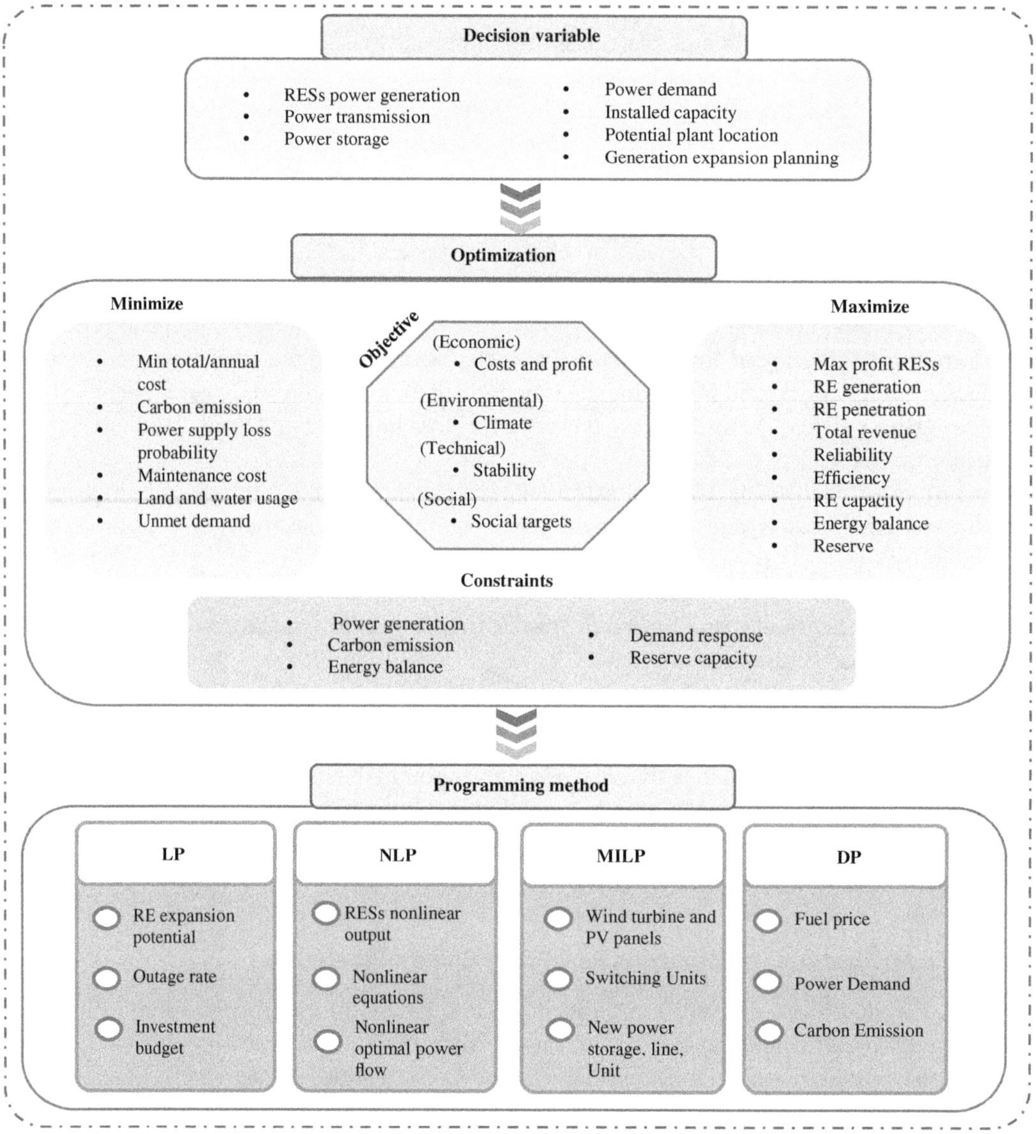

Figure 7.4 Detailed flow diagram of the optimization problem in RESs.

RESs comprising wind turbines, solar photovoltaic panels, and storage batteries. A technique is established to enhance the design of integrated renewable energy systems, focusing on reducing the net present cost and the levelized cost of electricity of these systems. In Table 7.5, some related references of ML applications in the optimization of RESs are summarized.

Table 7.4 Some indices and important features in RESs.

Type	Definition		Description
Parameters	Probability key parameters in the wind and solar resources in optimization problem		• Time, pressure, radiation, wind speed, humidity, radiation in a clear sky, blade pitch angle, gearbox oil temperature, rotor PRM, and generator winding temperature
Economic indices	$\text{LCOE} = \frac{\text{Total system life cycle cost}}{\text{Total annual energy generation}}$	(7.33)	• Leveled cost of energy • Necessary expenditure to generate a single unit of energy
Reliability indices	$\text{DPP} = \frac{\sum_{t=1}^{T} [\text{DP}(t)]}{\sum_{=1}^{T} [\, P_{load}\,(t)]}$	(7.34)	• Deficit power probability • The power deficiency to capacity power ratio
	$\text{LOLE} = \sum_{i=1}^{8760} \text{Toutage}\,(\,i\,)$	(7.35)	• Loss of load expected • Average hours when system demand exceeds generating capacity
	$\text{EENS} = \sum_{i=1}^{8760} \text{Eunserved}\,(i)$	(7.36)	• Expected energy not supplied • Energy is not delivered when system demand exceeds generation capacity
	$\text{LDR} = \frac{\sum_{t=1}^{7} (E_{load}(t) - E_{supply}\,(\text{t}))}{\sum_{t=1}^{7}\; E_{load}(t)} \%$	(7.37)	• Load dissatisfaction rate • The ratio of energy shortage at an hour to system energy demand
	$\text{LPSP} = \frac{\sum_{l=1}^{8760} \text{LPS}(t)}{\sum_{t=1}^{8760}\; P_{load}(t) \times \Delta t}$	(7.38)	• Loss of power supply probability
	$\text{ELF} = \frac{1}{T} \times \sum_{t=1}^{T} \frac{Q(t)}{P(t)}$	(7.39)	• Equivalent loss factor: indicates the probability of hybrid system failure to meet energy demand • The proportion of load loss to the required load
Social indices	$\text{HDI} = 0.091 \ln[P_{LY/C}] + 0.0724\text{HD}l_{y.z}$	(7.40)	• Human development index • Nation's overall performance assessment in the social and economic metrics
Environmental indices	$\%\text{RF} = \left(1 - \frac{\sum P \text{ Diesel generator}}{\sum P \text{ Renewable sources}}\right) \times 100$	(7.41)	• Renewable fraction • Electrical energy is provided to meet the necessary load demand in RE

Table 7.5 Summarization of references of ML application in optimization.

Reference	Goal/strategy	Algorithm and method	Results and description
[45]	Solar radiation estimation	SVM coupled with cuckoo search algorithm (CSA)	• Accuracy improvement of radiation estimation by using coupled CSA to SVM • Best parameters selection of SVM using CSA
[46]	RESs power generation forecasting	Advanced mode ANN with PSO, Jaya Algorithm, FTMA	• Error reduction to 71.261% and 80.514% in comparison with the fixed-sized dataset gradient descent-based methods
[47]	RESs power generation forecasting	EM and PSO	• Relying on ELM for prediction as a fast regression model
[48]	Short-term solar power forecasting	RNN and GA	• GA showed satisfying results
[42]	Solar energy management	ANN and electromagnetic field optimization EFO	• 16% higher accuracy compared to the shuffled frog leaping algorithm
[49]	Solar irradiation forecasting	LSTM and CNN	• Easy convergence • 37%–45% of forecast skill score over some models
[50]	Solar irradiation prediction	LSTM and Bayesian optimization	• Superior performance utilizing Bayesian optimization
[51]	Wind speed forecasting	CNN and Stochastic optimization	• The MAE value is 0.800227, and RMSE value is 0.999978
[52]	Wind energy prediction	Back-propagation neural network and GA	• This method reduces the battery storage system capacity and investment costs by 59.82%

7.4 Ensuring Grid Stability Through ML-Based Forecasting

The objective of stability analysis is to assess the electrical grid's ability to sustain stability during disruptions and to identify the nature of any instability. Timely implementation of emergency control measures is contingent upon the rapid identification of the specific type of instability. However, the extension of interconnected power grids and the incorporation of RE complicate the efficacy of old methods due to the system's intricate transitory features. The transient energy function (TEF) and the time-domain simulation (TDS) method are presently prevalent. Despite its high precision, TDS lacks computational efficiency for fulfilling the demands of online stability analysis. The TEF method does not necessitate time-intensive computations; nonetheless, deriving the energy function for large-scale power networks is challenging. The predominantly utilized second-order classical models in TEF, which overlook the intricate transient processes, may affect the precision of stability analysis. Security is characterized as the extent of risk associated with a power grid's capacity to endure imminent disturbances without disrupting customer service. Security analysis has two essential components: SSA and dynamic security assessment (DSA), differentiated by the inclusion of the post-disturbance dynamic transition process.

- SSA entails evaluating the capacity of a power system to endure a series of circumstances while adhering to system restrictions, including equipment rating limits, line power flows, and bus voltage. SSA emphasizes steady state while overlooking the transient dynamics of system variables.
- DSA assesses both steady-state and transient conditions. A dynamic secure system should satisfy security factors, including voltage and frequency changes, thermal overloading, and all types of stability, such as frequency stability, voltage stability, transient stability, and small-disturbance stability. DSA can be categorized into two distinct phases: pre-contingency and post-contingency DSA.

7.4.1 Grid Stability Forecasting

In recent years, the rising load demand and the growing integration of intermittent renewable energy have rendered stability a challenging issue. The stability issue may be classified into three categories: rotor angle stability, frequency stability, and voltage stability. Artificial intelligence has been utilized in all these domains because of its benefits in speed, precision, and flexibility. In this domain, the following concepts have a key role in assessing the power system stability assurance [53].

- **Evaluation of Rotor Angle Stability**: Rotor angle stability is the capacity of synchronous machines within an integrated power system to maintain synchronism following a disturbance. Conventional assessment approaches for large-disturbance rotor angle stability, known as transient stability analysis (TSA), possess inherent limitations. Upon inputting the necessary measurements into the ML model for online application, the TSA result may be promptly retrieved. In recent years, numerous ML-based methods have been employed for this purpose, including SVM, ELM, DBN, auto-encoders, LSTM, and CNN.
- **Evaluation of Voltage Stability**: Voltage stability denotes the capacity of a power system to sustain consistent voltages at all buses after a disturbance from a specified initial operating condition. As load demand continues to rise, power grids are compelled to function close to their transmission capacity constraints, hence jeopardizing voltage stability to a certain degree. The attributes of dynamic loads in industrial areas exacerbate the issue of voltage stability.

- **Evaluation of Frequency Stability**: Frequency stability denotes the capacity of a power grid to sustain a consistent frequency following a substantial system disturbance that causes a considerable disparity between load and generation. In contemporary smart grids, the incorporation of stochastic and volatile RE disrupts the balance between generation and consumption, while the utilization of power electronic devices diminishes system inertia. Consequently, frequency stability emerges as a significant and critical issue. In such conditions, frequency stability assessment serves as a vital method to assure stable system functionality.
- **Detection, Classification, and Localization of Faults**: Traditional fault detection algorithms, including impedance-based and traveling wave-based methodologies, are inadequate for accommodating the integration of distributed RESs and high-voltage direct current systems. Nevertheless, AI-driven approaches can circumvent intricate modeling and fault mechanism investigation.
- **Stability Control**: The SG faces fluctuating power demand and diverse operational problems, both of which affect its security. The smart grid operating states can be categorized based on their security degree, as illustrated in Figure 7.5. In each state, appropriate control measures are essential to maintain the security and stability of the grid. In the normal condition, all requirements are satisfied. Continuous minor fluctuations in load demand necessitate voltage control, damping control, generation control, and load frequency management to maintain or enhance voltage stability, rotor angle stability, and frequency stability, respectively. The alert condition is triggered when a credible contingency arises, leading the system to a condition where certain network equipment may become overloaded. In this condition, all constraints are satisfied; nonetheless, the margins of the reserve are inadequate. Preventive control techniques are essential to reinstate the system to its normal condition. SG can transition from the alert state to the emergency state if preventive control measures are either not implemented or ineffective, or it

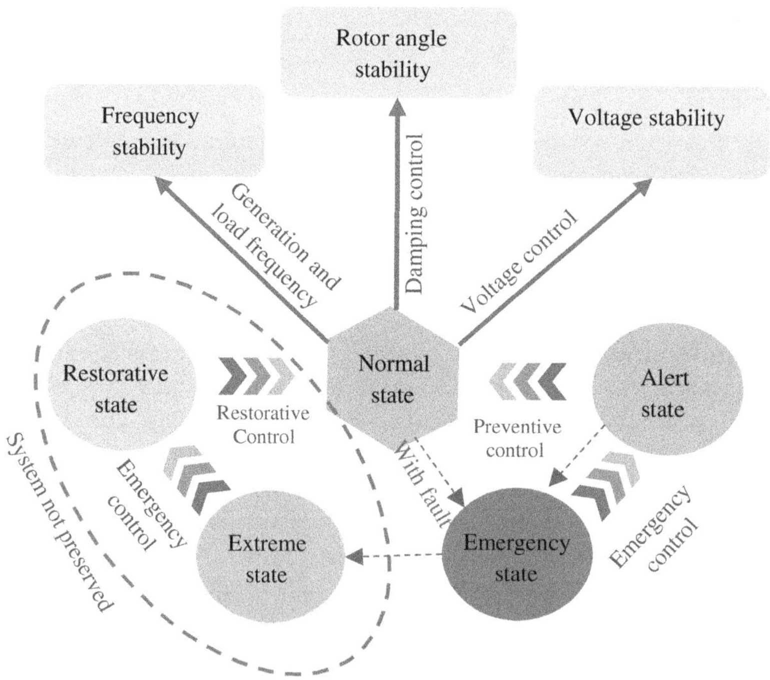

Figure 7.5 Operating states in the grid.

may shift immediately from the normal state due to severe contingencies (faults). In this condition, urgent control measures are required to revert the system to the alert state and then to the normal state. The system status could get worse and transition to an extreme condition if no emergency control measures are implemented. The principles of both inequality and equality are breached in this condition, rendering the system compromised. In this condition, efficient emergency protocols such as generator trip, load shedding, or controlled system segmentation are essential to revert the system to a restorative state. During the restorative phase, corrective control measures are implemented to reestablish all equipment and to reinstate the entire system load. The system could revert to either the alert state or the normal state.

Stability control mainly consists of generation control, voltage control, frequency control, preventive control, restorative control, and emergency control as follows:

- **Generation Regulation**: Due to the inadequate efficacy of conventional PID control for automatic generation control (AGC) in extensive interconnected power systems, smart generation control has emerged as a prevailing trend in recent years.
- **Voltage Regulation**: Automatic voltage control (AVC) pertains to the optimization of reactive power and voltage within the grid, utilizing real-time operational data gathered by the control system.
- **Damping Regulation**: The local PSS and wide-area damping controller that utilize the lead–lag link are overly dependent on precise system parameters and exhibit minimal adaptability to changes in operating conditions.
- **Preventive Regulation**: Preventive control is implemented to return the grid to a more stable and secure operational state when stability and security analyses indicate insufficient security under anticipated circumstances.
- **Emergency Control**: Emergency measures are implemented in a critical situation of the power system to restore it to the normal condition or to temporarily transition it to an alert state, or are enacted in an in-extremis scenario to revert it to its restorative state.
- **Restorative Regulation**: Restorative control comprises a sequence of systematic measures and actions implemented to return the system to a normal condition following emergency control.

In Table 7.6 some indices for stability assessment are provided [54]. In Figure 7.6, the overview of ML application to stability analysis is illustrated.

7.4.2 Grid Stability Through ML-Based Forecasting

Reference [55] introduced a fault classification technique utilizing CNN that processes characteristics derived from the Hilbert-Huang Transform in power distribution systems. Reference [56] suggested a schema for event-driven frequency and voltage stability evaluation and load shedding using SVM for stability forecasting. The study [57] presents a novel online DSA architecture utilizing semi-supervised learning. To avoid the computational expense of time-domain simulations, an algorithm known as tri-training is employed to identify the unlabeled data. The study [58] presents an innovative data-driven approach for pre-fault DSA of power systems. The study [59] presents an AdaBoost technique for the exact and accurate prediction of transient security assessments in power systems utilizing synchronized observations. A method for predicting TSA and instability modes utilizing a CNN is proposed in Ref. [60]. The method utilizes the bus voltage phasor, which is sampled by phasor measurement units during a brief observation window following a disturbance. It generates a prediction result in a timely manner, categorizing it as stable, aperiodic unstable,

Table 7.6 Some indices for stability assessment.

Equation	Description
$\text{VSLI} = \dfrac{4\left[V_s V_r \cos(\delta) - V_r^2 \cos(\delta)^2\right]}{V_s^2}$ V_s: the sending bus voltage magnitude V_r: the receiving bus voltage magnitude δ: voltage angle difference of two ends	Voltage Stability Load Index Stable: VLSI < 1; Unstable: VLSI > 1 Identifying critical buses
$\text{VSMI} = \dfrac{\delta_{r\max} - \delta}{\delta_{r\max}}$ $\delta_{r\max}$: Maximum allowable voltage angle difference	Voltage Stability Margin Index Stable: VSMI > 0; Unstable: VLMI<0 Stability margin evaluation and finding the weak location
$\text{LCPI} = \dfrac{4A\cos\alpha(P_r B\cos\beta + Q_r B\sin\beta)}{(V_s\cos\delta)^2}$ P_r and Q_r: active and reactive power β: angle associated with line impedance α: angle associated with load power factor	Line Collapse Proximity Index Stable: LCPI < 1; Unstable: LCPI > 1 Combine the active and reactive influence
$\text{P}-\text{index} = \dfrac{-2\dfrac{P_r}{V_r}\dfrac{dV_r}{dP_r}}{1-2\dfrac{H_T}{V_r}\dfrac{dV_r}{dP_r}}$	Power Index When $\dfrac{dV_r}{dP_r} = \infty$, the value would be 1 The ratio of power loss to power gained Dynamic voltage assessment and load shedding

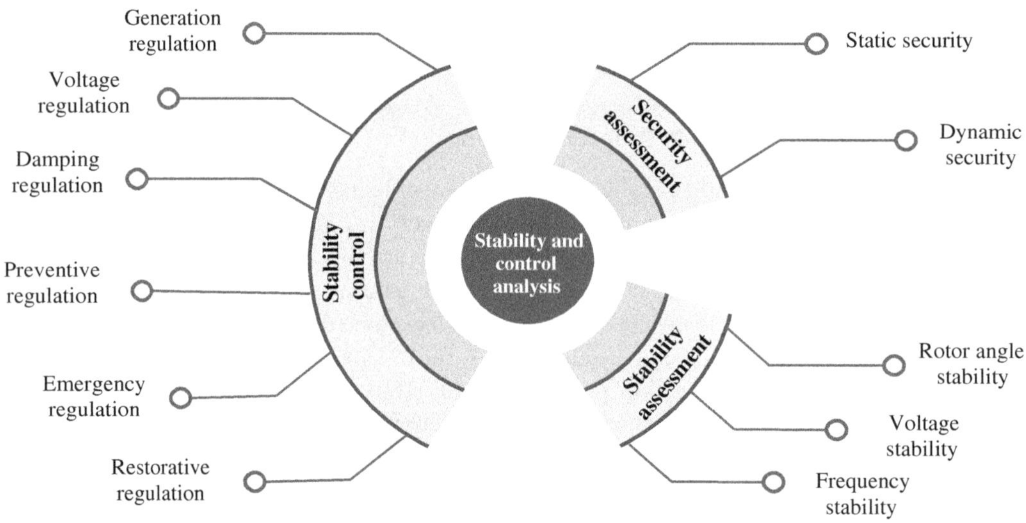

Figure 7.6 The overview of ML application to stability analysis.

Table 7.7 The summary of ML-based forecasting methods in RESs stability literature.

Reference	Strategy	Method	Results and description
[57]	Pre-contingency DSA	ELM	Tri-training labels unlabeled samples, while data editing mitigates the noise associated with inaccurately labeled samples.
[58]	Pre-contingency DSA	TL, ELM, RVFL	Feature-based TL enables the transfer of one DSA model to a previously similar fault.
[59]	Post-contingency DSA	SVM, AdaBoost	This methodology additionally assesses the coherence state of the generator and the synchronization of each generating unit's status.
[56]	TSP	SVM	Event-driven frequency and voltage stability evaluation
[60]	TSA	CNN	The approach can ascertain both the stability status and the mode of instability.
[61]	TSP	SVM	Multi-criteria decision-making was employed to ascertain the optimal timing for stability forecasting.
[62]	TSA	Data segmentation, AdaBoost	AdaBoost and data segmentation were utilized to address the issue of data imbalance.
[63]	Online DSA	Hybrid ELM-based ensemble model	Transient energy function, generator power outputs, and fault location are taken as input features for classification

or oscillatory unstable. Reference [61] introduces a novel method for predicting transient stability by utilizing multi-criteria decision-making (MCD) theory to concurrently assess both factors. The results indicate that they are contingent upon the prediction time; thus, addressing the MCD problem entails identifying the optimal time for stability prediction. The study [62] proposes a unique ensemble classification method based on data segmentation (DSEC) for real-time transient stability status prediction. The impact of the imbalanced data issue on the decision boundary and classification efficacy of real-time transient stability status predictions was examined comprehensively. A three-step DSEC approach is subsequently introduced. The summary of ML-based forecasting methods in RESs stability literature is provided in Table 7.7.

7.5 Challenges and Future Direction in ML-Based Approaches to RESs

Subsequent research may concentrate on enhancing forecasting methodologies and predictive modeling for renewable energy production. The incorporation of critical elements such as meteorological forecasting and real-time data from sensors and Internet of Things devices may significantly influence the optimization of RESs and grid operations management. As the energy sector evolves, it is imperative to confront the difficulties and capitalize on the possibilities arising from these developments. The future involves establishing a dynamic, adaptable, and intelligent

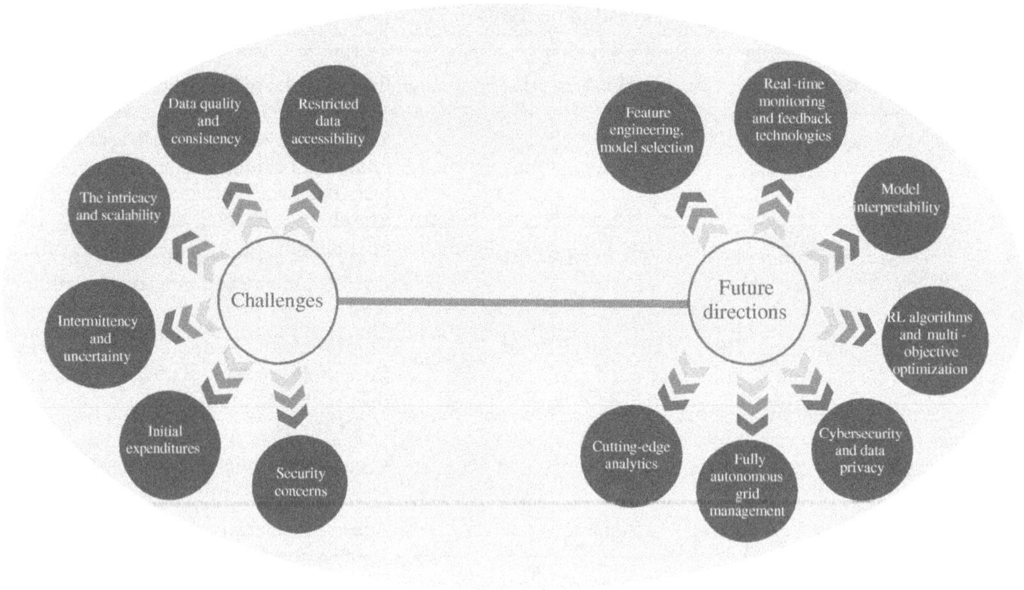

Figure 7.7 The overview of challenges and future directions in ML-based approaches to RESs.

ecosystem for a power grid that can efficiently utilize diverse RESs. The overview of challenges and future directions are illustrated in Figure 7.7

7.5.1 Challenges in ML-Based Approaches to RESs

Machine learning has demonstrated significant potential in enhancing and optimizing renewable energy systems; yet, the key limitations to its implementation in this field are listed in this section.

- **Restricted Data Accessibility**: ML models need substantial data for effective training. Nevertheless, for RESs, the volume of data may be insufficient or challenging to get, especially for nascent technologies that have not been in operation for extended durations.
- **Data Quality and Consistency**: Ensuring consistency and data quality presents a problem in RESs. The quality of data may be compromised by fluctuations in weather situations, equipment failures, and measurement inaccuracies. Weather data utilized for wind and solar energy forecasting may contain inaccuracies, and discrepancies among data from various sources might affect the precision of machine learning models.
- **The Intricacy and Scalability of Renewable Energy Systems**: RESs are inherently complicated, comprising numerous interdependent components, and complex systems frequently encompass numerous variables that are challenging to represent precisely. Machine learning models may find it challenging to accurately represent the intricate interactions among different components of the system, including the influence of shade on solar panels or the impact of wind turbulence on wind turbines. Consequently, creating machine learning models to enhance these systems necessitates a profound comprehension of their fundamental physics and engineering concepts.
- **Intermittency and Uncertainty of RESs**: The volatility of RESs presents a difficulty in sustaining a consistent power supply. RESs are characterized by significant uncertainty stemming from variables such as meteorological conditions, energy consumption, and storage capabilities.

Despite developments in this field, constraints in cost, capacity, and efficiency of storage systems persist, impeding the proper supply and demand balancing in real-world applications.

- **Significant Initial Expenditures and Extended Amortization Periods**: The implementation of advanced technologies, such as AMIs, necessitates substantial upfront investment along with prolonged payback durations, which may hinder general acceptance, particularly in underdeveloped areas or smaller markets.
- **Security Concerns**: ML models on renewable energy systems are experiencing many forms of attacks that potentially use machine learning in the domain of renewable energy systems, mainly presented below:
  - Adversarial attacks are the deliberate alteration of input data to induce misclassification or erroneous predictions by an ML algorithm. An attacker might alter sensor data from a wind turbine, resulting in the ML model predicting an erroneous wind speed or direction, which could lead to improper power generation and possible damage to the turbine.
  - Data poisoning attacks entail the introduction of harmful data into an ML model's training dataset to modify its behavior. This can be achieved by altering data inputs or generating fictitious data to influence the model's decision-making process. An attacker may introduce erroneous data into a solar panel efficiency dataset, causing the algorithm to derive an inaccurate efficiency curve, which would result in suboptimal energy generation. Similarly, in the context of RESs, an attacker might inject false information regarding weather conditions or energy production levels, leading to flawed predictions and potential harm to the renewable energy system.
  - Model stealing attacks entail the extraction of an ML model from a system, enabling the replication of the original model's behavior for ulterior goals. This can be achieved by reversing the model's parameters through the integration of data inputs and outputs. An attacker might alter the solar panel sensor data to circumvent the ML model's identification of a defective panel, resulting in less energy production and possible harm to the system.
  - DDoS attacks entail inundating a system with requests to render it inaccessible or inoperative [64]. This could be achieved by inundating an ML method with queries to impede its performance or cause it to fail. Within the framework of RES, an assailant may execute a denial-of-service attack on an ML model employed to regulate a renewable energy system, resulting in system failure.

7.5.2 Future Directions in ML-Based Approaches to RESs

The following future research directions and suggestions may enhance the comprehension and implementation of machine learning in requirements engineering processes, hence optimizing outcomes and efficiency. The recommendations are detailed below [1, 3].

- **Feature Engineering, Model Selection, and Tuning**: The optimization and selection of machine learning models must be meticulously evaluated to guarantee their suitability for the particular application and data attributes. The selection of model types, hyperparameters, and assessment criteria can profoundly influence model performance. Ensemble approaches enhance the performance and reliability of machine learning models by integrating different models to mitigate bias and volatility. In RES, ensemble approaches can amalgamate diverse data sources and model outputs to enhance system prediction and control. In RES, feature engineering, which involves the selection and transformation of pertinent input variables, can assist in identifying the critical variables that influence system performance and enhance their representation for model training.

- Advanced real-time monitoring and feedback technologies are essential for guaranteeing the precision and dependability of machine learning models in renewable energy systems. Ongoing assessment of system performance and comments regarding the model might enhance its accuracy and adaptability over time.
- Model interpretability is crucial for comprehending and elucidating the behavior of machine learning models in renewable energy systems. Methods such as feature importance analysis and model visualization facilitate the interpretation of results and yield insights for decision-making.
- Reinforcement learning algorithms and multi-objective optimization algorithms: Reinforcement learning algorithms facilitate the ability of renewable energy systems to acquire knowledge from experience and adjust to evolving conditions. Future studies may increasingly investigate the potential of reinforcement learning in renewable energy systems to enhance their performance and efficiency over time. Future studies may investigate the application of multi-objective optimization algorithms in RESs to concurrently optimize many objectives, including cost, performance, dependability, and sustainability.
- **Cybersecurity and Data Privacy**: With the increasing expansion of linked and data-dependent systems in SG, implementing robust cybersecurity frameworks is essential to safeguard the system from potential cyber threats. It is essential to use advanced security procedures to mitigate data privacy problems and develop safe data management processes.
- **Artificial Intelligence for Fully Autonomous Grid Management**: AI-based systems will autonomously monitor, evaluate, and regulate the operations of the grid without human involvement. This feature enables them to react to online data, anticipate possible problems, and dynamically enhance the performance of the grid.
- **Utilizing Cutting-Edge Analytics**: Enhancing the functionality of microgrids and virtual power plants could improve system stability and resilience. Simultaneously, research needs to examine the regulatory and economic implications of extensive integration of RESs. Recent developments in cybersecurity have facilitated the implementation of blockchain technology into the SG sector. By offering a safe, decentralized, and transparent platform for trading energy, it facilitates peer-to-peer (P2P) energy transfers between producers and consumers. Customers may engage in direct trading of energy credits with one another through this process.

7.6 Conclusion

This chapter thoroughly analyzed diverse machine learning algorithms in relation to RESs. The primary focus of the article was on the prospective uses of ML technologies, and the growing momentum of machine learning implications for RE underscores several application opportunities, especially in RESs. This chapter clarifies the significance of machine learning in renewable energy systems, along with a classification of several practical algorithms and their applications within the field. Furthermore, the role of ML in improving RESs generation is introduced by categorizing the ML applications and optimization methods in RESs. The presentation addresses grid stability using machine learning-based forecasting. Based on the insights presented in the chapter and the current research landscape for ML applications in RE, notable points arise:

- Machine learning and artificial intelligence act as potent techniques to address the constraints of requirements engineering. They offer several methods for tasks, including future conduct prediction of RESs sources, resource evaluation, optimization, and the analysis of grid stability. The emergence of DL algorithms enhances the reliability and precision of these duties.

Nonetheless, employing AI and deep learning entails substantial computing expenses and considerable resource implications, prompting an inquiry into how an RE sector can depend on such resource-intensive research methodologies.

- Given the nascent nature of these technologies, additional study is essential for their development and the enhancement of RESs stability, improving optimization efficiency and better applicability in RESs generation. However, they have a vital role in this study domain due to the intrinsic features that resources are unpredictable, expensive, and intricate.

The future of implementing machine learning in RE seems promising, particularly in light of the recommendations presented in recent research. Participation from all sectors is essential for advancing a sustainable and green energy future.

References

1 Allal, Z., Noura, H.N., Salman, O., and Chahine, K. (2024). Machine learning solutions for renewable energy systems: applications, challenges, limitations, and future directions. *Journal of Environmental Management* 354: 120392.

2 Rangel-Martinez, D., Nigam, K.D.P., and Ricardez-Sandoval, L.A. (2021). Machine learning on sustainable energy: a review and outlook on renewable energy systems, catalysis, smart grid and energy storage. *Chemical Engineering Research and Design* 174: 414–441.

3 Kiasari, M., Ghaffari, M., and Aly, H.H. (2024). A comprehensive review of the current status of smart grid technologies for renewable energies integration and future trends: the role of machine learning and energy storage systems. *Energies* 17 (16): 4128.

4 Ağbulut, Ü., Gürel, A.E., and Biçen, Y. (2021). Prediction of daily global solar radiation using different machine learning algorithms: evaluation and comparison. *Renewable and Sustainable Energy Reviews* 135: 110114.

5 Trizoglou, P., Liu, X., and Lin, Z. (2021). Fault detection by an ensemble framework of Extreme Gradient Boosting (XGBoost) in the operation of offshore wind turbines. *Renewable Energy* 179: 945–962.

6 Tama, B.A. and Lim, S. (2021). Ensemble learning for intrusion detection systems: a systematic mapping study and cross-benchmark evaluation. *Computer Science Review* 39: 100357.

7 Alazemi, T., Darwish, M., and Radi, M. (2024). Renewable energy sources integration via machine learning modelling: a systematic literature review. *Heliyon* 10 (4): e26088.

8 Josephine, P.K., Prakash, V.S., and Divya, K.S. (2021). Supervised learning algorithms: a comparison. *Kristu Jayanti Journal of Computational Sciences (KJCS)* 01–12.

9 Abisoye, B.O., Sun, Y., and Zenghui, W. (2023). A survey of artificial intelligence methods for renewable energy forecasting: methodologies and insights. *Renewable Energy Focus* 100529.

10 Usama, M., Qadir, J., Raza, A. et al. (2019). Unsupervised machine learning for networking: techniques, applications and research challenges. *IEEE Access* 7: 65579–65615.

11 Benti, N.E., Chaka, M.D., and Semie, A.G. (2023). Forecasting renewable energy generation with machine learning and deep learning: current advances and future prospects. *Sustainability* 15 (9): 7087.

12 Sarker, I.H. (2021). Machine learning: algorithms, real-world applications and research directions. *SN Computer Science* 2 (3): 160.

13 Alcañiz, A., Grzebyk, D., Ziar, H., and Isabella, O. (2023). Trends and gaps in photovoltaic power forecasting with machine learning. *Energy Reports* 9: 447–471.

14 Mienye, I.D. and Jere, N. (2024). A survey of decision trees: concepts, algorithms, and applications. *IEEE Access* 12: 86716–86727.

15 Dewangan, F., Biswal, M., Patnaik, B. et al. (2022). Smart grid stability prediction using genetic algorithm-based extreme learning machine. In: *Electric Power Systems Resiliency* (ed. R.C. Bansal, M. Mishra, and Y.R. Sood), 149–163. Academic Press.

16 Abualigah, L., Zitar, R.A., Almotairi, K.H. et al. (2022). Wind, solar, and photovoltaic renewable energy systems with and without energy storage optimization: a survey of advanced machine learning and deep learning techniques. *Energies* 15 (2): 578.

17 Shin, W., Han, J., and Rhee, W. (2021). AI-assistance for predictive maintenance of renewable energy systems. *Energy* 221: 119775.

18 Bhansali, Ashok, Namala Narasimhulu, Rocío Pérez de Prado et al. A review on sustainable energy sources using machine learning and deep learning models. *Energies* 16.17 (2023): 6236.

19 Khan, W., Walker, S., and Zeiler, W. (2022). Improved solar photovoltaic energy generation forecast using deep learning-based ensemble stacking approach. *Energy* 240: 122812.

20 Kisvari, A., Lin, Z., and Liu, X. (2021). Wind power forecasting – a data-driven method along with gated recurrent neural network. *Renewable Energy* 163: 1895–1909.

21 Wang, K., Qi, X., Liu, H., and Song, J. (2018). Deep belief network based k-means cluster approach for short-term wind power forecasting. *Energy* 165: 840–852.

22 Hussieny, O.A.A.-A., El-Beltagy, M.A., and El-Tantawy, S. (2020). Forecasting of renewable energy using ANN, GPANN and ANFIS (A comparative study and performance analysis). *2020 2nd Novel Intelligent and Leading Emerging Sciences Conference (NILES)*, Giza, Egypt (24–26 October 2020). IEEE.

23 Kumari, P. and Toshniwal, D. (2021). Extreme gradient boosting and deep neural network based ensemble learning approach to forecast hourly solar irradiance. *Journal of Cleaner Production* 279: 123285.

24 Tu, C.-S., Tsai, W.-C., Hong, C.-M., and Lin, W.-M. (2022). Short-term solar power forecasting via general regression neural network with grey wolf optimization. *Energies* 15 (18): 6624.

25 Li, N., He, F., and Ma, W. (2019). Wind power prediction based on extreme learning machine with kernel mean p-power error loss. *Energies* 12 (4): 673.

26 Mollaiee, A., Ameli, M.T., and Azad, S. (2022). Novel continuous learning scheme for online static security assessment based on the weather-dependent security index. *IET Generation, Transmission & Distribution* 16 (18): 3684–3705.

27 Paeizi, A., Ameli, M.T., and Azad, S. (2024). Distribution systems operation and planning: a literature review on smart resilience enhancement strategies. *Future Modern Distribution Networks Resilience* 111–139.

28 Ameli, M.T., Jalilpoor, K., Azad, S. et al. (2024). *Future Modern Distribution Networks Resilience: From Passive Operation to Strategic Active Paradigms*. Elsevier.

29 Dong, N., Chang, J.-F., Ai-Guo, W., and Gao, Z.-K. (2020). A novel convolutional neural network framework based solar irradiance prediction method. *International Journal of Electrical Power & Energy Systems* 114: 105411.

30 Xiao, L., Shao, W., Jin, F., and Zhuochun, W. (2021). A self-adaptive kernel extreme learning machine for short-term wind speed forecasting. *Applied Soft Computing* 99: 106917.

31 Hao, Y., Yang, W., and Yin, K. (2023). Novel wind speed forecasting model based on a deep learning combined strategy in urban energy systems. *Expert Systems with Applications* 219: 119636.

32 Kuo, P.-H. and Huang, C.-J. (2018). A green energy application in energy management systems by an artificial intelligence-based solar radiation forecasting model. *Energies* 11 (4): 819.

33 Lim, S.-C., Huh, J.-H., Hong, S.-H. et al. (2022). Solar power forecasting using CNN-LSTM hybrid model. *Energies* 15 (21): 8233.

34 Stoean, C., Zivkovic, M., Bozovic, A. et al. (2023). Metaheuristic-based hyperparameter tuning for recurrent deep learning: application to the prediction of solar energy generation. *Axioms* 12 (3): 266.

35 Shen, Z., Yao, Y., Wang, Q. et al. (2023). A novel micro power generation system to efficiently harvest hydroelectric energy for power supply to water intelligent networks of urban water pipelines. *Energy* 268: 126694.

36 Oladapo, B.I., Olawumi, M.A., and Omigbodun, F.T. (2024). Machine learning for optimising renewable energy and grid efficiency. *Atmosphere* 15 (10): 1250.

37 Suanpang, P. and Jamjuntr, P. (2024). Machine learning models for solar power generation forecasting in microgrid application implications for smart cities. *Sustainability* 16 (14): 6087.

38 Yu, S., You, L., and Zhou, S. (2023). A review of optimization modeling and solution methods in renewable energy systems. *Frontiers of Engineering Management* 10 (4): 640–671.

39 Hajipour, V., Niaki, S.T.A., Tavana, M. et al. (2023). A comparative performance analysis of intelligence-based algorithms for optimizing competitive facility location problems. *Machine Learning with Applications* 11: 100443.

40 Cai, W., Li, X., Maleki, A. et al. (2020). Optimal sizing and location based on economic parameters for an off-grid application of a hybrid system with photovoltaic, battery and diesel technology. *Energy* 201: 117480.

41 Tudorică, B.-G., Bucur, C., Panait, M. et al. (2024). Energetic equilibrium: optimizing renewable and non-renewable energy sources via particle swarm optimization. *Utilities Policy* 87: 101722.

42 Moayedi, H. and Mosavi, A. (2021). An innovative metaheuristic strategy for solar energy management through a neural networks framework. *Energies* 14 (4): 1196.

43 Farayola, A.M., Sun, Y., and Ali, A. (2018). ANN-PSO Optimization of PV systems under different weather conditions. *2018 7th International Conference on Renewable Energy Research and Applications (ICRERA)*, Paris, France (14–17 October 2018). IEEE.

44 Qiu, K. and Entchev, E. (2024). Modeling, design and optimization of integrated renewable energy systems for electrification in remote communities. *Sustainable Energy Research* 11 (1): 10.

45 Moazenzadeh, R., Mohammadi, B., Duan, Z., and Delghandi, M. (2022). Improving generalisation capability of artificial intelligence-based solar radiation estimator models using a bio-inspired optimisation algorithm and multi-model approach. *Environmental Science and Pollution Research* 29 (19): 27719–27737.

46 Zamee, M.A. and Won, D. (2020). Novel mode adaptive artificial neural network for dynamic learning: application in renewable energy sources power generation prediction. *Energies* 13 (23): 6405.

47 Golestaneh, F., Pinson, P., and Gooi, H.B. (2016). Very short-term nonparametric probabilistic forecasting of renewable energy generation – with application to solar energy. *IEEE Transactions on Power Systems* 31 (5): 3850–3863.

48 Ratshilengo, M., Sigauke, C., and Bere, A. (2021). Short-term solar power forecasting using genetic algorithms: an application using South African data. *Applied Sciences* 11 (9): 4214.

49 Kumari, P. and Toshniwal, D. (2021). Long short term memory–convolutional neural network based deep hybrid approach for solar irradiance forecasting. *Applied Energy* 295: 117061.

50 Zuo, H.-M., Qiu, J., Jia, Y.-H. et al. (2022). Ten-minute prediction of solar irradiance based on cloud detection and a long short-term memory (LSTM) model. *Energy Reports* 8: 5146–5157.

51 Huang, C.-J. and Kuo, P.-H. (2018). A short-term wind speed forecasting model by using artificial neural networks with stochastic optimization for renewable energy systems. *Energies* 11 (10): 2777.

52 Odero, H., Wekesa, C., and Irungu, G. (2022). Wind energy resource prediction and optimal storage sizing to guarantee dispatchability: a case study in the Kenyan power grid. *Journal of Electrical and Computer Engineering* 2022 (1): 4044757.

53 Shi, Z., Yao, W., Li, Z. et al. (2020). Artificial intelligence techniques for stability analysis and control in smart grids: methodologies, applications, challenges and future directions. *Applied Energy* 278: 115733.

54 Valuva, C., Chinnamuthu, S., Lone, T.k., and Kim, K.-C. (2023). A comprehensive review on the modelling and significance of stability indices in power system instability problems. *Energies* 16 (18): 6718.

55 Guo, M.-F., Yang, N.-C., and Chen, W.-F. (2019). Deep-learning-based fault classification using Hilbert–Huang transform and convolutional neural network in power distribution systems. *IEEE Sensors Journal* 19 (16): 6905–6913.

56 Singh, A.K. and Fozdar, M. (2019). Event-driven frequency and voltage stability predictive assessment and unified load shedding. *IET Generation, Transmission & Distribution* 13 (19): 4410–4420.

57 Liu, R., Verbič, G., and Ma, J. (2019). A new dynamic security assessment framework based on semi-supervised learning and data editing. *Electric Power Systems Research* 172: 221–229.

58 Ren, C. and Yan, X. (2019). Transfer learning-based power system online dynamic security assessment: using one model to assess many unlearned faults. *IEEE Transactions on Power Systems* 35 (1): 821–824.

59 Venkatesh, T. and Jain, T. (2019). Synchronised measurements based transient security assessment of power systems using AdaBoost classifiers. *IET Generation, Transmission & Distribution* 13 (17): 3980–3987.

60 Shi, Z., Yao, W., Zeng, L. et al. (2020). Convolutional neural network-based power system transient stability assessment and instability mode prediction. *Applied Energy* 263: 114586.

61 Hosseini, H., Naderi, S., and Afsharnia, S. (2019). New approach to transient stability prediction of power systems in wide area measurement systems based on multiple-criteria decision making theory. *IET Generation, Transmission & Distribution* 13 (21): 4960–4967.

62 Chen, Z., Han, X., Fan, C. et al. (2019). A data segmentation-based ensemble classification method for power system transient stability status prediction with imbalanced data. *Applied Sciences* 9 (20): 4216.

63 Singh, M. and Chauhan, S. (2023). A hybrid-extreme learning machine based ensemble method for online dynamic security assessment of power systems. *Electric Power Systems Research* 214: 108923.

64 Paeizi, A., Ameli, M.T., and Azad, S. (2023). Data analytics applications in digital energy system operation. In: *Energy Systems Transition: Digitalization, Decarbonization, Decentralization and Democratization* (ed. B. Mohammadi-Ivatloo and V. Vahidinasab), 25–52. Cham: Springer International Publishing.

8

Application of Artificial Neural Networks in Solar Photovoltaic Power Forecasting

Hamid Jabari[1], Afshin Ebrahimi[2], Ardalan Shafiei-Ghazani[1], and Farkhondeh Jabari[3]

[1] Faculty of Mechanical Engineering, Sahand University of Technology, Tabriz, Iran
[2] Faculty of Electrical Engineering, Sahand University of Technology, Tabriz, Iran
[3] Power Systems Operation and Planning Research Department, Niroo Research Institute (NRI), Tehran, Iran

8.1 RES Share in World Energy Transition

Renewable energy sources (RES) have a crucial role in global energy transition, aiming to mitigate climate change, enhance energy security, and supply rising energy demands. Until 2023, renewables account for over 30% of global electricity generation, a substantial increase from 19% in 2000 [1]. This growth is primarily attributed to the rapid expansion of solar and wind energy, which together contributed approximately 13.4% to global electricity generation in 2023. Although global energy-related emissions are projected to peak in 2024, they have continued to rise due to geopolitical tensions, particularly following the Russia–Ukraine conflict [2]. This situation has led to increased investments in fossil fuel projects and a resurgence in coal-fired power generation as countries seek energy security amidst rising prices for oil and gas. The International Renewable Energy Agency (IRENA) emphasizes that achieving net-zero emissions by 2050 requires cutting carbon dioxide emissions by approximately 37 GT from 2022 levels [3]. In 2022, global investments in energy transition technologies reached a record USD 1.3 trillion, yet fossil fuel subsidies surged simultaneously, complicating efforts to pivot towards cleaner energy sources. The current investment landscape indicates a significant gap; an additional USD 150 trillion is needed to meet the targets set by the Paris Agreement by 2050, averaging over USD 5 trillion annually. Looking ahead, projections indicate that RES could account for between 65% and 80% of global power generation by 2050, depending on various scenarios and policy frameworks [4]. The IRENA report suggests that from 2024 onward, renewables will grow by more than one percentage point annually, leading to a non-fossil share of 52% by mid-century. This transition will require substantial increases in renewable power capacity, necessitating annual additions of around 1066 GW from 2023 to 2050 under a scenario aligned with limiting global warming to 1.5 °C [3].

RES development is affected by several factors, such as technology, economy, policy, and social supports.

- Technology and innovation
  - **Innovation in Renewable Technologies**: Continuous improvements in solar, wind, and battery technologies are making RES more efficient and cost-effective [5]. For instance, advancements in photovoltaic cells and wind turbine designs have significantly increased energy output while lowering costs.

Data-Driven Energy Management and Tariff Optimization in Power Systems:
Shaping the Future of Electricity Distribution through Analytics, First Edition.
Edited by Hamidreza Arasteh, Pierluigi Siano, Niki Moslemi, and Josep M. Guerrero.

  - **Energy Storage Solutions**: Development of better energy storage systems, such as lithium-ion batteries, enables the effective use of intermittent energy sources like solar and wind power. This enhances the reliability and attractiveness of RES [6].
- Economic drivers
  - **Decreasing Costs**: The cost of generating energy from renewable sources has fallen dramatically over the past decade. According to recent studies, solar and wind energy are now among the cheapest forms of electricity generation globally [7].
  - **Investment Growth**: Increased investment in renewable technologies is being spurred by both public and private sectors, driven by potential for long-term returns and job creation in green industries [8].
- Policy support
  - **Government Incentives**: Many governments are implementing policies that favor renewable energy through subsidies, tax incentives, and renewable portfolio standards. These measures encourage both producers and consumers to adopt RES [9].
  - **International Agreements**: Global commitments to reduce carbon emissions, such as the Paris Agreement, are prompting countries to transition towards cleaner energy sources. This international pressure fosters national policies that support RES expansion [10].
- Social factors
  - **Public Awareness and Demand**: Growing awareness of climate change and environmental issues among consumers is increasing demand for clean energy solutions. Public support for sustainability initiatives is pushing governments and companies to prioritize RES [11].
  - **Corporate Sustainability Goals**: Many corporations are setting ambitious sustainability targets that include transitioning to renewable energy. This corporate shift not only helps reduce their carbon footprint but also influences market trends towards greener solutions.

The International Energy Agency forecasts that global solar photovoltaic (PV) capacity will reach approximately 15.5 terawatts (TW) by 2050 under a net-zero emissions scenario [12].

8.2 Applications of PV Panels in Energy Systems

PV panels play a crucial role in modern energy systems by converting sunlight directly into electricity. Their applications span various sectors and technologies, enhancing energy efficiency and sustainability. Below are some applications of PV panels:

- **Energy Generation**: PV panels are widely used in residential and commercial buildings to generate electricity. They can be installed on rooftops or integrated into building structures to supply power for daily use, offsetting electricity bills and contributing to grid stability. Grid-connected PV systems allow homeowners and businesses to sell excess electricity back to the grid, promoting renewable energy use and providing financial incentives for solar adoption [13]. In developing regions, particularly in remote areas lacking grid access, PV systems are increasingly deployed to provide electricity for lighting and basic appliances. For example, solar-powered LED lights replace kerosene lamps in rural India, improving safety and reducing costs over time [14].
- **Agricultural Uses**: PV panels are employed in solar-powered water pumps for irrigation, particularly in remote areas where traditional power sources are unavailable. This application is crucial for maintaining agricultural productivity during dry seasons [15]. Agrivoltaic combines solar energy generation with agricultural practices, allowing crops to grow beneath solar panels

while also generating electricity. This dual-use approach maximizes land efficiency. Photovoltaic panels can be used in aquaculture settings, where they reduce water evaporation and improve conditions for fish by providing shade [16].

- **Water Heating and Distillation**: PV systems can be utilized for water heating in the residential sector, swimming pools, and industrial processes [17]. This application is particularly beneficial in regions with high sunlight exposure. Solar energy can also be harnessed for water purification processes, providing clean drinking water in areas facing water scarcity [18].
- **Public Infrastructure Such as Street Lighting and Signage**: PV panels are integrated into street lights and road signs, providing a reliable power source for nighttime illumination without relying on grid electricity [19, 20]. This application enhances safety and reduces operational costs for municipalities [21].
- **Emergency Services**: Solar-powered systems are increasingly used in emergency telephones, traffic signals, and remote monitoring stations, ensuring functionality in off-grid locations [22, 23].

8.3 Disadvantages of PV Panels

Several limitations of PV technology are summarized as follows [24]:

- **High Initial Costs**: The most significant barrier to adopting solar PV technology is the high initial investment required for purchasing and installing solar panels. While prices have decreased over time, the upfront costs can still be prohibitive for many homeowners and businesses, particularly for larger installations.
- **Intermittency and Dependence on Weather**: Solar PV systems rely heavily on sunlight, which means their energy production is intermittent. They do not generate electricity at night and can be less efficient on cloudy or rainy days. This dependence on weather conditions can lead to inconsistent energy supply, necessitating additional solutions like battery storage to ensure reliability.
- **Space Requirements**: Installing solar panels requires significant space, which can be a limitation for properties with small roofs or limited land. Larger systems need ample area to accommodate enough panels to generate sufficient energy, making them unsuitable for densely populated or urban environments.
- **Energy Storage Costs**: To utilize solar energy effectively outside of daylight hours, users often need to invest in battery storage systems. These batteries can be expensive and add complexity to the overall system, increasing both the initial investment and ongoing maintenance costs.
- **Environmental Impact of Manufacturing**: While solar energy is clean during operation, the manufacturing process of solar panels involves significant resource consumption and can produce pollution. The extraction of raw materials and the energy-intensive production processes contribute to a carbon footprint that, while lower than fossil fuels, is still a concern.
- **Potential for Damage**: Solar panels are vulnerable to physical damage from extreme weather events such as hailstorms or high winds. Although warranties often cover such damages, repairs or replacements can still be costly and inconvenient.
- **Maintenance Requirements**: Although solar panels generally require low maintenance, they still need periodic cleaning and inspections to ensure optimal performance. Neglecting maintenance can lead to reduced efficiency over time.
- **Limited Applicability**: Not all roofs are suitable for solar panel installation due to factors like orientation, shading from trees or buildings, and roofing material compatibility.

8.4 Importance of PV Power Forecasting

Solar energy production is inherently variable due to factors like weather changes and time of day. Forecasting helps mitigate their intermittency by providing accurate predictions of solar output, which allows grid operators to balance supply and demand effectively. For instance, cloud cover can drastically reduce solar generation, sometimes by as much as 80% in minutes, necessitating quick operational adjustments to maintain grid stability [25].

Improved solar forecasts enable grid operators to optimize dispatch of solar and other energy sources, leading to a more flexible and resilient grid. This flexibility is crucial as the share of solar energy in the overall generation mix increases, allowing for better integration with other renewable sources and reducing reliance on fossil fuels [26, 27].

Accurate forecasting of PV power can lead to significant economic advantages by [28]:

- Reducing operational costs associated with managing reserves.
- Minimizing penalties for imbalances in electricity supply and demand.
- Enhancing their profitability by allowing better planning and management of resources.

As technologies like energy storage and electric vehicles become more prevalent, accurate solar forecasting becomes increasingly important. It allows for optimal charging strategies for batteries and helps integrate these technologies into the grid effectively, ensuring that they operate efficiently in conjunction with variable RES. Forecasts provide critical information for decision-makers regarding future power generation scenarios. This data is essential for negotiating contracts, managing financial risks, and planning maintenance schedules for solar facilities.

8.5 Proposed Algorithm for PV Power Prediction

The actual value of the PV power product (P_t) can be given by Eq. (8.1) [29].

$$P_t = N \times \eta \times A \times I_t \left(1 - 0.005 \times \left(T_t^{amb} - 25\right)\right) \tag{8.1}$$

where

N is the number of PV panels
η is the energy efficiency of each PV module
A is the absorber area of each PV module (m^2)
I_t is the solar irradiance in time t (W/m^2)
T_t^{amb} is the ambient air temperature in time t (°C)

In this research, an artificial neural network (ANN)-based optimization algorithm is used to forecast power, as illustrated in Fig. 8.1. The mean squared error (MSE) of the estimated values is calculated as an objective function and minimized using Eq. (8.2).

$$\text{MSE} = \frac{1}{T}\sum_{t=1}^{T}(P_t - \hat{P}_t)^2 \tag{8.2}$$

where P_t and $\hat{P}_t$ refer to the actual and predicted values of the PV electrical power output at hour t, respectively.

Both ambient air temperature and solar irradiation are selected as input data and should be normalized using Eq. (8.3).

$$x_t^{norm} = \frac{x_t - x_{\min}}{x_{\max} - x_{\min}} \tag{8.3}$$

where

x_t is the input data at hour t
x_t^{norm} is the normalized input data at hour t
$x_{\min}$ is the minimum values of input data
$x_{\max}$ is the maximum values of input data

Data normalization is a crucial preprocessing step in training ANNs, as it helps ensure that input features are on a similar scale with the following benefits:

- **Improves Convergence**: Normalized data helps gradient descent converge faster during training by preventing oscillations and ensuring a well-conditioned optimization process.
- **Mitigates Gradient Issues**: It addresses problems such as vanishing or exploding gradients, which can hinder model training.
- **Balances Feature Influence**: Different scales among features can lead to some features dominating others, which normalization helps mitigate.

Ambient air temperature and solar irradiation for 4662 hours (518 days) are used as input data. For one day, nine data points, each of which is related to weather conditions during a one-hour time interval, are used. In other words, each nine datasets represents inputs from 8:00 to 16:00 during one day. Then, input data related to seven days (63 data points) is selected as the test dataset. The solar PV power related to the test data is forecasted and then compared with the actual/observed values, aiming to minimize the MSE according to epochs. The number of epochs determines the number of times that the learning algorithm is run based on the training data.

Rectified Linear Unit (ReLU) is used as an activation function in neural networks due to its simplicity. In this chapter, ReLU as a piecewise linear function is applied to ANN. As modeled by Eq. (8.4), if input variable x is positive, ReLU will return it; otherwise, it will be zero.

$$\text{ReLu}(x) = \begin{cases} x & \text{if}\, x > 0 \\ 0 & \text{if}\, x \leq 0 \end{cases} \tag{8.4}$$

Batch size refers to the number of samples that are fed into the ANN at one time during training. It affects several aspects of model training:

- **Training Time**: Larger batch sizes typically result in faster epoch completion because they utilize parallel processing capabilities of modern hardware. However, this can lead to longer convergence times if the model struggles to generalize from broader updates.
- **Model Generalization**: Smaller batch sizes tend to yield better generalization on unseen data. They introduce more noise into the gradient estimation process, which can help avoid overfitting by allowing them to explore more diverse solutions in the loss landscape.
- **Memory Usage**: Larger batches require more memory, which can be a limiting factor depending on hardware capabilities. It's essential to find a balance that maximizes resource utilization without exceeding memory limits.

Dropout is a regularization technique used in ANNs to prevent overfitting. It involves randomly "dropping out" a fraction of neurons during training, which means that these neurons are temporarily removed along with their connections. A dense layer, also known as a fully connected layer, is a fundamental component of ANNs. This layer connects every neuron from the previous layer to each neuron in the dense layer, allowing for comprehensive information flow and interaction between all neurons.

The Adaptive Moment (Adam) optimizer is used in the training of ANNs due to its advantages:

- **Faster Convergence**: By adapting learning rates for each parameter, Adam often reaches optimal solutions more quickly than stochastic gradient descent.
- **Ease of Use**: The algorithm is straightforward to implement and requires minimal tuning of hyperparameters.
- **Robustness**: Adam performs well across different types of neural network architectures and is less sensitive to hyperparameter choices compared to other optimizers.

8.6 Numerical Results and Discussions

To demonstrate the accuracy and speed of the proposed ANN-based PV power forecasting strategy, a solar farm located in Ahwaz, Iran, is studied. The technical specifications of the PV panels are considered as follows:

- Number of PV panels = 20
- Energy efficiency of PV modules = 0.95
- Absorber area of PV modules = 5 m^2

Ambient air temperature and solar irradiation during 518 days (4664 data points) are selected as input data of the neural network, which can be found in [30]. For every day, nine data points (from $t = 8$ to $t = 16$), including air temperature and solar radiation, are considered. It should be mentioned that PV power output during seven days (63 data points) is forecasted and compared with actual values obtained from Eq. (8.1). The presented method is run to predict the output electrical power of the PV panel farm for one week. The forecasted and actual values of the PV power are compared in Fig. 8.2.

The MSE of the ANN-based optimization model is obtained as 0.0034. Moreover, its forecasted error is shown in Figure 8.3. It is obvious from Figures 8.2 and 8.3 that maximum forecasting error is less than 1 W. Using a Lenovo with a 2.10 GHz CPU and 4 GB RAM, the calculation time of the forecasting method (Eq. (8.1)–(8.4)) is almost 30 seconds. ANNs are computational models inspired by the human brain's neural networks, capable of learning complex patterns through training on historical data. They utilize various input features such as historical weather data (temperature, solar radiation) and past power output to predict future energy generation. The flexibility of ANNs allows them to adapt to different datasets and learning algorithms, improving their predictive capabilities. Additionally, tuning ANN parameters such as the number of hidden neurons can lead to more precise outputs while maintaining computational efficiency. Application of ANNs extends beyond large-scale solar farms; they are also being adapted for individual and small-scale solar power producers. This adaptability is crucial as more consumers become "prosumers," those who both produce and consume energy. Automated forecasting systems powered by ANNs can help these prosumers minimize costs and environmental impacts by optimizing their energy use based on accurate predictions.

8.7 Concluding Remarks

This chapter presents a comprehensive overview of the application of ANNs in the PV power forecasting approach. Temperature and solar irradiation are selected as input data. Root MSE of forecasted values is minimized as an objective function. Lessons learned from simulations and numerical results can be stated as follows.

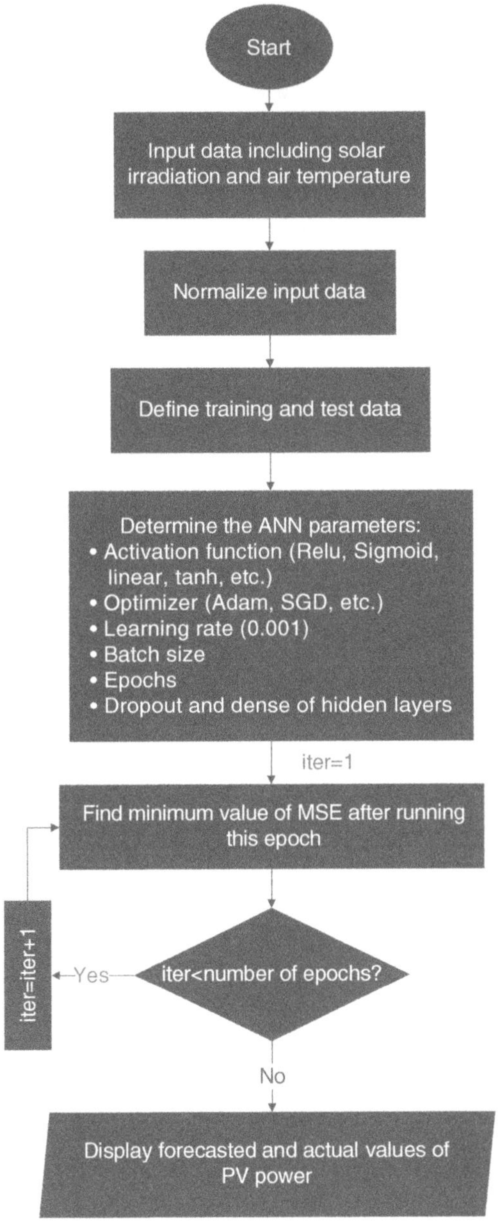

Figure 8.1 The proposed PV power forecasting algorithm.

Some key advantages of ANN in PV power forecasting are as follows:

- **High Accuracy**: Simulation has demonstrated that ANNs can achieve low MSE values, indicating high forecasting accuracy. This level of accuracy is crucial for optimizing energy management and grid stability.
- **Speed of Prediction**: ANNs are capable of processing large datasets quickly, allowing for real-time forecasting. This rapid processing is essential for applications where timely predictions can significantly impact operational decisions, such as energy trading and grid management.

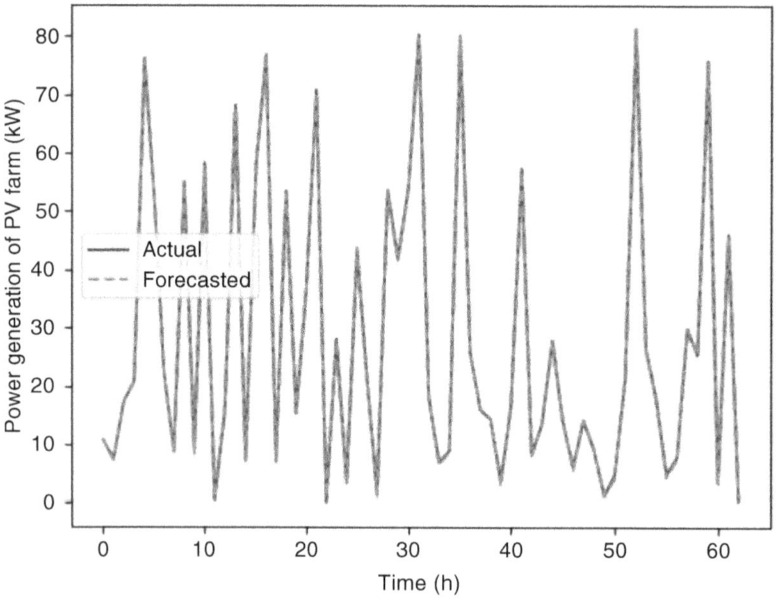

Figure 8.2 The forecasted and actual values of the PV farm for seven days (from 8:00 to 16:00).

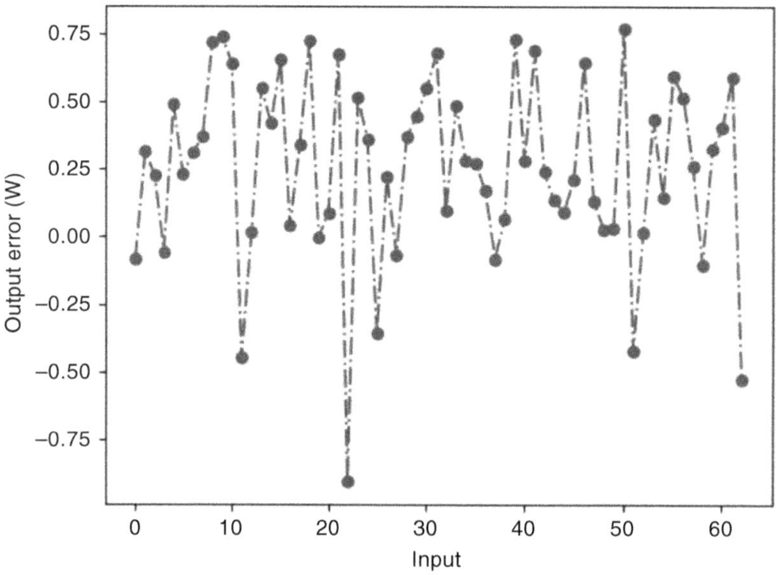

Figure 8.3 The forecasted error of the proposed algorithm for estimating the PV farm power during seven days (from 8:00 to 16:00).

- **Data Integration**: ANNs can effectively integrate various data sources, including numerical weather predictions and historical solar irradiance data. This capability enhances this model's robustness and reliability, as it can adapt to different environmental conditions and seasonal variations.

However, ANNs also exhibit several weaknesses that can impact their effectiveness in PV power forecasting:

- **Complexity and Interpretability**: ANNs can be complex and difficult to interpret. Their multi-layered architecture often makes it challenging for users to understand how input data influences predictions. This lack of transparency can hinder trust and acceptance among stakeholders in energy management systems.
- **Training Time**: Training ANNs requires significant computational resources and time, especially when dealing with large datasets typical in PV power forecasting. This lengthy training process can limit their adaptability to rapidly changing environmental conditions.
- **Overfitting**: ANNs are prone to overfitting, particularly when trained on limited or noisy datasets. Overfitting occurs when a model learns a noise in training data instead of an underlying pattern, leading to poor generalization to new, unseen data. This is a critical concern in PV forecasting, where accurate predictions are essential for grid management.
- **Sensitivity to Input Data Quality**: ANNs require a substantial amount of high-quality historical data for training. Inaccurate or incomplete data can lead to poor model performance. Moreover, ANNs may not effectively handle outliers or anomalies in data, which are common in real-world scenarios.

References

1 NA. (2024). World passes 30% renewable electricity milestone. https://ember-energy.org/latest-updates/world-passes-30-renewable-electricity-milestone/.

2 NA. (2024). Wind and solar added more to global energy than any other source. https://www.carbonbrief.org/analysis-wind-and-solar-added-more-to-global-energy-than-any-other-source-in-2023/.

3 NA. (2024). Climate action and energy transition. IRENA Members' survey on Nationally Determined Contributions. https://www.irena.org/-/media/Files/IRENA/Agency/Publication/2024/Jun/IRENA_Climate_action_Members_NDC_survey_2024.pdf.

4 NA. (2024). Global energy perspective 2024. https://www.mckinsey.com/industries/energy-and-materials/our-insights/global-energy-perspective/.

5 Elia, A., Kamidelivand, M., Rogan, F., and Gallachóir, B.Ó. (2021). Impacts of innovation on renewable energy technology cost reductions. *Renewable and Sustainable Energy Reviews* 138: 110488.

6 Sterner, M. and Stadler, I. (2019). *Handbook of Energy Storage: Demand, Technologies, Integration*. Springer.

7 Mehmood, U., Agyekum, E.B., Tariq, S. et al. (2022). Socio-economic drivers of renewable energy: empirical evidence from BRICS. *International Journal of Environmental Research and Public Health* 19 (8): 4614.

8 Bei, J. and Wang, C. (2023). Renewable energy resources and sustainable development goals: evidence based on green finance, clean energy and environmentally friendly investment. *Resources Policy* 80: 103194.

9 Abdmouleh, Z., Alammari, R.A., and Gastli, A. (2015). Review of policies encouraging renewable energy integration & best practices. *Renewable and Sustainable Energy Reviews* 45: 249–262.

10 Gentry, B.S. and Ronk, J.J. (2007). International investment agreements and investments in renewable energy. *From Barriers to Opportunities: Renewable Energy Issues in Law and Policy*, vol. 45.

11 Khambalkar, V.P., Katkhede, S.S., Dahatonde, S. et al. (2010). Renewable energy: an assessment of public awareness. *International Journal of Ambient Energy* 31 (3): 133–142.

12 Xu, C., Isabella, O., and Vogt, M.R. (2024). Future material demand for global silicon-based PV modules under net-zero emissions target until 2050. *Resources, Conservation and Recycling* 210: 107824.

13 Singh, G.K. (2013). Solar power generation by PV (photovoltaic) technology: a review. *Energy* 53: 1–13.

14 Goetzberger, A. and Hoffmann, V.U. (2005). *Photovoltaic Solar Energy Generation*. Springer Science & Business Media.

15 Marucci, A., Zambon, I., Colantoni, A., and Monarca, D. (2018). A combination of agricultural and energy purposes: evaluation of a prototype of photovoltaic greenhouse tunnel. *Renewable and Sustainable Energy Reviews* 82: 1178–1186.

16 Amaducci, S., Yin, X., and Colauzzi, M. (2018). Agrivoltaic systems to optimise land use for electric energy production. *Applied Energy* 220: 545–561.

17 Sharon, H., Vivar, M., and Fuentes, M. (2022). A review on role of solar photovoltaic (PV) modules in enhancing sustainable water production capacity of solar distillation units. *Journal of Environmental Management* 320: 115781.

18 Dev, R., Kashyap, Y., Tewari, K., and Pal, P. (2023). Solar distillation and water heating systems integration with photovoltaic technology. In: *Renewable Energy: Accelerating the Energy Transition* (ed. R. Goyal, S. Patel, and A. Sharma), 139–165. Springer.

19 Danner, M.T. (2014). Design, prototyping and testing of a high-performance solar-powered street light.

20 Shah, D.K. (2004). Study on implementation of solar photovoltaic traffic lighting system for Gwarko, Lalitpur, Nepal. Tribhuvan University.

21 Jiang, W., Wang, T., Yuan, D. et al. (2024). Available solar resources and photovoltaic system planning strategy for highway. *Renewable and Sustainable Energy Reviews* 203: 114765.

22 Li, Q., Li, T., and Zanelli, A. (2024). Performance evaluation of flexible photovoltaic panels for energy supply in post-disaster emergency shelters. *Journal of Building Engineering* 98: 111285.

23 Franceschi, J., Rothkop, J., and Miller, G. (2014). Off-grid solar PV power for humanitarian action: from emergency communications to refugee camp micro-grids. *Procedia Engineering* 78: 229–235.

24 Bondarets, V. and Nykyporets, S. (2023). Solar cells: advantages and disadvantages. ВНТУ.

25 Antonanzas, J., Osorio, N., Escobar, R. et al. (2016). Review of photovoltaic power forecasting. *Solar Energy* 136: 78–111.

26 Kouro, S., Leon, J.I., Vinnikov, D., and Franquelo, L.G. (2015). Grid-connected photovoltaic systems: an overview of recent research and emerging PV converter technology. *IEEE Industrial Electronics Magazine* 9 (1): 47–61.

27 Hamid, A.K., Mbungu, N.T., Elnady, A. et al. (2023). A systematic review of grid-connected photovoltaic and photovoltaic/thermal systems: benefits, challenges and mitigation. *Energy & Environment* 34 (7): 2775–2814.

28 Das, U.K. et al. (2018). Forecasting of photovoltaic power generation and model optimization: a review. *Renewable and Sustainable Energy Reviews* 81: 912–928.

29 Jabari, H., Ghazani, A.S., and Jabari, F. (2024). Thermodynamic modeling and optimal short-term scheduling of a green H_2-fueled combined heat and power generation system for industrial zones. *International Journal of Hydrogen Energy* 84: 235–245.

30 https://www.meteoblue.com/en/weather/.

9

Power System Resilience Evaluation Data Challenges and Solutions

Mohammad Reza Sheibani, Habibollah Raoufi, and Javad Nezafat Namini

Power Systems Operation and Planning Research Department, Niroo Research Institute (NRI), Tehran, Iran

9.1 Introduction

Resilience is a relatively new concept that has gained significant attention in recent years due to the increasing frequency of disasters and their highly destructive impacts on power systems. This interest has been reflected in the works of many academic and industrial researchers, particularly within the fields of power system planning and operation studies.

A disaster is defined as an event that occurs infrequently but has a highly destructive and significant impact on various sectors of society, including power systems [1, 2]. In [3, 4], disasters are referred to as high-impact low-frequency (HILF) events, which can have severe consequences for power systems, yet they happen rarely and, in some cases, may never occur at all.

In many references, the concept of probability is utilized in naming disasters, with terms such as high-impact low-probability events [5–11], high-consequence low-probability events [4, 12], or severe events with high fatalities and low probability [13] being employed. However, using the concept of probability to describe these occurrences is not particularly accurate; the mathematical concept of probability is challenging to apply to events that have a low frequency of occurrence. Indeed, measuring the likelihood of these events and their impacts is difficult, and the frequency and severity of past events may not adequately indicate the potential impacts of future occurrences. It seems that for disasters, using the concept of frequency is more appropriate than using probability because the frequency of a disaster can be easily derived from historical event data.

Power systems are exposed to disasters from physical, cyber, and cyber-physical vulnerabilities [14]. Disasters can not only damage physical and cyber equipment but also harm the personnel associated with them [15]. In the event of a disaster, due to the centralized structure of power systems, even a relatively minor impact on one or more critical areas of the system can lead to significant outages affecting a large portion or even the entirety of the system [1].

Disasters can be categorized into two main types: natural and man-made. Natural disasters arise from natural factors such as weather and geological phenomena, including earthquakes, tsunamis, storms, tornadoes, blizzards, ice storms, floods, wildfires, volcanic eruptions, severe droughts, heat waves, and pandemics [1, 16–20]. Recent statistics indicate a rise in the frequency, intensity, and costs associated with weather-related natural disasters due to extensive climate changes [21–23]. For example, Hurricane Sandy struck the United States in October 2012, with estimated damages exceeding $70 billion. This hurricane affected over 8 million electricity customers (equating to more than 20 million people) and resulted in power outages lasting nearly 10 days.

Data-Driven Energy Management and Tariff Optimization in Power Systems: Shaping the Future of Electricity Distribution through Analytics, First Edition.
Edited by Hamidreza Arasteh, Pierluigi Siano, Niki Moslemi, and Josep M. Guerrero.

While the transmission network sustained minimal damage, the distribution network experienced significant impacts [1]. Man-made disasters, on the other hand, are caused by deliberate human actions and include warfare, coordinated criminal activities, and terrorism (both physical and cyberattacks) [15]. A notable instance occurred in December 2015 when Ukraine's power grid was targeted by a cyberattack, leaving approximately 225,000 people without electricity for up to six hours [24].

Natural disasters can be divided into two categories: predictable and unpredictable. Predictable disasters primarily include weather-related events such as storms, tornadoes, floods, and blizzards. These events can often be anticipated by analyzing meteorological data in advance, allowing for appropriate measures to be taken to mitigate their impact. In contrast, unpredictable disasters, such as earthquakes and major wildfires, cannot be forecasted beforehand. The approach to managing these types of events differs significantly; preparedness must be tailored based on the historical occurrence and severity of such events in a given area. Therefore, when addressing natural disasters, factors such as type, predictability, intensity, and duration must be considered. Additionally, outages caused by natural disasters can last from several days to several months [1, 25].

Although it is clear that resilience is a concept related to the impact of disasters on power systems, there is no universally accepted definition for physical-cyber resilience in power systems [14]. However, power system resilience can be defined as the ability of the system to efficiently withstand disasters (HILF events) in such a way that minimizes outages imposed on customers, maintains essential social services, and facilitates rapid recovery to normal conditions as much as possible.

In the literature on power systems, there is extensive discussion regarding the necessity of assessing and measuring resilience. According to various sources, while evaluating the resilience of power systems is a complex and challenging topic, it is essential for effectively assessing the efficacy of resilience enhancement measures and refining them.

Since resilience is a qualitative concept, efforts must be made to quantify it using various tools. This is accomplished through resilience metrics, which serve as mathematical metrics for measuring the resilience of power systems against disasters [5, 7, 26]. Proper assessment of the resilience level in power systems using resilience metrics leads to effective and rational actions for strengthening that resilience [6]. The appropriate evaluation of system resilience and improvements requires the development of resilience metrics applicable in planning, operation, and policymaking within power systems [13]. Given that various factors influence system resilience, accurately measuring it can be difficult. A straightforward approach in this regard is to combine different factors into a composite metric. However, challenges arise concerning how to select factors, measure them, combine them, and weight them [27].

Providing high-accuracy data for measuring resilience in power systems is one of the challenges that can affect the accuracy and reliability of resilience metrics. In this book chapter, the data needed to measure the resilience of power systems are discussed along with the challenges facing resilience studies in terms of data and the proposed solutions to solve these challenges.

To this end, in the following, the resilience metrics will first be introduced and classified. Following that, the necessary data for resilience studies in power systems will be examined. Then, by highlighting the data-related challenges encountered at various stages of resilience studies, proposed solutions will be presented.

9.2 A Review of Power System Resilience Metrics

The most comprehensive conceptual framework for classifying resilience metrics of power systems has been presented in [28], as illustrated in Figure 9.1. In this framework, resilience metrics are divided into two categories at the first layer: "performance-based" and "non-performance-based."

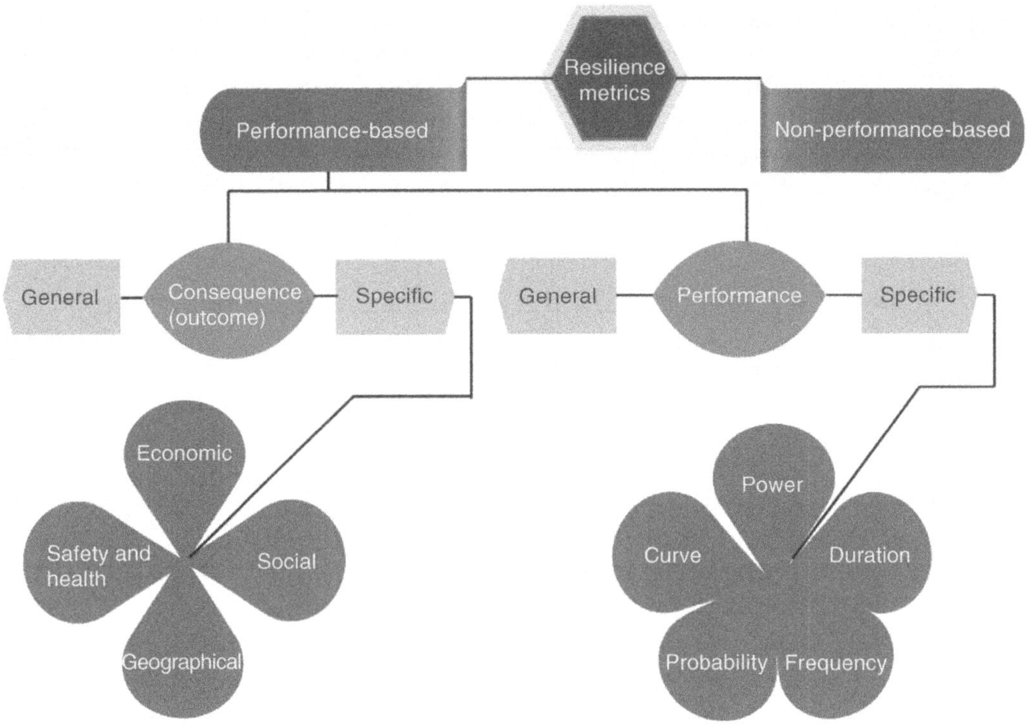

Figure 9.1 Conceptual framework for classification of power system resilience metrics.

The performance of a power system is a metric that directly reflects the output of the power system [4, 6, 7]. This performance can be derived from data obtained through analyzing the past behavior of the system during a disaster event, or it can be predicted by modeling and simulating how a disaster impacts the system over a specified time frame, particularly through sequential Monte Carlo simulations [29]. Performance-based metrics depend on the system's performance and are calculated based on it. Once the system's performance is determined over a time interval, various types of performance-based metrics can be computed. In contrast, non-performance metrics do not rely on the system's performance and are calculated based on several factors that influence the system's resilience before, during, and after a disaster.

In the proposed conceptual framework, the non-performance metrics are not further categorized; however, performance-based metrics are subdivided into several additional layers. Performance-based indicators can be broadly classified into two main categories: "performance metrics" and "outcome metrics." Performance metrics present the system's performance value (direct system output) as a metric, while outcome metrics represent the system's impact on various aspects of society. Both performance and outcome metrics are further divided into "general" and "specific" categories. Specific metrics are unambiguously defined based on specific quantities. In contrast, general metrics are defined with some ambiguity, using broad concepts such as "performance" and "impact." A general metric, once clarified and all related quantities are specified, transforms into a specific metric and fits into one of the classifications presented in Figure 9.1. In this framework, general metrics are not further categorized; however, specific metrics (both performance and outcome) have additional subdivisions. The specific performance metrics are divided into five groups: power, duration, number, probability, and curve. The specific outcome metrics are categorized into four groups: economic, social, geographical, and safety and health. These classifications are described in detail in [28].

Given the significant importance of performance-based metrics and their widespread use in the literature on power systems, Tables 9.1 and 9.2 present key examples of various specific performance and outcome metrics, along with the necessary data for their calculation. It is important to note that the items listed in these tables represent the simplest forms of these indicators. As indicated in [28], these metrics can also be utilized in more complex forms, such as normalized, weighted, expected, worst-case scenarios, etc. In such cases, additional data will be required for their calculation depending on the nature of the relationships involved.

9.3 The General Framework for the Resilience Assessment of the Power System

Based on the studies conducted in the previous section, the framework presented in Figure 9.2 can be considered as a general framework for the resilience assessment of the power system. Data gathering is considered the initial step. In this step, comprehensive data including past events, infrastructure details, and environmental factors is collected.

In the resilience assessment studies, a variety of data is required, collected from different sources, so there is a need to analyze this data, verify it, and correct it if necessary. Disaster modeling based on the input data of the problem will be the next step, for which different methods can be used.

Several mathematical and statistical methods are employed in disaster modeling, such as

- **Probability Analysis**: Predicts the likelihood of different catastrophe scenarios
- **Monte Carlo Simulation**: Provides a range of possible outcomes by simulating a large number of random samples
- **Stochastic Event Generation**: Creates databases of scenario events with specific strengths, locations, and probabilities

After the disaster modeling, it is necessary to examine how the disaster will affect the power system performance. The methods used to determine the impact are selected based on the type of metric desired and the necessary outputs of calculating the metric. Finally, the resilience metric is determined based on the results of the studies conducted in the previous steps. In the following, the above issues are examined with a focus on the numerical challenges and solutions ahead.

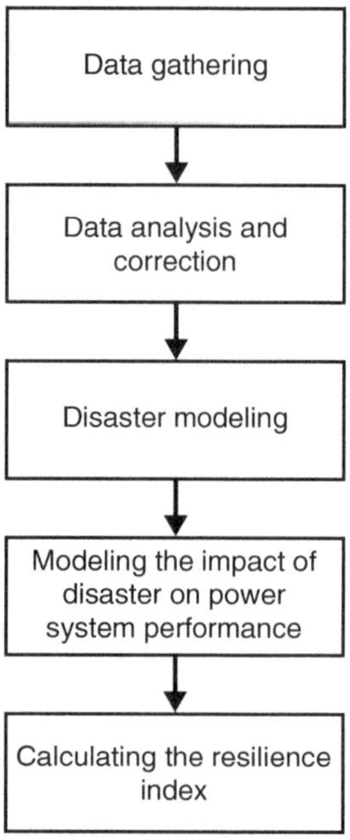

Figure 9.2 General framework for the resilience assessment of the power system.

9.4 Data Required for Power System Resilience Studies

Many factors influence the vulnerability of power systems to outages following natural disasters in a given geographical area. The required data for statistical models can be divided into the following categories:

- Data of natural origin
- Basic data of the power system
- Data on failure and restoration rates

In the following, the data categories previously mentioned will be investigated in detail.

Table 9.1 Specific performance-based resilience metrics.

Metric type	Example	Required data
Power	Not supplied load	Interrupted load statistics
	Supplied load	Total and interrupted load statistics
	Not supplied energy	Interrupted load statistics and duration of interruption in a time period
	Supplied load	Total energy consumption, interrupted load, and duration of load interruption in a time period
	Restored load	Curve of the restored load in a time period
	Restored Energy	Curve of the restored load in a time period
	Generation and demand mismatch	System generation and demand statistics
	Nonconnected generation capacity	Unavailable generation capacity statistics
	Connected generation capacity	Total and unavailable generation capacity statistics
Duration	Load or energy interruption duration	Interruption duration statistics
	Restoration duration	Restoration curve
	Drop state duration after disaster	Grid performance curve in the considered time period
	Duration before performance drop (between disaster start time and performance drop)	Grid performance curve in the considered time period
Frequency	Consumers' outage statistics over a period of time	Number of consumer outages
	Grid-connected consumer statistics	Grid-connected consumer numbers
	Out-of-service transmission lines statistics	Number of out-of-service transmission lines
Probability	The probability of various disaster scenarios occurring, the amount of energy lost per scenario	The probability that the amount of supplied load exceeds a certain amount
	The probability of various disaster scenarios occurring, the amount of energy lost per scenario	The probability that the amount of unsupplied load is greater than or equal to a certain value
	Possibility of no load shedding	Probability distribution curves for total energy produced, total energy consumed, and total energy provided by the energy storage
	Possibility of load shedding	The probability of various disaster scenarios occurring

(Continued)

Table 9.1 (Continued)

Metric type	Example	Required data
Curve	The area under the actual performance curve	Performance curve of the grid in the desired time period
	The area between the actual and ideal performance curves	Performance curve of the grid in the desired time period
	The area between the actual and worst-case related performance curves	Performance curve of the grid in the desired time period
	Ramping rate of the performance curve in the dropping or restoration period	Performance curve of the grid in the desired time period

9.4.1 Data of Natural Origin

Data of natural origin refers to information related to environmental factors that can impact power systems. This includes:

- **Meteorological Data**: Information on weather patterns, including temperature, precipitation, wind speed, and storm events.
- **Geophysical Data**: Data related to geological events such as earthquakes, landslides, or floods.
- **Historical Disaster Records**: Documentation of past natural disasters and their consequences on power systems in specific geographical areas

Weather data can significantly influence the determination of failure causes and recovery durations within power networks. Since resilience modeling must be dynamic, real-time weather data is essential for accurately predicting natural disasters. For instance, data related to severe storms, including wind intensity, gusts, precipitation levels, humidity, and temperature, are critical. This information is continuously updated in geographical and meteorological databases. Consequently, distribution network operators can query weather conditions through meteorological databases to enhance their preparedness and response strategies.

In situations where obtaining accurate information is challenging, such as data related to flood events, simulated data for flooding and associated strong winds is often generated and utilized. However, the adequacy of this simulated data in accurately reflecting infrastructure failures remains a significant challenge.

9.4.2 Basic Data of the Power System

The foundational data used in power system resilience studies can be categorized as follows.

- **Network Data**
  - Initial and terminal codes
  - Conductor type
  - Length
  - Construction date
  - Pole data
  - Maintenance and inspection records
  - Geographical location data

Table 9.2 Specific outcome performance-based resilience metrics.

Metric type	Example	Required data
Economic	Energy or power not supplied cost	Statistics on disconnected load, duration of load interruption, and cost per unit of load interruption or load energy interruption
	Restoration cost	Cost of equipment, employees, and other items used for restoration
	Increased operation cost	Operating cost statistics before and after the disaster
	Gross regional product reduction	Gross regional product before and after the disaster
	Costs imposed on industries, consumers, and the government	Statistics on the damage to economic activities in various sectors and their costs
Social	Reduced number of working hours for workers during recovery period	Statistics on workers' working hours before and after the disaster
	Number of citizens able to work	Statistics of citizens ready to work before and after the disaster
	Affected population	Statistics on the number of people without electricity
	Number or percentage of households with electricity after the disaster	Statistics of households with electricity before and after the disaster
	Number or percentage of households with electricity restored for the restoration period	Statistics on households with electricity before and after the disaster, statistics on how the load on the network was restored over a period of time
	The probability that, within a given time period, the number or percentage of households with electricity after a disaster will exceed a certain threshold.	Statistics on households with electricity before and after the disaster, statistics on how the load on the network was restored over a period of time
Geographical	The affected geographical area	Area of the interrupted zone
Safety and health	Human casualties	Death toll after the disaster
	Number of available hospital beds	Statistics on the number of available hospital beds after the disaster

- **Substation Data**
  - Substation name
  - Transformer serial number
  - Busbar configuration
  - Auxiliary system data (such as remote control, protection mechanisms, and compensation systems)
  - Construction data
  - Inspection records
  - Data required for main diagram representation
- **Transformer Data**
  - Position in network

  - Manufacturer serial number and year of production
  - Year of repair
  - Technical specifications: Detailed specifications including:
    - Nominal power ratings
    - Impedances
    - Dimensions
- **Switching Equipment Data**
  - Position in network
  - Identifier symbol
  - Type code
  - Nominal voltage and current ratings
  - Nominal tripping and closing currents
  - Relay settings and transformation ratios
  - Year of production and maintenance operations
  - Statistical values (such as operation data)
  - Equipment status
- **Load Data**
  - Consumers' location
  - Number of consumers
  - Consumers' code
  - Codes of consumers' groups and tariffs
- **Other Data**
  - Technical characteristics of conductors
  - Grounding data
  - Cable connections
  - Economic data

In power system data collection, it is essential to categorize the data. The data is classified based on the following criteria:

- Types of consumers (residential, commercial, industrial, and agricultural customers)
- Types of power outages (unplanned outages, scheduled outages, and outages due to disasters)
- Weather conditions (climatic conditions of the area, including factors such as temperature, humidity, precipitation, and other atmospheric elements that may affect power system performance)
- Network configuration (network topology, voltage levels, and type of equipment)

These categorizations help in more accurate data analysis and better understanding of power system performance.

9.4.3 Data on Failure and Restoration Rates

The electric power industry collects data related to equipment failure rates, network recovery time, and the impact of outages on consumers. The main objective of gathering this information is to manage power outages in the power system. Examples of data related to failure rates and network recovery include:

1) Type of fault
2) Fault location
3) Geographical location of the network where the fault occurred

4) Recovery time
5) Type of protective equipment
6) Number of customers affected by the outage

These data points are crucial for effective power outage management and system reliability analysis.

One of the important components of failure data is the location and position of the fault, including latitude and longitude. Access to such data is determined through the location of protective equipment operation. This data is used to identify the relationship between weather parameters and failures.

Using the precise geographical location of the incident along with the duration of the outage and the performance of protective equipment can be utilized for studying the temporal and spatial dimensions of severe natural events. This approach allows for the following functions:

- Correlation analysis
- Predictive modeling
- Risk assessment
- Resource allocation
- Infrastructure planning

9.5 Data Analysis and Correction

Data analysis and validation are fundamental steps in resilience studies. In resilience studies, a large volume of data is considered as input. The accuracy of this data significantly impacts the outcome of resilience assessments. Therefore, validating input data and applying corrections where possible are crucial and influential processes. On the other hand, in resilience studies, the possibility of encountering incorrect and low-accuracy data will be greater than in other studies. Because these studies are conducted in the event of a disaster in the power system, in these conditions, the infrastructure for generating and transmitting information may also experience accidents and problems. Therefore, in this situation, verification will be much more important.

Various methods can be employed for data validation, including

- Statistical analysis to identify anomalies
- Consistency checks across different data sets
- Historical data comparisons
- Expert review and validation

Where possible, identifying and correcting data errors is essential. This process may involve

- Cross-referencing data from multiple sources
- Applying statistical methods to identify outliers
- Using domain expertise to assess data plausibility

The integration of machine learning methods into power system resilience studies has become increasingly important due to the complexity of modern power grids, which incorporate renewable energy sources and advanced technologies. Data validation plays a crucial role in ensuring the reliability and effectiveness of machine learning models applied in this context. In this situation data validation is essential for several reasons [30–32]:

- **Quality Assurance**: Ensures that the data used for training machine learning models is accurate, consistent, and complete. In power systems, data can be affected by various factors such as sensor errors, communication failures, or environmental conditions, leading to inconsistencies that can skew model predictions.

- **Model Performance**: Proper validation techniques help assess how well a model generalizes to unseen data. This is particularly critical in power systems, where operational decisions based on faulty predictions can lead to significant disruptions or failures.
- **Regulatory Compliance**: Many power systems are subject to regulatory standards that require robust data management practices. Effective validation processes help meet these requirements and maintain stakeholder trust.

Several techniques are commonly employed in the validation of data for machine learning applications in power system resilience:

- **Cross-Validation**: This method involves partitioning the dataset into subsets, training the model on some subsets while validating it on others. K-fold cross-validation is a popular approach where the dataset is divided into *kk* parts, allowing each part to serve as a validation set at some point.
- **Data Cleaning**: Identifying and correcting inaccuracies or inconsistencies in the dataset is crucial. This may involve removing outliers, filling missing values, or aggregating data from different sources to ensure uniformity.
- **Synthetic Data Generation**: In scenarios where real-world data is scarce or incomplete, synthetic data can be generated using simulation models. This approach can help augment training datasets and improve model robustness.

Despite its importance, several challenges exist in the data validation process for machine learning in power systems:

- **Data Volume and Variety**: The sheer volume of data generated by modern power systems can make validation cumbersome. Moreover, the variety of data types (e.g., time-series data from sensors, categorical data from operational logs) requires diverse validation strategies.
- **Dynamic Nature of Power Systems**: Power systems are subject to rapid changes due to fluctuating demand and supply conditions. This dynamism necessitates continuous validation processes to keep models relevant and effective.
- **Integration with Traditional Methods**: Many existing power system protection and control strategies are based on traditional mathematical models. Integrating machine learning methods with these established practices while ensuring robust validation remains a significant challenge.

9.6 Disaster Forecasting in Power System Resilience Studies

Disaster forecasting in power system resilience studies involves predicting potential disruptions to power systems caused by natural disasters, cyberattacks, or other significant events. This predictive capability is essential for enhancing the resilience of power systems, allowing for better preparedness and response strategies.

The importance of disaster forecasting includes the following items:

- **Proactive Risk Management**: By forecasting potential disasters, utilities can implement proactive measures to mitigate risks, such as reinforcing infrastructure or developing contingency plans. This preemptive approach can significantly reduce the impact of disruptions on service delivery.
- **Resource Allocation**: Accurate forecasting enables more effective allocation of resources. Utilities can prioritize areas that are most likely to be affected by disasters, ensuring that emergency response teams and equipment are strategically positioned.

- **Improved Recovery Strategies**: Understanding the potential impacts of various disaster scenarios allows for the development of tailored recovery strategies. This includes planning for rapid restoration of services and minimizing downtime.

A number of techniques for disaster forecasting are presented below.

- **Statistical Methods**: Traditional statistical models analyze historical data to identify patterns and trends associated with disasters. These methods can provide insights into the frequency and severity of past events, aiding in future predictions.
- **Machine Learning Approaches**: Machine learning techniques are increasingly used due to their ability to process large datasets and identify complex patterns that traditional methods might miss. Machine learning models can be trained on historical disaster data and real-time sensor information to improve forecasting accuracy.
- **Simulation Models**: Simulation-based approaches create virtual environments to model the behavior of power systems under various disaster scenarios. These simulations help in understanding potential vulnerabilities and the effectiveness of different response strategies.
- **Geospatial Analysis**: Geographic Information Systems are utilized to analyze spatial data related to disaster risks, such as flood zones or earthquake-prone areas. This spatial analysis aids in identifying critical infrastructure at risk and planning mitigation measures accordingly.

The challenges in disaster forecasting can be summarized as follows:

- **Data Quality and Availability**: The effectiveness of forecasting models heavily relies on the quality and completeness of data. Inconsistent or missing data can lead to inaccurate predictions, undermining preparedness efforts.
- **Dynamic Nature of Threats**: The evolving nature of threats, including climate change impacts leading to more frequent extreme weather events, complicates forecasting efforts. Models must be continuously updated to remain relevant.
- **Complex Interdependencies**: Modern power systems consist of interconnected components that can influence each other during disasters. Understanding these interdependencies is crucial for accurate forecasting but can be challenging due to their complexity.

9.7 Modeling the Impact of Disaster on Power System Performance

A wide range of methods have been used to model power system recovery after natural disasters. Methods such as engineering fragility curve fitting methods, deterministic resource constraint methods, probabilistic methods such as Monte Carlo simulation, optimization methods, discrete-state, discrete Markov processes, network-based methods, and statistical regression methods have been presented.

Disaster modeling in power system resilience studies involves the assessment and analysis of how power systems respond to various disruptive events, such as natural disasters, cyberattacks, or equipment failures. This modeling is crucial for understanding vulnerabilities and enhancing the robustness and recovery capabilities of power systems.

Key components of disaster modeling are as follows:

- **Risk Assessment**: Evaluating potential risks associated with different types of disasters is fundamental. This includes identifying high-risk areas and understanding the likelihood and impact of specific events on power infrastructure.
- **Modeling Techniques**: Various modeling techniques are employed to simulate disaster scenarios, including

  - **Probabilistic Models**: These models assess the likelihood of different failure modes and their impacts on system performance.
  - **Agent-Based Models**: These simulate interactions between individual components of the power system, allowing for a detailed examination of system dynamics during disasters.
  - **Machine Learning Approaches**: Recent studies highlight the use of machine learning algorithms to analyze historical data and predict system behavior under stress, enhancing the predictive capabilities of disaster models [31].
- **Phases of Disaster Response**: Disaster modeling typically encompasses several phases [30, 33]:
  - **Pre-Disruption Phase**: This involves anticipation and preparation strategies to mitigate impacts before a disaster occurs. It includes resource allocation and contingency planning.
  - **During Disruption Phase**: This phase focuses on the system's ability to withstand and absorb shocks, maintaining operational continuity as much as possible.
  - **Post-Disruption Phase**: After a disaster, the focus shifts to recovery strategies, assessing damage, and restoring services efficiently.

Applications of disaster modeling are as follows.

- **Infrastructure Resilience**: By simulating various disaster scenarios, utilities can identify weaknesses in their infrastructure and implement improvements to enhance resilience.
- **Emergency Response Planning**: Disaster models inform emergency response strategies by predicting how quickly systems can recover and what resources will be needed during restoration efforts.
- **Policy Development**: Insights from disaster modeling can guide policymakers in developing regulations that promote resilience in power systems, ensuring that utilities are prepared for potential disruptions.

Challenges in disaster modeling are the same as the disaster forecasting challenges, including the data limitation, complex interdependencies, and dynamic nature of threats.

Machine learning-based methods are one of the available solutions for solving the challenges facing modeling and resilience studies, which are examined below based on static and dynamic machine learning methods.

9.8 Static Model in Machine Learning

The majority of work conducted in resilience studies has focused on modeling the occurrence of severe weather conditions and the resulting initial failures. These models are static in nature, with their primary focus on combining weather variables and failure data. Some models in this area begin with a single node (such as distribution network lines or other components) and examine the effects of weather variables on its failure rate. For example, the failure rate, defined as the average number of failures per mile per hour of distribution network line length, has been presented as a quadratic curve in relation to storm intensity [34]. For the other example, power system equipment failure curves as a function of wind speed have been investigated for modeling failure rates. These approaches aim to quantify the relationship between severe weather conditions and power system component failures, providing a foundation for reliability assessments and resilience planning in electrical grids. The static nature of these models allows for a focused analysis of how specific weather variables contribute to system failures, which is crucial for understanding and predicting the impact of extreme weather events on power infrastructure.

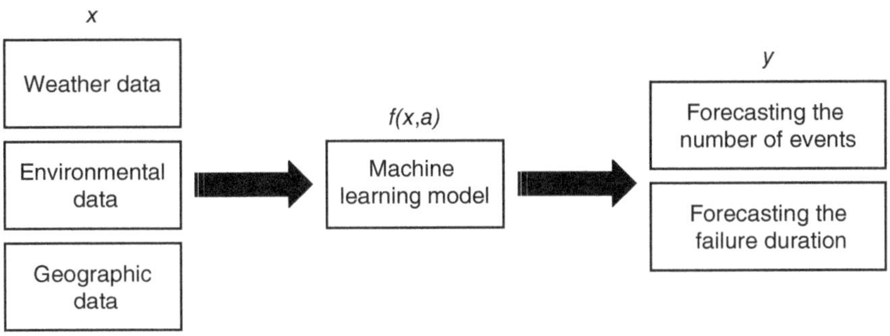

Figure 9.3 Static model in machine learning mechanism.

These models, although diverse, can utilize a machine learning mechanism as illustrated in Figure 9.3. For this problem, an n-dimensional vector space is considered as the input variable, comprising $x \in R^n$ where $n \geq 1$.

The modeling is such that the static path map is specified as $f(x;\, a): x \rightarrow y$ between the initial variables x and the problem output $y \in R^m$ for $m \geq 1$. Here y can represent the failure including the number of failures or the duration of the failure or the probability of failure. Also, for unknown parameters that affect the problem model, a parameter $a \in R^l$ is introduced for $l \geq 1$.

For input and output variables, a database is required as a matrix $D = \{x^{(k)},\, y^{(k)}\}$ for input variables x and the problem objective y, which is the failure. It should be noted that the variable k represents the sample number. In other words, $x^{(k)}$ is the concept of the k-th sample of the input variable to the machine learning system. The main goal of using a machine learning system is to obtain a failure map or f corresponding to the data in the database D. That is, the machine learning system can estimate f, which is a map from input x to y. This is the main context of using a machine learning system in the discussion of resilience. This model is static when neither f nor the parameters a or the input and output $(x;\, y)$ are constant variables. The input variables x mainly represent weather variables, including wind speed and intensity and precipitation [35].

As mentioned, a machine learning system is needed to fit the impact of natural disasters on the failure rate. A number of data fitting methods are examined below.

- **Generalized Linear Models (GLM)**: GLM is a generalization of linear regression used for data that is not normally distributed. For example, the number of failures, which is a discrete quantity, can be predicted with the help of a GLM.
  In general, a GLM consists of three parts:
  - **Random Component**: The dependent variable (response) is Y, which assumes that the observations related to the data are independent of each other, have the same distribution, and belong to the exponential family.
  - **Systematic Component**: In principle, the linear predictor is dependent on the variable X in the form of the relation $\eta = X\beta$, where β is a $p \times 1$ vector of unknown parameters and X is an $n \times p$ matrix consisting of independent variables.
  - **Link Function**: The link function relates the linear predictor, i.e., η, to the mathematical expectation of the response variable, i.e., μ, and is a monotonic ascending and differentiable function.

- **Generalized Additive Models (GAM)**: The GAM consists of a random component, a cumulative component, and a link function. GAM differs from GLM in that the cumulative predictor is replaced by a linear predictor.
- **Accelerated Failure Time Model (AFT)**: The AFT model has been used to estimate the duration of power outages. This model is a parametric model in survival analysis for survival time data. These data are commonly known as survival data and are asymmetrically distributed, often with a positive skew. Therefore, it is unreasonable to assume a normal distribution for them.
- **Tree-Based Data Mining Models**: Tree-based data mining models are widely utilized techniques for classification and regression tasks in data mining and machine learning. These models leverage hierarchical structures to make predictions based on input features, making them intuitive and easy to interpret.

Classification and regression trees, Bayesian additive regression trees (BART), and multivariate adaptive regression splines are examples of tree-based data mining models that can be used for resilience assessment studies. Among these methods, it has been observed that the BART method is very efficient in predicting the duration of failure in the power system in proportion to weather variables.

In the static model, it is assumed that the occurrence of a fault is independent of time and location. The assumption of independence is invalid in practice because locations in close proximity may experience similar weather effects. Furthermore, in some specific geographical locations, the probability of weather-related failures is higher than in other areas. For this reason, time and location dependence must be considered, which is discussed below.

9.9 Spatiotemporal Random Process

When spatial and temporal scales are considered in modeling failure rates and resilience, the modeling process becomes time- and location-dependent and requires the use of time- and location-dependent stochastic analysis. Many studies have considered the dynamic effect of time and location in presenting resilience models. Some of these models are listed below:

9.9.1 Dynamic Model for Chain Failures

Dynamic models consider the progression of the failure in time and location. In static systems, static failure-based analysis shows the vulnerability of the power grid, and they do not describe the progression of failure and outage in the system. This challenge is solved by using dynamic models. In dynamic analysis, dynamic activities such as network power flow with redispatch (in order to consider the distribution of network load after the failure and considering overload in the network) are performed.

Dynamic modeling is divided into two groups. The first group uses graph theory analysis, and the second group uses power network analysis with economic dispatch after each failure.

9.9.2 Nonstationary Failure-Recovery-Impact Processes

Nonstationary failure-recovery-impact processes refer to dynamic systems where the rates of failure and recovery are not constant over time. These processes are crucial for understanding how systems behave under varying conditions, especially in environments where external factors can significantly influence operational stability. The formulation of this method is bottom-up, starting with the failures of the distribution network infrastructure, then the recovery services are considered, and finally the impact of the failure on the consumers will be addressed. It should be noted that in this process, the recovery services are included in this model through the duration of the failure.

9.10 Lessons Learned and Concluding Remarks

In this chapter, the challenges related to data for assessing resilience in power systems are examined. By identifying the challenges facing these studies, proposed solutions are presented to address them and better and more accurately assess resilience in the power system as a critical infrastructure.

Resilience assessment studies in power systems require a large and diverse amount of data, which will lead to complications due to the large number and variety of necessary data, including meteorological and environmental conditions, basic power system data, and event data. The large amount of data increases the possibility of errors in recording, storing, and communicating them. On the other hand, the data related to disasters is not available in some cases. The reason for this could be that the disaster did not occur for the desired system. Therefore, analyzing available data and making the necessary corrections to them is one of the steps that can create desirable input data for resilience assessment. Data volume and variety, the dynamic nature of the power system data, and integration with traditional methods are the main challenges facing the validation and analysis studies of the power system.

Research in this domain suggests several avenues for enhancing data validation processes:

- **Advanced AI Techniques**: Leveraging advanced artificial intelligence (AI) methodologies such as deep learning could improve the accuracy of predictions and enhance the robustness of validation techniques by automatically identifying patterns in complex datasets.
- **Real-Time Validation Frameworks**: Developing frameworks that allow for real-time data validation could significantly improve the responsiveness of power systems to emerging threats or operational challenges.
- **Collaboration Across Disciplines**: Interdisciplinary collaboration among engineers, data scientists, and regulatory bodies can lead to more comprehensive validation strategies that address both technical and compliance-related aspects effectively.

Predicting a disaster and determining its impact on the power system is one of the important steps in resilience assessment studies. The integration of advanced technologies such as AI and big data analytics holds promise for improving disaster forecasting in power systems. AI algorithms can enhance predictive capabilities by analyzing vast amounts of real-time data from smart grids and IoT devices, leading to more accurate forecasts. Disaster forecasting is a critical component of enhancing power system resilience. By employing a combination of statistical methods, machine learning, simulation models, and geospatial analysis, utilities can improve their preparedness for potential disruptions, ultimately leading to more reliable energy systems.

Studying the impact of disasters on the power system and how to respond to them determines the level of resilience of the power system and, on the other hand, can determine future solutions to increase resilience. Challenges in disaster modeling are the same as the disaster forecasting challenges. These challenges are the data limitation, complex interdependencies, and dynamic nature of threats.

Machine learning-based methods are one of the available solutions for solving these challenges that can determine the impact of methods to improve resilience in the power system.

9.11 Future Work

As a suggested research work in continuation of the studies conducted in this book chapter, it is suggested to conduct additional studies to determine appropriate resilience metrics, taking into account the impact of the number of data required as well as the error of the necessary predictions.

Since the impact of a disaster on a power system is a time-dependent phenomenon, any metric that can better model this impact is more appropriate in terms of the concept of resilience. In this regard, metrics that fully consider how the system performance curve changes over time (such as the common area-based metrics) have the most conceptual validity, and metrics that are calculated based on only one point of the system performance curve (such as unsupplied demand) have the least conceptual validity. It is clear that considering the entire performance curve requires more data, and therefore data errors can have a greater impact on the evaluation result.

Future power system simulations are usually performed without modeling some aspects (such as the interdependence of critical infrastructures in society or disaster management activities in different sectors) or by simplifying some aspects (such as how power system recovery activities are carried out and their impacts). This causes the simulated performance curve to be in error compared to what will actually happen in the power system. The greater the simplifications and omissions, the greater the error in the system performance curve. The greater the simplifications and omissions, the greater the error in the system performance curve.

Considering the above, examining resilience assessment metrics by considering the required data from past performance as well as forecast errors is important and necessary, requiring in-depth and detailed studies.

References

1 IEC. (2014). Microgrids for disaster preparedness and recovery – with electricity continuity plans and systems. International Electrotechnical Commission (IEC), Switzerland.

2 Sheibani, M.R. and Moshari, A. (2020). Operation planning of a microgrid considering the resiliency in the presence of energy storage systems. *2020 10th Smart Grid Conference (SGC)*, Kashan, Iran (16–17 December 2020), pp. 1–6. IEEE.

3 NERC. (2012). Severe impact resilience: Considerations and recommendations. North American Electric Reliability Corporation (NERC).

4 Jimenez-Estevez, G., Navarro-Espinosa, A., Palma-Behnke, R. et al. (2017). Achieving resilience at distribution level: learning from isolated community microgrids. *IEEE Power and Energy Magazine* 15 (3): 64–73.

5 Panteli, M., Mancarella, P., Trakas, D. et al. (2017). Metrics and quantification of operational and infrastructure resilience in power systems. *IEEE Transactions on Power Systems* 32 (6): 4732–4742.

6 Li, Z., Shahidehpour, M., Aminifar, F. et al. (2017). Networked microgrids for enhancing the power system resilience. *Proceedings of the IEEE* 105 (7): 1289–1310.

7 Liu, X., Shahidehpour, M., Li, Z. et al. (2017). Microgrids for enhancing the power grid resilience in extreme conditions. *IEEE Transactions on Smart Grid* 8 (2): 589–597.

8 Gholami, A., Shekari, T., Aminifar, F. et al. (2016). Microgrid scheduling with uncertainty: the quest for resilience. *IEEE Transactions on Smart Grid* 7 (6): 2849–2858.

9 Farzin, H., Fotuhi-Firuzabad, M., and Moeini-Aghtaie, M. (2016). Enhancing power system resilience through hierarchical outage management in multi-microgrids. *IEEE Transactions on Smart Grid* 7 (6): 2869–2879.
10 Espinoza, S., Panteli, M., Mancarella, P. et al. (2016). Multi-phase assessment and adaptation of power systems resilience to natural hazards. *Electric Power Systems Research* 136: 352–361.
11 Panteli, M., Trakas, D.N., Mancarella, P. et al. (2016). Boosting the power grid resilience to extreme weather events using defensive islanding. *IEEE Transactions on Smart Grid* 7 (6): 2913–2922.
12 Manshadi, S.D. and Khodayar, M.E. (2015). Resilient operation of multiple energy carrier microgrids. *IEEE Transactions on Smart Grid* 6 (5): 2283–2292.
13 Bie, Z., Lin, Y., Li, G. et al. (2017). Battling the extreme: A study on the power system resilience. *Proceedings of the IEEE* 105 (7): 1253–1266.
14 Arghandeh, R., von Meier, A., Mehrmanesh, L. et al. (2016). On the definition of cyber-physical resilience in power systems. *Renewable and Sustainable Energy Reviews* 58: 1060–1069.
15 The_National_Academies (2012). *Terrorism and the Electric Power Delivery System*. Washington, DC: The National Academies Press.
16 Gholami, A., Aminifar, F., and Shahidehpour, M. (2016). Front lines against the darkness: enhancing the resilience of the electricity grid through microgrid facilities. *IEEE Electrification Magazine* 4 (1): 18–24.
17 Abi-Samra, N., McConnach, J., Mukhopadhyay, S. et al. (2014). When the bough breaks: managing extreme weather events affecting electrical power grids. *IEEE Power and Energy Magazine* 12 (5): 61–65.
18 Marnay, C., Aki, H., Hirose, K. et al. (2015). Japan's pivot to resilience: How two microgrids fared after the 2011 earthquake. *IEEE Power and Energy Magazine* 13 (3): 44–57.
19 Ansari, B. and Mohagheghi, S. (2015). Optimal energy dispatch of the power distribution network during the course of a progressing wildfire. *International Transactions on Electrical Energy Systems* 25 (12): 3422–3438.
20 Arab, A., Khodaei, A., Han, Z. et al. (2015). Proactive recovery of electric power assets for resiliency enhancement. *IEEE Access* 3: 99–109.
21 Bajwa, A.A., Mokhlis, H., Mekhilef, S. et al. (2019). Enhancing power system resilience leveraging microgrids: a review. *Journal of Renewable and Sustainable Energy* 11 (3): 035503.
22 MEA. (2014). Maryland resiliency through microgrids: task force report. Maryland Energy Administration.
23 Arghandeh, R., Brown, M., Del Rosso, A. et al. (2014). The local team: leveraging distributed resources to improve resilience. *IEEE Power and Energy Magazine* 12 (5): 76–83.
24 The_National_Academies (2017). *Enhancing the Resilience of the Nation's Electricity System*. Washington, DC: National Academies Press.
25 Abbey, C., Cornforth, D., Hatziargyriou, N. et al. (2014). Powering through the storm: MIcrogrids operation for more efficient disaster recovery. *IEEE Power and Energy Magazine* 12 (3): 67–76.
26 Gholami, A., Shekari, T., Amirioun, M.H. et al. (2018). Toward a consensus on the definition and taxonomy of power system resilience. *IEEE Access* 6: 32035–32053.
27 The_National_Academies (2012). *Disaster Resilience: A National Imperative*. Washington, DC: The National Academies Press.

28 Raoufi, H., Vahidinasab, V., and Mehran, K. (2020). Power systems resilience metrics: a comprehensive review of challenges and outlook. *Sustainability* 12 (22): 9698.

29 Panteli, M. and Mancarella, P. (2017). Modeling and evaluating the resilience of critical electrical power infrastructure to extreme weather events. *IEEE Systems Journal* 11 (3): 1733–1742.

30 Zahraoui, Y., Korotko, T., Rosin, A. et al. (2024). AI applications to enhance resilience in power systems and microgrids – a review. *Sustainability* 16 (12).

31 Porawagamage, G., Dharmapala Kalana, K., Chaves, J.S. et al. (2024). A review of machine learning applications in power system protection and emergency control: opportunities, challenges, and future directions. *Frontiers in Smart Grids* 3: https://doi.org/10.3389/frsgr.2024.1371153.

32 Almasoudi, F.M. (2023). Enhancing power grid resilience through Real-Time fault detection and remediation using advanced hybrid machine learning models. *Sustainability* 15 (10): 8348.

33 Lin, J.H. and Wu, Y.K. (2024). Review of power system resilience concept, assessment, and enhancement measures. *Applied Sciences* 14 (4): 1428.

34 Brown, R.E. (2008). *Electric Power Distribution Reliability*, 2nde. CRC Press.

35 Ji, C., Wei, Y., and Poor, H.V. (2017). Resilience of energy infrastructure and services: modeling, data analytics, and metrics. *Proceedings of the IEEE* 105 (7): 1354–1366.

10

Nonintrusive Load Monitoring in Smart Grids Using Deep Learning Approach

Sobhan Naderian[1] *and Hamidreza Arasteh*[2,3]

[1] *System Power and Energy (SPE) Department, James Watt School of Engineering (JWSE), University of Glasgow, Glasgow, United Kingdom*
[2] *Power Systems Operation and Planning Research Department, Niroo Research Institute, Tehran, Iran*
[3] *Center for Research on Microgrids (CROM), Huanjiang Laboratory, Zhuji, Shaoxing, Zhejiang, China*

10.1 Introduction

Nonintrusive load monitoring (NILM) is a well-known analyzing method that disaggregates the electricity usage data based on different resolutions (seconds, minutes, etc.) to the specified appliance usage without measuring each appliance's electricity consumption [1]. Due to the potential of NILM on energy disaggregation in smart grids and its improvements and robustness, several studies focused on it during the past decade [2–4]. Early research on NILM began in 1985 at the Massachusetts Institute of Technology under George Hart, who introduced the concept of nonintrusive appliance load monitoring (NIALM or NILM) [5]. A hybrid approach that combines artificial intelligence methods (machine learning and neural networks) with signal processing has been used in NILM to identify specific appliance usage patterns from electricity consumption data collected in residential areas. Another important reason for NILM is its potential to reduce carbon footprints in electricity distribution networks, as it improves energy efficiency [6]. Furthermore, with the increasing affordability and efficiency of deep learning computation, NILM is expected to play a significant role in smart grids [7]. The need for innovative technologies to reduce carbon emissions and apply new machine learning methods has been discussed in reference [8].

A hybrid method based on subtractive clustering and the maximum likelihood classifier was proposed in reference [9], but it only focused on the resistive loads. The application of a temporal multi-label classification method has been discussed in reference [10], where statistical measures like average and variance are used to detect specific appliances. A new hybrid approach based on wavelet transform and decision trees has been explored by the authors of reference [11], with the key strength being the design of custom wavelet functions tailored to power system applications, as opposed to standard wavelet functions meant for image processing. The use of the Stockwell transform to extract current waveform features and appliance identification through the 0-1 knapsack method has been presented in reference [12], and experimental results were validated in this study.

A novel event-driven framework for NILM based on convolutional neural networks (CNN) for residential appliances was introduced in reference [13], with its main advantage being the ability to detect the on-off states of each appliance. An algorithm for identifying specific appliance signatures

Data-Driven Energy Management and Tariff Optimization in Power Systems: Shaping the Future of Electricity Distribution through Analytics, First Edition.
Edited by Hamidreza Arasteh, Pierluigi Siano, Niki Moslemi, and Josep M. Guerrero.

using the Karhunen-Loéve expansion was proposed by the authors of reference [14], where the active power signal is decomposed into subspace components to recognize appliances and estimate energy consumption. The combination of several time-domain features for appliance detection and problem-size reduction was proposed in reference [15], where a decision tree classifier was used to identify specific appliances through feature-reduced approaches.

Different NILM methods have been proposed, focusing on data analysis and pattern recognition, with meaningful features extracted from large energy consumption datasets. For this purpose, deep learning methods, artificial neural networks (ANN), and pattern recognition techniques have been applied to appliance recognition [16–20]. A new NILM method using CNN with differential input was proposed by the authors of reference [21], where transient parameters were mapped by a five-layered CNN from differentiated data that was fed into the network and combined with supplementary input. The use of CNNs in NILM is a significant strength of this study. In reference [22], a new approach using the pinball quantile loss function to guide deep neural networks for NILM was proposed, enhancing the efficiency and accuracy of deep networks like CNN and RNN.

Recent studies have been thoroughly reviewed in reference [23], where new algorithms for better energy management and potential future applications of NILM were suggested. The authors of reference [15] introduced a method based on multi-descriptor fusion to identify appliance signatures, proposing a dimension reduction technique that applies to fused time-domain features. A bagging tree classifier was implemented to classify appliances in this research. A microgrid energy management system using NILM was introduced in reference [24], where deep neural networks processed smart meter data to identify appliance signatures. Event-based NILM using adaptive weighted recurrence graph blocks and CNN-based event-driven NILM methods was proposed by reference [25]. A similar appliance identification technique using deep neural networks based on feature mapping was carried out in reference [26], using an open-source dataset of four fluorescent lamps to implement NILM with a combined method of locality-constrained linear coding and machine learning classifiers.

Simultaneous appliance energy disaggregation was developed in reference [27], using mixed integer programming to model the on-off status of appliances, with event-driven methods reducing computational complexity. A real-time NILM system for simultaneous appliance identification was proposed in reference [28], using a deep dictionary learning model for real-time load monitoring. A new framework based on two-dimensional (2D) phase encoding of power waveforms was proposed by the authors of reference [29]. This method, known as 2D phase encoding of power signals (2D-PEP), creates a 2D representation of power waveforms, and its effectiveness and accuracy were validated using datasets such as REEND, UK-DALE, and WHITED. Implementation of phase domain rather than time domain is a key strength of this study.

Based on the reviewed research, identifying specific appliances in smart grids is essential as it provides precise appliance-level energy usage data, benefiting end-users and enabling more flexible and accurate demand-side management. This study presents a new framework for NILM, combining CNN and long short-term memory (LSTM) to disaggregate electricity consumption data and output appliance-level data. To achieve this, the hybrid approach is trained using supervised learning methods. The main contributions of this method can be summarized as follows:

- Development of a hybrid deep learning model based on LSTM and CNN, capitalizing on the strengths of both networks. This hybrid method provides high performance in appliance recognition. CNN layers reduce the problem's complexity, while LSTM layers generate more accurate estimates for each appliance's usage data.

- Validation of the proposed NILM method using the REFIT dataset, with an eight-second sampling frequency, to assess its accuracy and efficiency.
- A detailed comparison of the proposed method's accuracy, number of appliances considered, and the size of deep neural networks, as compared to previous NILM methods.

This study is organized as follows: Section 10.2 details the proposed deep neural network, Section 10.3 describes the NILM methodology, and Section 10.4 presents the results based on the REFIT dataset and compares them with earlier methods.

10.2 Deep Learning Neural Networks

10.2.1 RNN

Recurrent neural networks (RNNs) are a type of ANN where the connections between nodes form a feedback loop, enabling the network to exhibit temporal dynamic behavior. This allows RNNs to use their internal memory to process input sequences of varying lengths, making them suitable for tasks such as time series analysis, handwriting recognition, and speech processing. Due to their feedback mechanism, RNNs are particularly effective for time series analysis [30, 31], which makes them a practical choice for NILM compared to feedforward neural networks. In NILM, appliances are turned on and off in a random manner, generating dynamic and varying electricity usage data, and RNNs can efficiently track these changes due to their ability to handle dynamic conditions.

However, RNNs face challenges with short-term memory when processing long input sequences. When the sequence length increases, RNNs tend to forget earlier data points, making it difficult for them to recall initial information. To address this issue, LSTM networks were developed. LSTMs are designed to handle long sequences of data and include feedback connections. They can be fed with either individual data points or entire sequences, making them highly effective for tasks requiring long-term memory retention.

10.2.2 LSTM

The sequential model for LSTM has been demonstrated in Figure 10.1, which slides on the predefined sequence of input data to generate an output sequence of data points or output time series. Each cell is fed by the sequence of input time-series data points. Upcoming output data points and the cell's output are concatenated together to generate a new input data point for the next data point. This repetitive procedure will be carried out to cover whole data points. F. Gers introduced the LSTM network in 1999, which is the new type of RNN that consists of four main parts, namely, input gate, output gate, input candidate gate, and forget gate [32]. The forget gate plays a key role in LSTM to forget the former non-important cell's state and remember the crucial state as expressed:

$$f_t = \sigma(W_f \cdot [h_{t-1}, x_t] + b_f) \tag{10.1}$$

the σ or sigmoid is an activation function widely used for adding nonlinear features to a model as follows:

$$\sigma(X) = \frac{1}{1 - e^{-x}} \tag{10.2}$$

to generate new cell state C_t', the input gate and input candidate gate are activated simultaneously, and

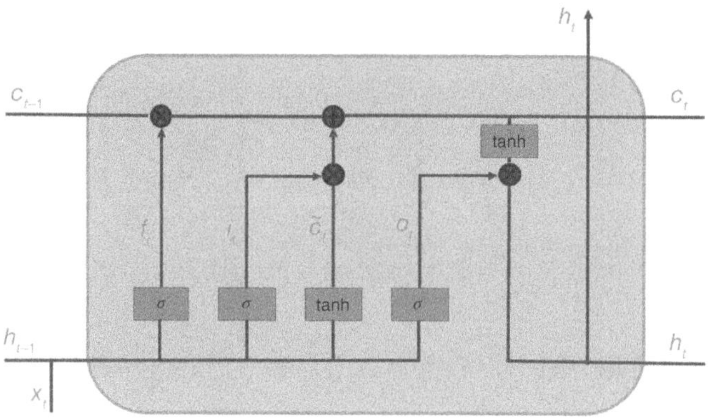

Figure 10.1 Internal structure of the LSTM cell.

Then repeat this process to cover whole data points. The input gate and input candidate gate use sigmoid and hyperbolic tangent functions, respectively, to carry out the new cell's state procedure and generate C_t' as follows:

$$i_t = \sigma(W_i \cdot [h_{t-1}, x_t] + b_t) \tag{10.3}$$

$$C_t' = \tanh(W_c \cdot [h_{t-1}, x_t] + b_c) \tag{10.4}$$

The tanh function is a hyperbolic tangent function that maps input to a continuous number between −1 and 1.

10.2.3 CNN

CNN have been widely used in scientific and computer science research topics such as image analysis. CNN consists of several kernels that recognize different features of fed data. Detection of specified appliance consumption data points has been carried out by filters. Also, discrimination between appliance consumption data points and irrelevant data has been carried out by filters. During the training phase of the CNN network filters' characteristics are set automatically [21].

Three basic layers create CNN networks: input layer, hidden layer, and output layer. Hidden layers are located between input and output layers, and convolution is carried out by built-in activation functions inside them. In CNN networks, activation functions could be different functions like tanh and ReLU. ReLU is widely used in layers that fulfill multiplication. There are three various types of layers for CNN: convolutional layer, pooling layer, and fully connected layer. CNN architecture presented in Figure 10.2.

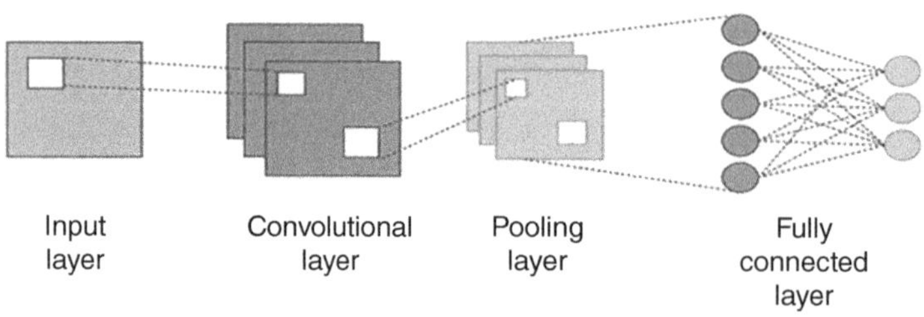

Figure 10.2 The architecture of CNN.

10.2.4 Convolutional Layer

The convolutional layer is the most important component of a CNN. It builds a network of connections and weights based on input data to learn from and generate a feature map. In this layer, different feature maps work to establish meaningful relationships from the input data. Each feature map is made up of neurons that help in creating the appropriate map from the data passed in from the previous layers. The input data is convolved with a kernel function to produce new features, which are then passed through a nonlinear activation function. Commonly used activation functions include Sigmoid, tanh, and ReLU [33].

10.2.5 Pooling Layer

To extract features from the input data, pooling layer could also be advantageous for lowering the feature map dimension and enhancing the effectiveness of the feature extraction process. Pooling layers have been placed between two convolutional layers for optimum efficiency. The moving step of each kernel is a criterion for the size of each feature map. Average pooling [34] and max pooling [35] are widely used for pooling operations. To find more complicated patterns out of input data, cascading pooling layers and convolutional layers have been used in past studies.

10.2.6 Fully Connected Layer

In order to generate understandable output in CNN, a classifier layer should be implemented. Fully connected layers carry out the classifier role in CNN by converting input data from prior layers into meaningful data for the output. Softmax regression is widely used for the problems involved in classification due to its significance for generating proper probability distribution for outputs. Hybrid classification models based on support vector machine (SVM) and CNN have been widely implemented [35].

10.3 The Proposed Method

10.3.1 Pre-Processing and Preparing Data

A sequence-to-sequence model is implemented to train and validate the proposed approach. A sliding window of specified input data points (main electricity usage data) is used and fed into the model to efficiently predict the output of a specific appliance's energy consumption. The pre-processing analysis method establishes a framework for organizing both input and output consumption data points. To normalize the input and output data during pre-processing, the following method is applied:

$$\mathrm{DP}_{i_{Norm}} = \frac{\mathrm{DP}_i - \mathrm{DP}_{avg}}{\mathrm{DP}_{std\_dev}} \tag{10.5}$$

where

$\mathrm{DP}_{i_{Norm}}$ stands for normalized data points
DP_i is actual data point
DP_{avg} and DP_{std_dev} are considered for mean and standard deviation of data points, respectively.

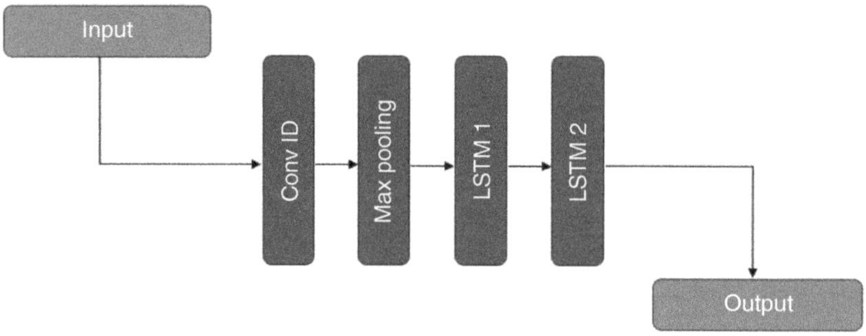

Figure 10.3 The architecture of the proposed deep neural network.

10.3.2 Proposed Method Architecture

Figure 10.3 represents the proposed model for the NILM based on CNN and LSTM networks. One dimension convolving layer and max pooling layer make a CNN layer at the first two layers. As discussed before, CNN networks are quite strong due to the convolutional layer; it enables the CNN to learn about hidden features in the input data points. One of the most important issues of convolutional layers is the output feature map's large dimensions and complexity. To alleviate this concern, one dimension convolving layer is cascaded by a MaxPooling layer. The MaxPooling layer is capable of decreasing the complexity two times and minimizing the size of the problem significantly. Two LSTM layers and a 1-D CNN layer are cascaded and followed by two dense layers to generate accurate output data and recognize the specified appliance.

10.3.3 Proposed Method's Parameters

The overall parameter selection process involves testing various architectures to achieve the most accurate and time-efficient results. Additionally, some components of the architecture have been adapted from reference [37]. Table 10.1 provides details about each layer of the proposed model. Data from each appliance is fed into the model, with a total of 100,000 data points considered for each appliance, divided into training and testing sets (70% for training and 30% for testing). This process is repeated 500 times to identify the consumption patterns of specific appliances from the aggregated data. For each appliance, a sliding window of 100 sample sequences is applied to the aggregated data to produce the necessary input data points. The electricity usage data of each appliance serves as the output for the network during training, and the same sliding window method is used on the appliance's own consumption data.

Two data matrices are created for each appliance: an input matrix with dimensions $M \times 100$ (where $M = N$ 2212 Window_Size $+ 1$, with N being the total number of data points) that includes only aggregated data, and an output matrix of the same dimensions that contains only the power consumption data of the appliance. A kernel size of 48 is used in the convolutional layer, followed by a max pooling layer with a pool size of 3. The LSTM layers consist of 128 and 256 neurons, cascaded with two fully connected layers to generate the output for each appliance. The mean absolute error is utilized as the loss function to measure accuracy.

Table 10.1 Parameters of the proposed deep neural network.

Layer	Detail
Convolution layer	Kernels: 48
MaxPooling	Pool size: 3
LSTM1	Number of nodes: 256
LSTM2	Number of nodes: 128
Fully connected layer	Number of nodes: 128
Fully connected layer	Number of nodes: 100

10.3.4 Performance Evaluation

To measure the accuracy and efficiency, four metrics have been proposed on the output data. Root mean square error (RMSE) and average normalized error (ANE) have been calculated for the specified appliance as follows:

$$\text{RMSE} = \sqrt{\frac{1}{N}\sum_{t=1}^{N}(x_t - \hat{x}_t)} \tag{10.6}$$

where

x_t is actual value
$\hat{x}_t$ is the predicted value of the specified appliance at time t
N is the length of data points

Average normalized error (ANE) is calculated by the following expression:

$$\text{ANE} = \frac{\left|\sum_{i=1}^{N}\overline{p}_i - \sum_{i=1}^{N}\hat{p}_i\right|}{\sum_{i=1}^{N}\overline{p}_i} \tag{10.7}$$

ANE can measure the total energy difference between actual and predicted electricity consumption data points for five different types of appliances where $\hat{p}_i$ is the predicted power by the model for all considered appliances at time i and $\overline{p}_i$ is the actual power consumed by all mentioned appliances at the same time. To measure accuracy and F1 score metrics and assess the performance of the method, the following expressions are considered [32]:

$$\text{Accuracy} = \frac{\text{TP} + \text{TN}}{\text{TP} + \text{TN} + \text{FP} + \text{FN}} \tag{10.8}$$

$$\text{F1 score} = 2 \times \frac{\text{Precision} \times \text{Recal}}{\text{Precision} + \text{Recal}} \tag{10.9}$$

where Precision $= \frac{\text{TP}}{\text{TP+FP}}$ and Recal $= \frac{\text{TP}}{\text{TP+FN}}$

TP, NP, FP, and FN represent total repetition of true positives, true negatives, false positives, and false negatives, respectively. The number of time slices in which the appliance n was either correctly classified as being on (TP), classified as being on while it was actually off (FP), classified as on while it was actually on (FN), or correctly classified as being on (TN). Mentioned metrics have been calculated as follows:

$$\mathrm{TP}^{(n)} = \sum_t \mathrm{AND}\left(x_t^{(n)} = \mathrm{on}, \hat{x}_t^{(n)} = \mathrm{on}\right) \tag{10.10}$$

$$\mathrm{FP}^{(n)} = \sum_t \mathrm{AND}\left(x_t^{(n)} = \mathrm{off}, \hat{x}_t^{(n)} = \mathrm{on}\right) \tag{10.11}$$

$$\mathrm{TN}^{(n)} = \sum_t \mathrm{AND}\left(x_t^{(n)} = \mathrm{off}, \hat{x}_t^{(n)} = \mathrm{off}\right) \tag{10.12}$$

$$\mathrm{FN}^{(n)} = \sum_t \mathrm{AND}\left(x_t^{(n)} = \mathrm{on}, \hat{x}_t^{(n)} = \mathrm{off}\right) \tag{10.13}$$

10.4 Results and Discussion

This study implements five deep learning models to disaggregate the electricity usage of five appliances: fridge, microwave, dishwasher, washing machine, and kettle. The REFIT dataset (specifically, data from house 2) is utilized to train and test each model to accurately predict the electricity consumption of the specified appliances. Table 10.2 presents the accuracy of the proposed method for each appliance in detail, showing that the hybrid model achieves notable accuracy in recognizing the energy consumption patterns of the appliances.

To disseminate the results of the proposed model more effectively, the selected outputs of each deep learning model for the specified appliances are displayed in Figure 10.4, with blue representing actual power consumption and red indicating predicted power consumption. The fridge, which exhibits a regular consumption pattern, shows a high level of accuracy in its power usage identification. In contrast, the other appliances operate randomly and without any set patterns. The washing machine, which has two operational processes – heating and washing – demonstrates varying accuracy for each process, with the complex washing pattern being identified with some error. Both the microwave and kettle are typically used for short durations of one to two minutes, leading to random usage patterns; however, the proposed method still provides significantly accurate predictions for these appliances.

To evaluate the performance of the proposed method in comparison with previous studies, various criteria are considered, including the implemented method, dataset, number of appliances, and accuracy. Table 10.3 displays the output results for this comparison. The accuracy results from earlier studies focused on NILM were sourced from references [13, 38–41]. The findings in Table 10.3 indicate that the proposed algorithm significantly outperforms the results reported

Table 10.2 Output accuracy measures for each appliance.

Metrics	Fridge	Washing machine	Dishwasher	Microwave	Kettle	Overall
ANE	0.037	0.0207	0.0303	0.7176	0.2110	**0.2034**
RSME	19.066	297.896	201.992	97.849	666.783	**256.717**
Accuracy (%)	94.21	95.140	97.790	99.15	93.381	**95.934**
F1 (%)	90.11	96.15	94.59	47.61	76.190	**80.936**

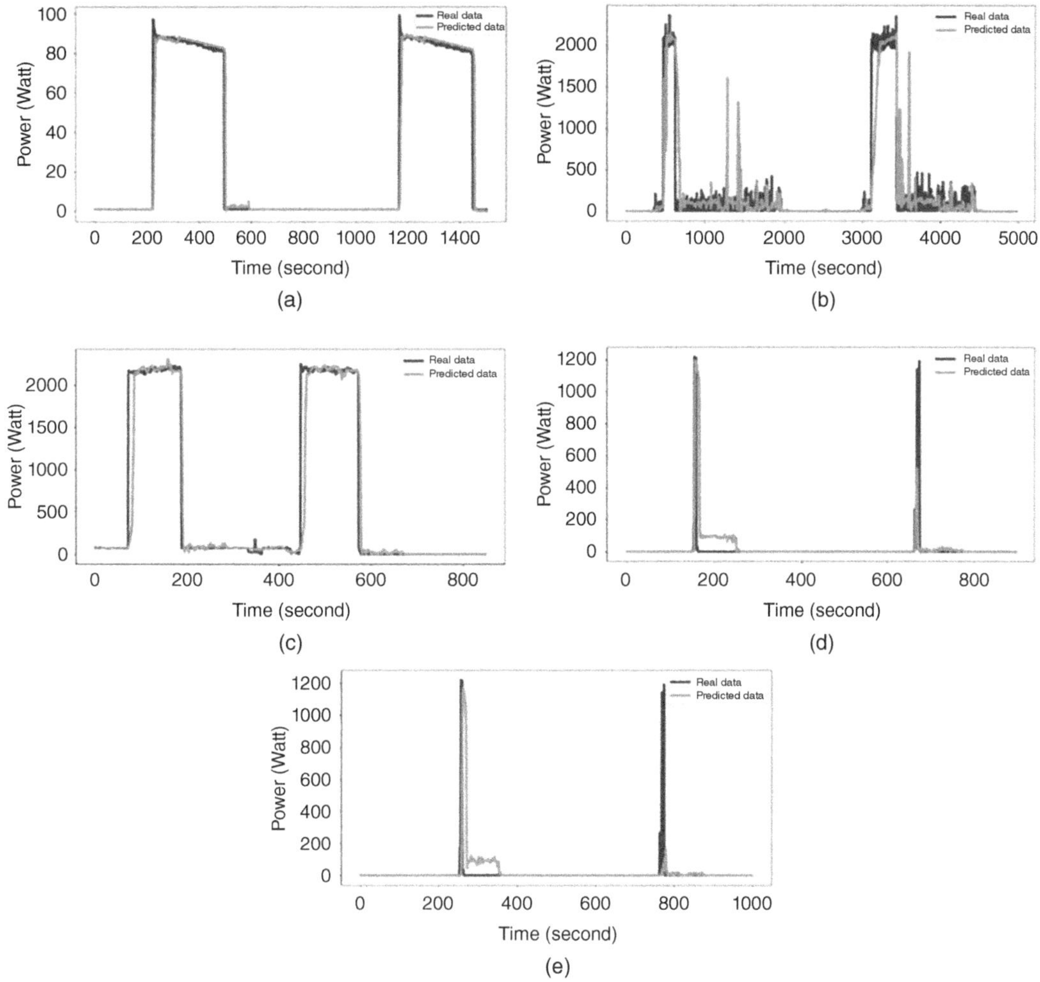

Figure 10.4 Output results for (a) fridge, (b) washing machine, (c) dishwasher, (d) microwave, and (e) kettle (black lines are real data and gray lines are predicted data).

Table 10.3 Comparison of the results of the proposed method and past studies.

Study	Method	Dataset	# Appliances	Accuracy (%)
[38]	SDPR	REFIT	5	78
[39]	SSGSP	REFIT	6	69.33
[40]	HMM	REDD	5	88.6
[41]	GSP	REFIT	6	71
[13]	CNN	UK-DALE	7	93.8
Proposed method	LSTM-CNN	REFIT	5	95.93

in references [38–41] and is slightly more accurate than those in reference [13]. The slightly lower accuracy in reference [13] can be attributed to its training on a larger number of appliances compared to the proposed method. Both the studies referenced [38–41] and the proposed method utilized the same dataset, but they employed different and enhanced variations of graph signal processing and semi-definite programming relaxation for NILM, resulting in accuracies ranging from 69.33% to 78%. In contrast, the accuracy of the proposed method is considerably higher. References [13, 40] were trained on five and seven appliances, respectively, using hidden Markov models and CNNs, but their accuracies were lower in comparison to the proposed method. While reference [13] demonstrated promising accuracy for more appliances (i.e., seven), it leveraged CNN networks to achieve those results.

10.5 Challenges and Future Trends

Similarly, with other complex challenges in energy research, implementing NILM models capable of accurately disaggregating energy usage patterns, including simultaneous appliance operations, remains a difficult task due to high uncertainty and complexity. Additionally, energy consumption patterns differ widely across individual appliances and users, making it hard to design a general model that fits all scenarios. Deep learning models have enhanced the generalization of NILM models. However, most deep learning-based NILM approaches rely on single-task learning, where a particular model is trained for each appliance, leading to high computational costs. This approach overlooks the concurrent operation of multiple devices and their interdependencies. Future work proposes multi-task learning for simultaneous appliance detection and power estimation to address these limitations.

Additionally, exploring hardware implementations could facilitate real-time applications of this approach. Investigating NILM for the same appliances, particularly in scenarios involving simultaneous on/off events for multiple appliances, would further enhance the model's applicability and provide valuable insights into energy management strategies.

10.6 Conclusion

This study proposed a novel hybrid deep learning approach that combines LSTM and CNN architectures for NILM applications in smart grids. The proposed network is trained on data from five appliances within the REFIT dataset, demonstrating its effectiveness in accurately disaggregating the energy consumption patterns of each appliance. One of the key strengths of this approach lies in its accuracy and the ability to capture the unique characteristics of different appliances. The integration of both CNN and LSTM networks significantly enhances overall accuracy. While CNNs can extract spatial features from the input data, LSTMs are proficient in processing temporal sequences, allowing the hybrid model to effectively analyze complex energy consumption patterns over time.

The proposed method is evaluated against established accuracy metrics and compared with results from previous studies, showing significant performance. This comprehensive evaluation highlights the potential of the hybrid model to advance NILM applications in smart grids. Future research opportunities could include evaluating the proposed method across various datasets to test its robustness and adaptability.

References

1 Naderian, S. (2022). Machine learning approach for non-intrusive load monitoring in smart grids: new deep learning method based on long short-term memory and convolutional neural networks. *2022 Eighth Iranian Conference on Signal Processing and Intelligent Systems (ICSPIS)*, Behshahr, Iran, Islamic Republic of (28–29 December 2022), pp. 1–7. IEEE. https://doi.org/10.1109/ICSPIS56952.2022.10044076.

2 Cominola, A., Giuliani, M., Piga, D. et al. (2017). A hybrid signature-based iterative disaggregation algorithm for non-intrusive load monitoring. *Applied Energy* 185: 331–344.

3 Naderian S. A novel hybrid deep learning approach for non-intrusive load monitoring of residential appliance based on long short term memory and convolutional neural networks. arXiv preprint arXiv:2104.07809. 2021.

4 Bassamzadeh, N. and Ghanem, R. (2017). Multiscale stochastic prediction of electricity demand in smart grids using Bayesian networks. *Applied Energy* 193: 369–380.

5 George William Hart (1992). Nonintrusive appliance load monitoring. *Proceedings of the IEEE* 80 (12): 1870–1891.

6 Carmel, C.K., Shrimali, G., and Albert, A. (2013). Disaggregation: the holy grail of energy efficiency. *Energy Policy* 52: 213–234.

7 Townson, B. (2016). Nilm: Vehicle or destination. *International Workshop on Non-Intrusive Load Monitoring*, Vancouver, British Columbia, Canada (14–15 May 2016).

8 Nalmpantis, C. and Vrakas, D. (2019). Machine learning approaches for non-intrusive load monitoring: from qualitative to quantitative comparation. *Artificial Intelligence Review* 52 (1): 217–243.

9 Henao, N., Agbossou, K., Kelouwani, S. et al. (2015). Approach in nonintrusive type I load monitoring using subtractive clustering. *IEEE Transactions on Smart Grid* 8 (2): 812–821.

10 Basu, K., Debusschere, V., Bacha, S. et al. (2014). Nonintrusive load monitoring: a temporal multilabel classification approach. *IEEE Transactions on Industrial Informatics* 11 (1): 262–270.

11 Gillis, J.M., Alshareef, S.M., and Morsi, W.G. (2015). Nonintrusive load monitoring using wavelet design and machine learning. *IEEE Transactions on Smart Grid* 7 (1): 320–328.

12 Lin, Y.-H. and Tsai, M.-S. (2013). Development of an improved time–frequency analysis-based nonintrusive load monitor for load demand identification. *IEEE Transactions on Instrumentation and Measurement* 63 (6): 1470–1483.

13 Yang, D., Gao, X., Kong, L. et al. (2020). An event-driven convolutional neural architecture for non-intrusive load monitoring of residential appliance. *IEEE Transactions on Consumer Electronics* 66 (2): 173–182.

14 Dinesh, C., Nettasinghe, B.W., Godaliyadda, R.I. et al. (2015). Residential appliance identification based on spectral information of low frequency smart meter measurements. *IEEE Transactions on Smart Grid* 7 (6): 2781–2792.

15 Himeur, Y., Alsalemi, A., Bensaali, F., and Amira, A. (2020). Effective non-intrusive load monitoring of buildings based on a novel multi-descriptor fusion with dimensionality reduction. *Applied Energy* 279: 115872.

16 Chang, H.-H., Lin, L.-S., Chen, N., and Lee, W.-J. (2013). Particle-swarm-optimization-based nonintrusive demand monitoring and load identification in smart meters. *IEEE Transactions on Industry Applications* 49 (5): 2229–2236.

17 Sung, M. and Ko, Y. (2015). Machine-learning-integrated load scheduling for reduced peak power demand. *IEEE Transactions on Consumer Electronics* 61 (2): 167–174.

18 Devlin, M.A. and Hayes, B.P. (2019). Non-intrusive load monitoring and classification of activities of daily living using residential smart meter data. *IEEE Transactions on Consumer Electronics* 65 (3): 339–348.

19 Rafiq, H., Zhang, H., Li, H., and Ochani, M.K. (2018). Regularized LSTM based deep learning model: first step towards real-time non-intrusive load monitoring. *2018 IEEE International Conference on Smart Energy Grid Engineering (SEGE)*, Oshawa, Ontario, Canada (12–15 August 2018), pp. 234–239. IEEE.

20 Moradzadeh, A., Mohammadi-Ivatloo, B., Abapour, M. et al. (2021). A practical solution based on convolutional neural network for non-intrusive load monitoring. *Journal of Ambient Intelligence and Humanized Computing* 12 (10): 9775–9789.

21 Chen, Y., Zhang, G., Bai, M. et al. (2019). Automatic waveform classification and arrival picking based on convolutional neural network. *Earth and Space Science* 6 (7): 1244–1261.

22 Gomes, E. and Pereira, L. (2020). Pb-nilm: pinball guided deep non-intrusive load monitoring. *IEEE Access* 8: 48386–48398.

23 Gopinath, R., Mukesh Kumar, C., Joshua, P.C., and Srinivas, K. (2020). Energy management using non-intrusive load monitoring techniques–state-of-the-art and future research directions. *Sustainable Cities and Society* 62: 102411.

24 Çimen, H., Çetinkaya, N., Vasquez, J.C., and Guerrero, J.M. (2020). A microgrid energy management system based on non-intrusive load monitoring via multitask learning. *IEEE Transactions on Smart Grid* 12 (2): 977–987.

25 Faustine, A., Pereira, L., and Klemenjak, C. (2020). Adaptive weighted recurrence graphs for appliance recognition in non-intrusive load monitoring. *IEEE Transactions on Smart Grid* 12 (1): 398–406.

26 Gopinath, R., Kumar, M., and Srinivas, K. (2020). Feature mapping based deep neural networks for non-intrusive load monitoring of similar appliances in buildings. *Proceedings of the Seventh ACM International Conference on Systems for Energy Efficient Buildings, Cities, and Transportation*, Japan (18–20 November 2020), pp. 262–265.

27 Hua, D., Huang, F., Wang, L., and Chen, W. (2021). Simultaneous disaggregation of multiple appliances based on non-intrusive load monitoring. *Electric Power Systems Research* 193: 106887.

28 Liu, Y., Liu, W., Shen, Y. et al. (2021). Toward smart energy user: real time non-intrusive load monitoring with simultaneous switching operations. *Applied Energy* 287: 116616.

29 Himeur, Y., Alsalemi, A., Bensaali, F., and Amira, A. (2021). An intelligent nonintrusive load monitoring scheme based on 2d phase encoding of power signals. *International Journal of Intelligent Systems* 36 (1): 72–93.

30 Funahashi, K.-i. and Nakamura, Y. (1993). Approximation of dynamical systems by continuous time recurrent neural networks. *Neural Networks* 6 (6): 801–806.

31 Hermans, M. and Schrauwen, B. (2013). Training and analysing deep recurrent neural networks. *Advances in Neural Information Processing Systems* 26.

32 Petersen, N.C., Rodrigues, F., and Pereira, F.C. (2019). Multi-output bus travel time prediction with convolutional LSTM neural network. *Expert Systems with Applications* 120: 426–435.

33 Nair, V. and Hinton, G.E. (2010). Rectified linear units improve restricted Boltzmann machines. *Proceedings of the 27th International Conference on International Conference on Machine Learning (ICML2010)*, Haifa, Israel (21–24 June 2010).

34 Wang, T., Wu, D.J., Coates, A., and Ng, A.Y. (2012). End-to-end text recognition with convolutional neural networks. *Proceedings of the 21st International Conference on Pattern Recognition (ICPR2012)*, Tsukuba, Japan (11–15 November 2012), pp. 3304–3308. IEEE.

35 Boureau, Y.L., Ponce, J., and LeCun, Y. (2010). A theoretical analysis of feature pooling in visual recognition. *Proceedings of the 27th International Conference on Machine Learning (ICML-10)*, Haifa, Israel (21–24 June 2010), pp. 111–118.

36 Stankovic, L., Wilson, C., Liao, J. et al. (2015). Understanding domestic appliance use through their linkages to common activities. *Eighth International Conference on Energy Efficiency in Domestic Appliances and Lighting*, Switzerland (26–28 August 2015).

37 Ullah, I., Hussain, M., Aboalsamh, H. et al. (2018). An automated system for epilepsy detection using EEG brain signals based on deep learning approach. *Expert Systems with Applications* 107: 61–71.

38 He, K., Stankovic, L., Liao, J., and Stankovic, V. (2016). Non-intrusive load disaggregation using graph signal processing. *IEEE Transactions on Smart Grid* 9 (3): 1739–1747.

39 He, K., Jakovetic, D., Zhao, B. et al. (2019). A generic optimisation-based approach for improving non-intrusive load monitoring. *IEEE Transactions on Smart Grid* 10 (6): 6472–6480.

40 Makonin, S., Popowich, F., Bajić, I.V. et al. (2015). Exploiting hmm sparsity to perform online real-time nonintrusive load monitoring. *IEEE Transactions on Smart Grid* 7 (6): 2575–2585.

41 Zhao, B., He, K., Stankovic, L., and Stankovic, V. (2018). Improving event-based non-intrusive load monitoring using graph signal processing. *IEEE Access* 6: 53944–53959.

11

Power System Cyber-Physical Security and Resiliency Based on Data-Driven Methods

Hamed Delkhosh, Mahdi Ghaedi, and Maryam Azimi

Faculty of Electrical and Computer Engineering, Tarbiat Modares University, Tehran, Iran

11.1 Introduction

Various efforts to modernize electric power grids are underway around the globe, integrating cutting-edge technologies to address the challenges faced by existing infrastructure. However, with each technological innovation, a new layer of complexity emerges, accompanied by its specific vulnerabilities. In this regard, the digitalization of the power grids has highlighted the importance of cyber adversaries aiming to disrupt power supply on a large scale, causing substantial technical, economic, and social costs. Recognizing the inevitability of such vulnerabilities, there is a shift in focus within the field of cyber-physical power systems (CPPSs). Beyond security measures against mild and frequent cyber and physical incidents, the emphasis is now on fostering resilience to ensure uninterrupted supply in the face of high-impact and low-probability (HILP) cyber-physical threats. The coordination of cyberattacks and physical deliberate or nondeliberate attacks or incidents highlights the need to study the resilience of CPPSs. Hence, there is a pressing need for an extensive exposition of diverse modeling and analytical techniques. Specifically, as the integration of CPPSs deepens and considerable measurement data becomes more accessible, it becomes crucial to introduce and apply various data analytic methods within these systems.

The objective of the present chapter is to present a thorough categorization of techniques for analyzing and evaluating CPPSs, focusing particularly on data-driven methods aimed at evaluating and enhancing both security and resilience. After a concise introduction, in the second section, the fundamental concepts of CPPSs are discussed while outlining the major challenges they face, focusing on security and resiliency. In the third section, data analytic techniques are discussed within two categories: basic (supervised and unsupervised learning) and advanced (dimensionality reduction, feature engineering, reinforcement learning, and integrated models). The fourth section delves into different modeling techniques for CPPSs. Their applications, advantages, and disadvantages are explored based on four main categories: direct modeling, testbeds, game-theoretic, and machine learning. In the fifth section, significant attention is dedicated to a key weakness of CPPSs—their vulnerability facing different threats. In this section, various cyber-physical threats, including physical, cyber, and coordinated attacks, are briefly described to provide a comprehensive understanding of their implications. In the sixth section, various types of defensive measures against these threats are discussed based on two general categories: preventive measures and mitigation actions. Preventive measures encompass methods that utilities must undertake before the attack occurs (under normal conditions of CPPSs). This includes supply

Data-Driven Energy Management and Tariff Optimization in Power Systems: Shaping the Future of Electricity Distribution through Analytics, First Edition.
Edited by Hamidreza Arasteh, Pierluigi Siano, Niki Moslemi, and Josep M. Guerrero.

chain security, access control, personnel training, resource allocation, infrastructure hardening, and moving target defense (MTD). During and after an attack, mitigation actions are necessary to alleviate the impacts and restore the CPPS to its normal state. In this regard, it is essential to engage in intrusion detection and data recovery during the attacks, followed by restoration and forensic analysis afterward. This chapter thoroughly reviews all attack and defense methods, with a focus on data-driven methods accompanied by recent examples in the literature. The content presented in this chapter is concluded, and the related references are provided at the end.

11.2 Fundamental Concepts

To provide a general overview of the topic, it is necessary to first provide preliminaries about the concept of CPPSs and their challenges regarding security and resiliency, especially focusing on their differences.

11.2.1 Cyber-Physical Power System (CPPS)

Nowadays, most of the key infrastructures, such as energy, transportation, and water supply, are highly coupled together. This is commonly known as interdependency among these sectors. The recent great advances in digital technologies have enabled the integration of cyber components with these physical grids, developing them into cyber-physical systems. This evolving field combines the physical and cyber worlds while facilitating smarter monitoring, automated controlling, and real-time optimization for physical processes. The different applications of such combinations may be found in all aspects of modern human life, for example, transportation, health care, and also energy grids (power, gas, etc.).

The modern power grids are also in the form of a CPPS. They are the new and promising forms of modern power grids, allowing the integration of physical grids (including various components such as generators, transmission lines, and substations) with cyber networks based on digital technologies (including sensors, communication networks, and various software). A CPPS allows continuous data collection and control, which offers a large potential for effective monitoring and automated decision-making [1]. Integrating different cyber and physical components may lead to interoperability issues, where various components with different ages and from different manufacturers cannot work together effectively. This issue is not within the scope of the present chapter. Considering the inevitability of forming a CPPS, three key challenges arise:

1) CPPSs are complex networks that are strongly vulnerable to various natural disasters, deliberate attacks, and human errors. This makes it essential to continuously check the vulnerability, predict the risks, analyze the security and reliability, and improve the system's resiliency against all kinds of deliberate and nondeliberate threats. In risk and vulnerability analysis [2], the destructive consequences of different hazards (most vulnerable points) are quantified in advance, respectively with and without considering the probability. In security and reliability analysis [3], the power network is usually designed to be normal considering different common contingencies. In resiliency analysis [4], low probability-high impact events are studied in both cyber networks and physical grids, which can even cause cascading failure. All of these fields are well-developed studies for the traditional power grids, which must be modified and extended in order to be appropriate for the emerging modern CPPSs. This chapter mainly deals with examining and improving the grid's resiliency and security against various cyber-physical threats to the CPPS, especially with a focus on data-driven methods.

2) CPPSs highly rely on the two-way flows of electricity and information, and thereby their provided interdependent networks. Therefore, any fault or malfunction in either network can strongly affect the overall performance of the other network. This issue necessitates proper interdependent models, the different concepts of which are also discussed in this chapter.
3) By acknowledging the key role of human factors, including the stress level, experience, and knowledge of operators in effective management of the grids, and particularly highlighting the growing participation level of consumers along with the extensive integration of smart meters, demand response programs, and social media, CPPSs have evolved into CPSPS [5]. A CPSPS encompasses the physical grid, cyber networks (including the social layer), and control centers of the power system, as illustrated in Figure 11.1. In the physical layer, both measurement and status data are received through remote terminal units based on smart meters from throughout the grid. This information will be communicated between the control center and the physical layer through the cyber layer. In the control layer, all acquired data are analyzed, and the proper control actions, which are taken either through automatic decision-making or human operators, are finally exchanged to the physical layer. CPSPSs must also deal with the way that humans (operators, defenders, attackers, insider threats, and prosumers) interact with

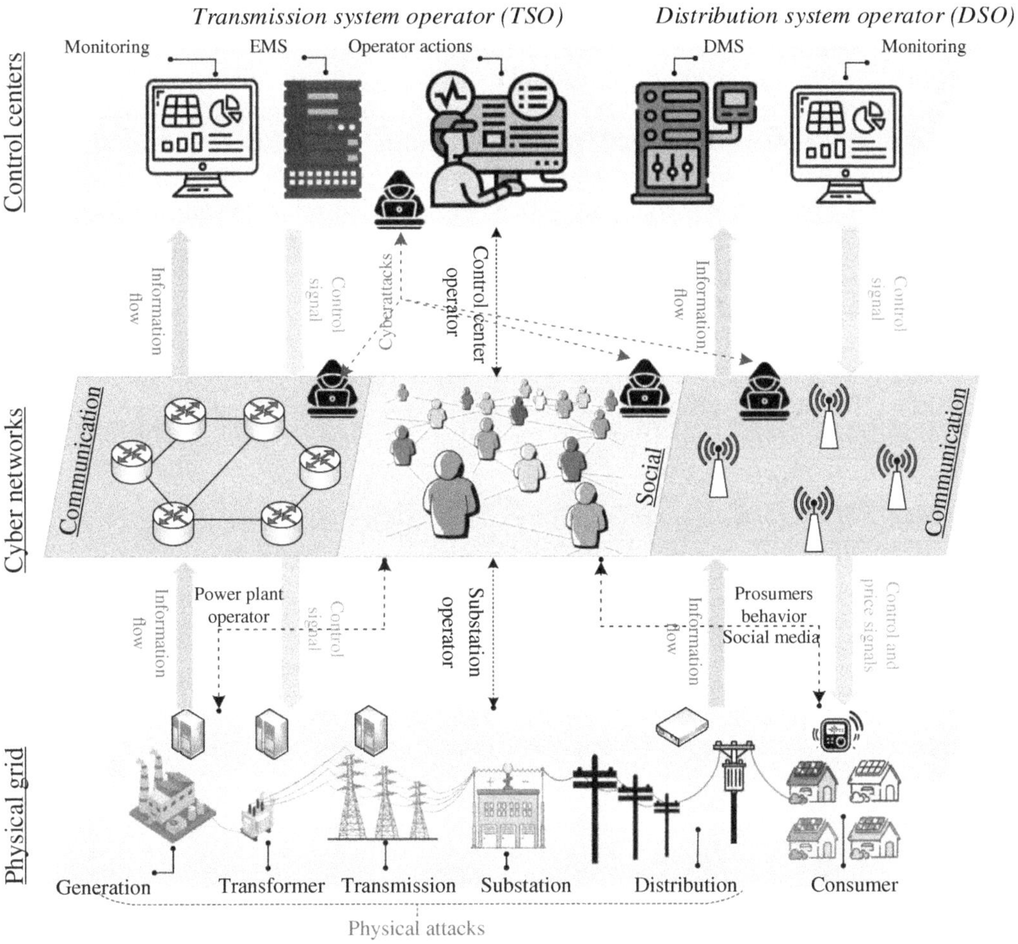

Figure 11.1 Structure of a typical cyber-physical-social power system (CPSPS).

the CPPS, emphasizing their behavior and preferences. For example, insider threats can be considered to intentionally impact the likelihood of malicious attacks being successful, depending on their motivation, collective behavior, and different access levels [6]. It is worth noting that CPSPSs and their related contexts are outside the scope of the present chapter and are concisely mentioned in various sections that are strongly related to the security and resiliency of the CPPS.

11.2.2 Security and Resiliency

Resiliency is defined as the grid's capability to successfully prepare for, withstand, and recover from unexpected HILP disruptions. These disruptions usually refer to $N - k\ (k > 2)$ contingencies, which can cause a cascading failure. Note that the grid security or reliability studies are usually executed based on $N - 1$ or $N - 2$ analysis, neglecting the cascading failure and recovery states. Evaluating the effects of a broader range of high-impact cyber and physical threats (including deliberate coordinated attacks) on the CPPS needs to be studied in the resiliency context. In this regard, the present chapter is mainly focused on the resiliency of the CPPS against various attacks, different types of which are explained and classified in Section 11.5. Nevertheless, it must be noticed that there is no standard terminology in the CPPS literature, i.e., the cybersecurity and cyber-resiliency terms are often used interchangeably.

The concept of resiliency is depicted in Figure 11.2 in a schematic manner, consisting of three phases. It begins with the pre-event phase (Phase 1) with a normal operational state ($t0$–$t1$). Various types of preventive measures that can be implemented in this phase are briefly discussed in Section 11.6.1. Following an event in $t1$, the grid enters a post-event alert state. In the case of resiliency studies and HILP events, the alert is not corrected promptly, and the cascading failure might happen, leading to an abnormal outage or emergency state (Phase 2). Note that the corrective actions are usually adopted in Phase 2 to alleviate the total damage after the HILP event. However, at the end of Phase 2, the system operates at a new normal condition ($t2$–$t3$), where the power grid's performance is drastically dropped. Herein, the vulnerability assessment is mainly specified based on quantifying the consequence of events considering the load shedding and other topological metrics of the degraded network. Besides the preventive and corrective actions,

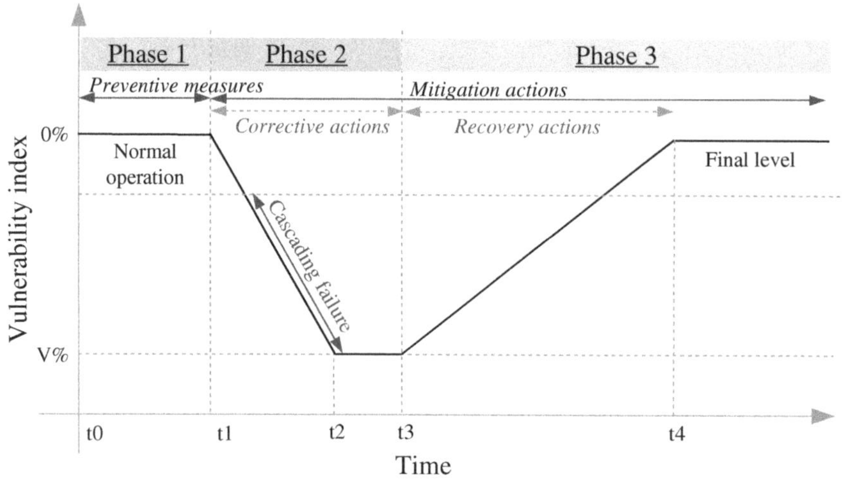

Figure 11.2 Three-phase performance of a CPPS under HILP disruptions.

resiliency also deals with the importance of recovery actions (Phase 3), answering how fast the system will recover from the failure. During this phase ($t3$–$t4$), traditionally the critical loads are sequentially and optimally restored following the repair of damaged infrastructure. In the evolving landscape of resiliency in CPPS, data and cyber node recovery are also of great importance, sometimes having higher priority compared to the physical facilities recovery. Mitigation actions, including both corrective (during the event) and recovery (after the event), are discussed in Section 11.6.2.

Cyber-physical threats continue to evolve, making it essential to continuously prioritize the vulnerabilities while taking different mitigation strategies [7]. Note that some mitigation actions are also required after the recovery actions, such as forensic analysis. To quantify resiliency, a set of different metrics specifically measure how fast and how low power system performance drops in Phase 2, how extensive Phase 2 is, and how promptly the power grid restores (the average time to recovery) in Phase 3 to the normal state before the event. Commonly, the area of the semi-trapezoid in Figure 11.2, calculated as the integral of the trapezoid for the duration of Phases 2–3, is used as the resiliency metric. Sometimes simpler metrics, such as the degradation after Phase 2, are utilized as the index. Numerous studies employ data-driven methods (machine learning algorithms) for resiliency assessment and improvement [8].

11.3 Role of Data Analytics

With the increasing volume and complexity of data in future power systems, the role of data analytics will be expanded for various applications, such as cybersecurity and resiliency purposes [9]. The raw data from smart meters, IoT devices, various field sensors, and utilities is initially gathered, cleaned, pre-processed, and transferred for further investigations. Herein, traditional monitoring and control centers usually have executed their duties based on various extracted charts and tables from this raw data. Nevertheless, a deeper level of understanding can be derived using analytical techniques, statistical analysis, and machine learning (ML)-based algorithms. Interpreting the data can provide helpful insights and valuable guidance for better decision-making throughout the power grid (from the generation sector to the customer level). This includes playing a crucial role in preventing, detecting, and mitigating cyber-physical threats. As shown in Figure 11.3, basic methods and advanced techniques of ML are briefly reviewed in the following to provide the base knowledge for the explanations of Sections 11.5 and 11.6.

11.3.1 Basic Methods

The basic methods of ML, i.e., supervised and unsupervised learning, are briefly described here.

11.3.1.1 Supervised Learning (SL)

Supervised learning (SL) algorithms are specifically designed to learn from the labeled databases. This means that the training data consists of a feature vector (input) paired with the available output. In SL, the goal is to find the patterns within datasets in order to establish the relationships between inputs and outputs. SL is classified into regression and classification for the continuous (e.g., predicting price) and discrete (e.g., distinguishing attack and normal conditions) output cases, respectively. Different labeled datasets, such as access controls, activity logs, network traffic, and especially various field measurements, enable the power grid to detect abnormal behaviors, diagnose cyber intrusions of insider threats, and design mitigation mechanisms for cyber threats.

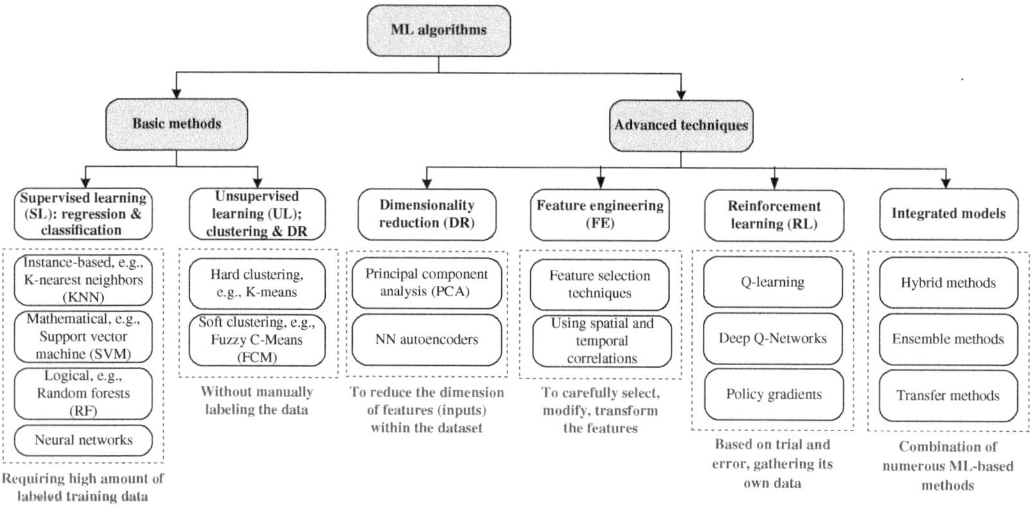

Figure 11.3 Overview of various machine learning techniques.

Various SL methods have been utilized in the literature for different applications in the cyber security and resiliency field. To name a few, polynomial regression, K-nearest neighbors, support vector machine, random forests (RF), and different types of neural networks (NNs) can be mentioned. Nevertheless, SL requires a large amount of labeled training data, acquiring which can be time-consuming and expensive, especially considering the data privacy concerns. Note that in basic ML-based methods, if the data pattern changes once the model has been trained, the model must be updated through retraining. Advanced techniques discussed in Section 11.3.2 can resolve some of these issues.

11.3.1.2 Unsupervised Learning (UL)

Unsupervised learning (UL) algorithms allow the model to be learned without manually labeling the data. The core goal behind UL is to find hidden patterns in the input data without human intervention and guidance. Herein, the training data only consists of a feature vector with no explicit label or output. There are two main types of UL approaches, i.e., clustering and dimensionality reduction (see Section 11.3.2.1 for further information). Clustering methods, such as K-means hard clustering and Fuzzy C-means soft clustering, are used in the cyber-physical field (especially for intrusion detection). Hard clustering techniques specifically determine the belongingness of the data to a cluster (or group), while soft clustering only provides the probability of data assignment to multiple clusters. The main difference between the clustering algorithms and the classification methods is that they do not need any normal and attack condition labels on the database, specifically for the intrusion detection application. Despite its wide application, UL is mainly sensitive to noisy data or outliers, which can cause various issues such as overfitting. To take advantage of both SL and UL approaches, a small set of labeled data can be leveraged within the unlabeled dataset, forming methods often called semi-supervised learning [10].

11.3.2 Advanced Techniques

In this section, some of the advanced techniques of ML are briefly introduced.

11.3.2.1 Dimensionality Reduction (DR)

Dimensionality reduction (DR) is a subset of UL methods that can effectively reduce the dimension of features (inputs) within the dataset while preserving all important underlying patterns. The model is more simplified to learn with fewer dimensions, leading to a faster training time, better memory utilization, and a more accurate fit. DR can also be used for filtering out noisy and irrelevant features so as to be less prone to overfitting. Different techniques like principal component analysis and NN autoencoders are commonly used to achieve DR goals. Specifically speaking in the field of cybersecurity and resiliency of the power system, a large volume of noisy and private data is gathered from a variety of sources. Notice that the higher the quality of the dataset is, the more reliable and insightful the results are. By leveraging DR and improving the quality of the dataset, the most important features can be identified within the dataset, helping timely detection of attacks, responding to them, and making informed decisions to mitigate their effects [11]. Moreover, DR can resolve the data privacy concern by effectively recognizing the patterns with compact and reduced features instead of all features.

11.3.2.2 Feature Engineering

ML models mainly face the issue of datasets with a high dimension, i.e., various and numerous features or inputs. This is especially true in large-scale power grids, which can result in ineffective performance of data-driven methods. To handle this issue, Feature Engineering (FE) not only allows the decision-maker to carefully select, modify, and transform the existing features but also helps to create new ones to improve both the accuracy and learning power of ML models. This can also help in dimensionality reduction and handling the irrelevant and missing features. Generally speaking, defining special features related to the specific characteristics of a problem can enhance the performance of various data-driven methods. Especially in the cybersecurity and resiliency field, using statistical measures instead of all measurements can be helpful for assessing the overall vulnerability and detecting cyber-physical threats based on drastically reduced dimension of features. On the other hand, both spatial and temporal characteristics of raw features can facilitate diagnosing the sequence of events, attacker and defender interactions, and relationships between different detective measures. All in all, proper FE can strongly help in designing preventive measures and mitigation actions, confronting cyber-physical threats [12].

11.3.2.3 Reinforcement Learning

In reinforcement learning (RL), a model is trained through trial and error without labeled (in SL) or unlabeled (in UL) input data. In this approach, an agent lively interacts with its defined environment, taking different actions and receiving rewards or penalties. During the training, the agent adapts its strategies, relying on its own gathered data, compared to the basic ML methods that utilize the historical database. The main advantage of the model-free RL is its applicability in all complex, dynamic, and uncertain environments. On the other hand, its training can be highly time-consuming due to a large number of iterative interactions with the environment. Q-learning, Deep Q-Networks, and policy gradients are the most commonly used RL algorithms in the field of cybersecurity [13]. In addition to the application in intrusion detection, RL algorithms can also be utilized to identify the best defensive and mitigation strategies considering the dynamic behavior of the adversarial attackers. In addition, RL has significant potential in identifying and prioritizing high-risk vulnerabilities without requiring any historical dataset. Moreover, some of the best preventive measures (e.g., resource allocation) can be proactively determined by RL algorithms considering the long-term rewards, based on both immediate and future threats. RL can also be

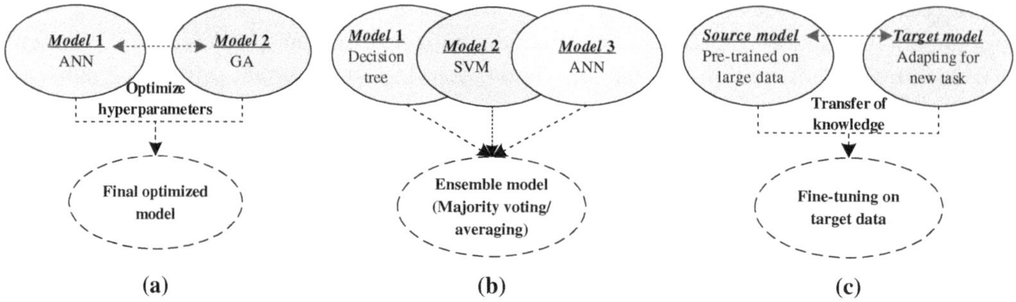

Figure 11.4 Three different architectures for various integrated ML methods. (a) hybrid method, (b) ensemble method, and (c) transfer method.

used for considering complex interactions of various players (e.g., attacker and defender) in the context of CPSPSs, which will be stated in Section 11.4.

11.3.2.4 Integrated Models

Integrated models involve the combination of numerous ML-based methods, capitalizing on the strengths of various algorithms, especially useful when tackling high-dimensional and complex problems. These models can be classified into three groups, as shown in Figure 11.4: hybrid, ensemble, and transfer methods. In hybrid methods, multiple different heterogeneous models, like optimization techniques (e.g., Genetic Algorithm (GA)) and neural networks, are combined while utilizing each algorithm's strengths and compensating for their weaknesses. This could be implemented by utilizing GA for hyperparameter optimization of artificial NN.

Ensemble methods, in contrast, focus on combining the predictions of these individual models using techniques like majority voting and decision templates for consensus. Transfer learning involves applying the output (insight) gained from one model as the input of another method to enhance the performance. Integrated models can also be challenging along with providing powerful performance. This is especially considerable in terms of increased model complexity, substantial computational costs, and intricate hyperparameter tuning. Additionally, these methods are highly dependent on data quality and availability, highlighting the role of effective FE, which can be quite challenging as well. In the recent literature of the cybersecurity and resiliency field, such as [14], it is tried to design various combinations and types of integrated models for different preventative and mitigative purposes.

11.4 Interdependency Modeling

All types of critical infrastructures within different countries (e.g., electricity, gas, water, transportation, and communications) are highly linked, interconnected, and interdependent. In CPPS, interdependency specifically involves the multifaceted relationship between the power grid and its highly linked supporting cyber networks. In CPPS, physical grids rely on the interdependent cyber network to supply their required information for monitoring (decision-making) and control (implementing the actions). On the other hand, usually, it is assumed that the cyber components rely on the interdependent physical grid (and sometimes independent batteries) to supply their required energy. Developing a comprehensive and effective tool that can accurately model interdependent CPPS is not completely addressed yet. As a main difference complicating the interdependency modeling, the power grid is commonly represented as a continuous-time system, while the cyber network has a discrete behavior. Depending on the type of cyber-physical attacks, the coupling between these two interdependent networks is primarily examined from two distinct points of view, namely, failure propagation and impact evaluation.

The majority of events like physical attacks can be modeled as component outages, regardless of their origins, while altering the topology of the attacked network. In this regard, the failure propagation perspective handles the cyber-physical coupling in a manner that deals with how the outages (removing buses/nodes) in one network are propagated into the other one. In addition, the flow of information and different types of routing schemes can also be combined into such CPPSs interdependent models. Two different models are mainly considered for the failure propagation approach, as illustrated in Figure 11.5. In the one-to-one model (see Figure 11.5a), one element in each network is directly coupled with a single element on the other one. In a one-to-many model (see Figure 11.5b), each element in one system is linked to numerous elements in another system. These two failure propagation approaches are considered in various interdependent CPPS methods. It must be noted that the one-to-one model is founded on a relatively practical assumption based on the supporting cyber networks of the power grids in real large-scale systems. For more information in this context, interested readers can refer to [15].

However, all events, such as cyber integrity attacks, do not necessarily lead to topological changes. For instance, both data (information and command) latency or manipulation do not alter the cyber network topology, while they can still have a significant impact on both state estimation (SE) and optimal power flow (OPF) modules of the control centers. The impact evaluation perspective (see Figure 11.5c) manages the cyber-physical coupling dealing with how the events in the cyber network impact the operational performance of the power grid. Nevertheless, the nature or origins of the events in the cyber network are neglected in this point of view. Therefore, the effect of cyber networks in terms of data delay, loss, and manipulation on the performance of different components of the power grid is evaluated. The impact evaluation approach emphasizes the power grid modeling, considering the effects of cyber networks. So, it is only proper when the power grid is the main focus and it is not appropriate for comprehensive CPPS modeling.

Properly modeling the CPPS has become a prominent area of contemporary research in both academia and industry sectors. This issue becomes more complicated considering the social aspect of the problem, changing the CPPSs to CPSPSs as previously described. In the following sections, different methods of CPPSs and CPSPSs modeling are briefly discussed, considering the two previously mentioned perspectives for handling the interdependency. This includes direct modeling methods, developed testbeds, game-theoretic approaches, and ML-based models.

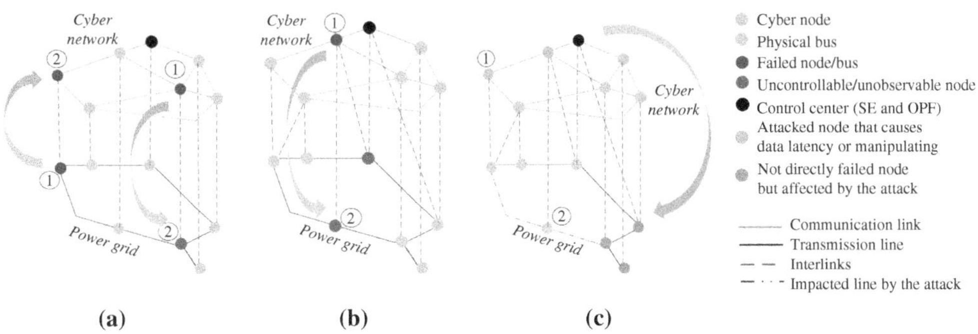

Figure 11.5 Different types of cyber-physical interdependency. (a) One-one failure propagation, (b) one-many failure propagations, and (c) impact evaluation.

11.4.1 Direct Modeling

Direct modeling attempts to model cyber networks and physical grids directly based on either mathematical or topological methods. In mathematical methods, the continuous behavior of the power grid is usually represented by differential equations, while the discrete behavior of the cyber network is commonly captured using difference equations. In the context of the power grid, the model must capture the variation of physical quantities (like voltage and power) based on equations such as Kirchhoff's laws and power system dynamics. In the context of the cyber network, the model demonstrates how various states change over a set of discrete sampling times. Highlighting the dynamics of both cyber networks and power grids, different interfaces are required to effectively interconnect these coupled systems. These traditional methods, also referred to as hybrid methods in the literature, often result in precise mathematical models that are accurate yet hard to solve. Such complicated models can be handled in the testbeds (see Section 11.4.2) based on various specific software and hardware for accurately and separately executing the calculations and managing the interactions of the cyber network and power grid.

On the other hand, topological methods try to handle individual network calculations and interdependency modeling in simplified forms. As a key example of these methods, the graph-based models are widely applied for the failure propagation viewpoint of interdependency. In this context, CPPS is usually modeled by two coupled distinct graphs, each with its different nodes and links, denoting different components and their connections (either one-to-one or one-to-many). After graph-based visualization of CPPS, complex networks can be used to estimate the vulnerability facing various cyber-physical threats. This can help to identify the critical components in both the cyber network and power grid and provide the proper suggestion for hardening CPPS. The performance and vulnerability of CPPS are usually measured by complex network metrics such as degree, betweenness, and closeness, only relying on the network's topological aspects. By utilizing a more realistic assumption regarding the operation mechanism of the CPPS (e.g., consecutive SE and OPF calculations), the flow patterns (data in the cyber network and energy in the power grid) can also be used to weigh these vulnerability indices. Such methods are usually called as functional, which commonly calculates the vulnerability based on DCOPF considering the cascading failure. Some of the other topological methods, like finite state machine and Petri net, are also widely utilized to model the dynamic behavior of interdependent systems under different types of deterministic and stochastic events by focusing on state changes. However, the related papers have properly highlighted the practicality of these methods for large-scale systems. For more information, interested readers can refer to [16].

11.4.2 Testbeds

The CPPS testbeds are designed as controlled environments including cyber and physical components along with the control procedure that can be continuously followed for safely exploring the effects of different events (e.g., cyber-physical threats), accurately evaluating the system resilience, and testing defending solutions without risking real-world systems. Testbeds can be divided into three different groups: physical, software-based, and hardware in loop (HIL). In the physical testbeds, the power grid and the cyber network are modeled using real physical equipment. Despite the high accuracy and development costs of the obtained results, it has limitations for some specific studies and large-scale CPPSs. The software-based testbeds allow relatively accurate analysis of a large-scale CPPS at lower costs, facing numerous repetitive experiments. They usually consist of advanced and specific software for various sectors, for example,

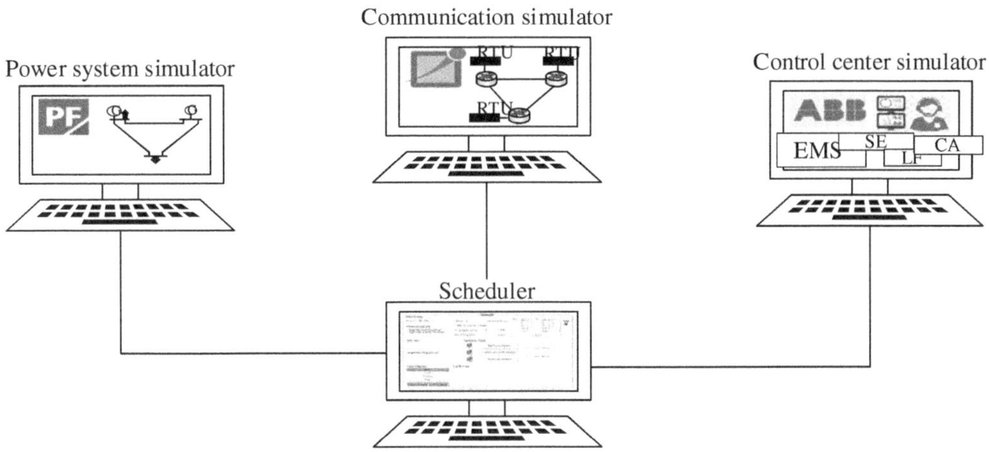

Figure 11.6 A typical software-based cyber-physical testbed with centralized scheduling.

power grids, cyber networks, and control centers. They can be regarded as the implemented form of direct mathematical methods for interdependency modeling. However, depending on the simulated environments, the accuracy of their results can vary over a wide range. The HIL testbed is a promising form of testbed working as an intermediary solution, allowing for the real-time integration of real physical components with simulation software. Note that in different forms of testbeds, time synchronization and data management among different consisting elements is a sophisticated task, which can be managed in a centralized or distributed manner [17]. An example of a software-based testbed for the CPPS comprising three different layers (physical, communication, and control) is shown in Figure 11.6, describing the centralized scheduling.

11.4.3 Game-Theoretic

Game-theoretic methods are primarily used for the impact evaluation of threats considering the intricate and complex interactions of various players, especially the defender and attacker. These agents interact live with the others, as well as the environment. Depending on the perspective taken (attacking/defending) and whether focusing on the preventive (before the attack) or mitigation (after the attack) defense strategies, the modeling is commonly grouped into Defense-Attack (D-A), Attack-Defense (A-D), and Defense-Attack-Defense (D-A-D), as illustrated in Figure 11.7. D-A is implemented for the sake of designing the most effective preventive defense strategies considering the most impactful attack. A-D is executed from the attacker's perspective to identify the most destructive attacks considering the optimal mitigating actions of the defender. D-A-D is a tri-level model for the defender considering both preventive and mitigative defenses.

It is worth noting that all players make their decisions either simultaneously (i.e., static games) or in a sequence at different points in time (i.e., dynamic games) with and without knowledge of the other agents' choices. Game-theoretic methods are mainly handled by either multi-level optimizations or classical equilibrium-based calculations. In the first one, the nonconvex optimization is usually managed by Karush-Kuhn-Tucker (KKT) reformulation, decomposition techniques, and heuristic algorithms like column-and-constraint generation (CCG). In the second one, the equilibrium points are calculated as the solutions where the agent cannot gain a higher payoff given the impacts of all other agents. Moreover, the effectiveness of game-theoretic modeling for vulnerability analysis and optimal defense design highly relies on the underlying assumptions on

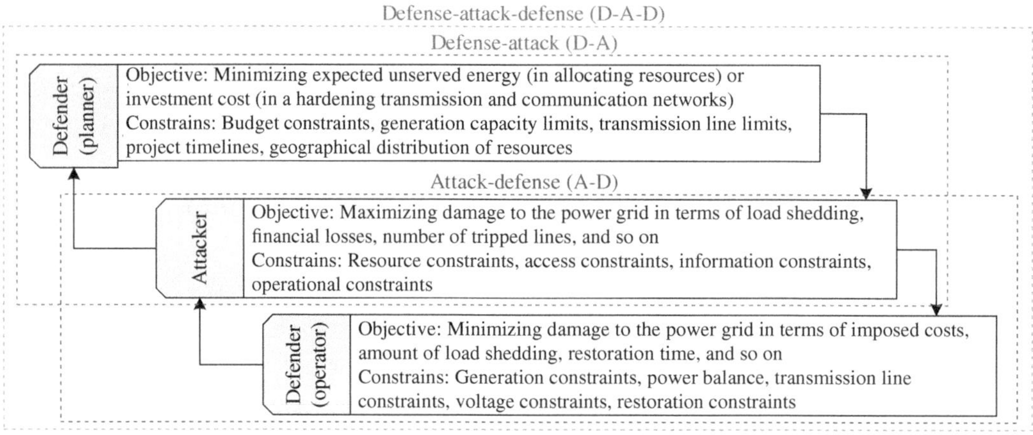

Figure 11.7 Multi-level models architecture (D-A, A-D, and D-A-D) for attacker–defender interactions.

the agent behaviors, for example, common simplifying and non-accurate assumptions on totally rational behavior of the human agents. For more information regarding game-theoretic methods for defender and attacker interactions, references such as [18] are useful.

11.4.4 Machine Learning

ML is one of the most recent effective tools for interdependency modeling. In this context, different data-driven methods can be helpful for modeling the failure propagation or cascading failure in the CPPS. Reviews show that most of the previously introduced methods oversimplify the failure propagation, leading to conservative prediction of vulnerability. Model-free ML-based techniques can capture the complex interdependency in CPPS, allowing accurate vulnerability assessment in cascading failures [19]. Regarding the impact evaluation perspective, RL can be effectively applied considering the attacker's and defender's behavior and interactions [20]. In this regard, a digital twin, which is a digital replica of an interdependent system virtually reflecting the real-time system's behavior similar to testbeds, can also be regarded as a helpful tool for interdependency modeling.

As previously mentioned, the interdependent CPSPS emphasizes the direct and indirect roles and effects of human decision-making, expertise, and behavior on the power grid. Moving toward future distribution networks is strongly related to the growing participation of prosumers, primarily facilitated by a wide integration of both smart meters and social media. This social coupling covers a broad range of problems, including event (anomaly or attack) detection, adopting new technologies (e.g., smart meters and electrical vehicles), and adopting demand-side management strategies. To leverage the obtained helpful social data in real time, it is essential to process and transform it into information while properly utilizing different data-driven techniques. In this regard, various aspects of the data (e.g., pricing, privacy, and integrity) have been widely studied [21] to evaluate the value of data, preserve the privacy of consumers, ensure data trustworthiness, and promote data sharing. Emphasizing the key role of human operators and other stakeholders in effectively managing (defender) or even intentional and unintentional threatening (attacker and insider threats) of the power grid at the transmission level, CPSPSs are studied considering the human factors such as the stress, experience, access level, and knowledge [22].

11.5 Cyber-Physical Threats

As one of the primary threats to the power grid, cyber and physical attacks have escalated with the development of information technologies. As illustrated in Figure 11.8, numerous key cyberattacks with different motives (e.g., economic, social, and political objectives) have already been launched on CPPSs of different nations. In recent years, various sectors (academia, industry, and government) have increasingly focused on the security and resiliency of this key infrastructure so as to deal with these threats. In this regard, it is important to quantify the impacts of various cyber-physical threats in advance, which can be helpful for designing both optimal preventive measures and mitigation actions. In this context, mainly the term "vulnerability analysis" is used for identifying, analyzing, and measuring the weaknesses of CPPS [23]. Understanding concepts of various types of deliberate cyber-physical threats [24], briefly described in the following (see Figures 11.9 and 11.10 for more details), lays the groundwork for effective vulnerability analysis.

Two approaches are used in the literature for vulnerability analysis against cyber-physical threats. The first, commonly considered the worst-case attacks (most destructive), neglects the

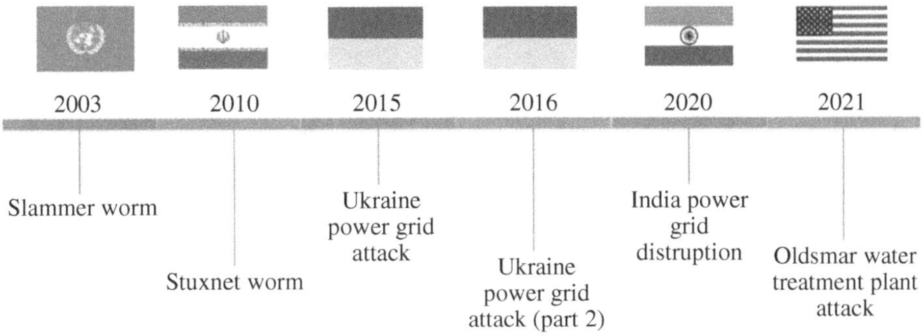

Figure 11.8 Timeline of some of the most relevant and recent cyber events on power grids.

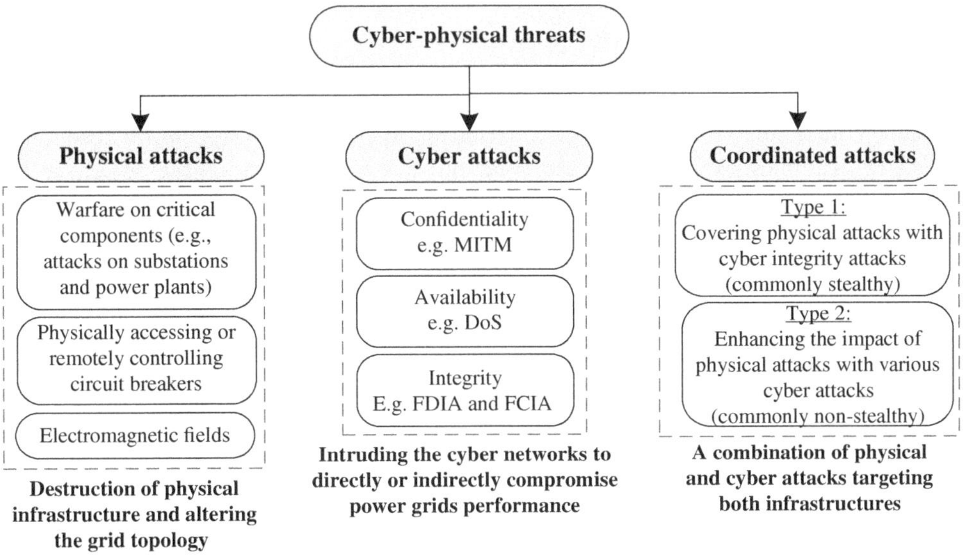

Figure 11.9 Overview and classification of different cyber-physical threats.

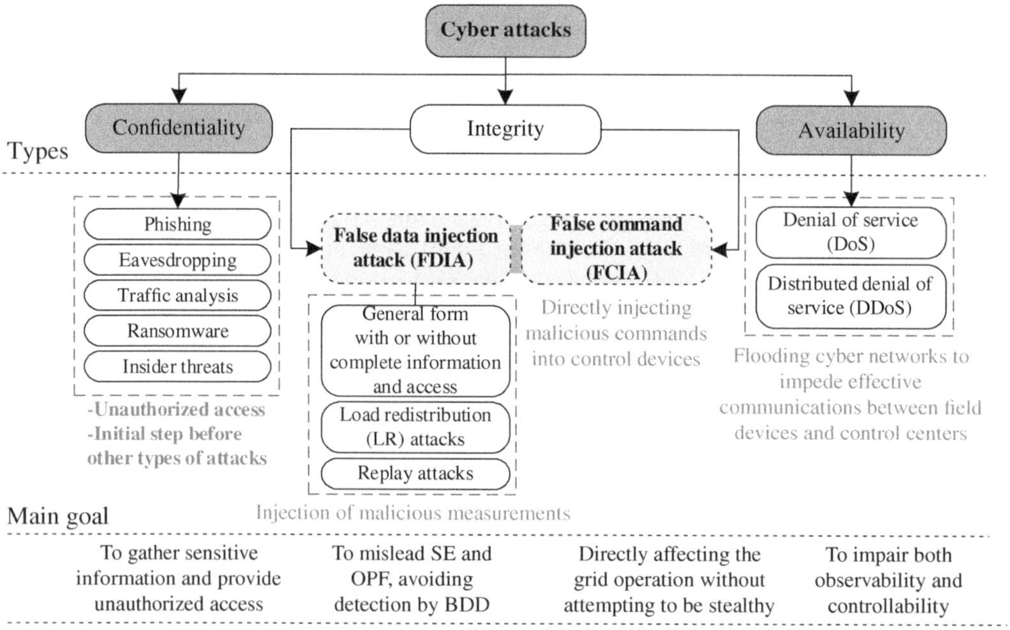

Figure 11.10 Overview and classification of different cyberattacks.

cascading failure. This approach often involves solving a bi-level optimization problem from the attacker's perspective (previously described as A-D interdependency model), taking the defender's responses (corrective actions) into account. In this approach, only the initial stage of the cascading failure is evaluated for the vulnerability assessment, while ignoring the failure propagation. The second approach enumerates one or a set of random attacks while relying on simulation or ML-based methods for cascading failure modeling [25]. Therefore, the conditions of CPPS can be evaluated after each specific attack, considering failure propagation. Even in most of the related papers using this approach, the attack (incident) is already known or randomly selected, compared to the optimally found threat in the previous approach.

Note that the developed studies for both approaches commonly assume specific types of cyber-physical threats and are limited in considering various attack types, especially simultaneously. From the vulnerability indices viewpoint, various perspectives have been introduced in the literature. As two common examples, the economic impacts of the attack on the power market (e.g., operation cost and attacker's profit) [26] or damages to the static security (e.g., overloading of lines and other components or supply-demand imbalance) [27] can be mentioned.

11.5.1 Physical Attacks

Physical events on the grid, including disconnection of transmission lines and transformers, various short circuits, and disruption of power plants, are commonly investigated in the context of power system security and resiliency in traditional studies. Physical disruptions (either non-deliberate events or deliberate attacks) usually alter the system topology and affect its power balance, leading to drastic and widespread changes in indications (demonstrating equipment status) and measurements, making them easy to detect. Coordinated cyber-physical attacks can improve the imposed effects of the physical attack and raise their undetectability chance. This is why most of the papers in the CPPS field have especially focused on the coordination of these events (deliberate physical attacks) with various forms of cyber threats (see Section 11.5.3 for more details).

In the context of deliberate physical attacks, the transmission lines are more at risk due to their wide geographical distribution. The lines are susceptible to physical attacks by either remotely controlling or physically accessing their circuit breakers, leading to short-term disruptions. The attacker commonly attempts to identify the most critical lines before implementing the attack, aiming to maximize the total amount of load shedding, repair/replacement costs, and probability of cascading failures. Substations in strategic locations and large power plants are also important targets susceptible to warfare (e.g., missile attacks). Such drastic attacks can put the stable and optimal operation of the power grid at risk for a long-term period. In fact, they impose significant economic and social impacts until the repair and replacement of the equipment are completed. The key facilities of the power system can also be indirectly targeted by the electromagnetic fields, causing excessive voltages or electromagnetic pulses leading to disabled operation.

Facing physical attacks, vulnerability analysis is mainly performed based on the cascading failure approach. The development of CPPS and its cyber-physical interdependency has made it possible to collect real-time measured data and apply immediate control commands (e.g., generation dispatch and load shedding) to alleviate the destructive effects of the attack. However, this interdependency has complicated the cascading failure modeling and, consequently, the vulnerability analysis. In CPPSs, an event on the physical grid affects the cyber network and vice versa (as explained in Section 11.4). As a complicating aspect, after an initial failure in some links and routers in the cyber network, the data flow is redistributed to other devices, leading to further impacts similar to cascading failure. In this regard, it is needed to take into account the role of routing strategy, channel bandwidth, and data latency [28].

The second perspective of vulnerability analysis (a multi-level optimization for finding the worst attack) could also be utilized for physical attacks involving attacker-defender interactions. Taking various uncertainties (such as renewables integration, load demand, and market prices) into account, the optimal attack can be modeled as a tri-level optimization, in which the attacker with and without the uncertainty is included in the second and the first levels, respectively, and the defender is represented in the third level [29].

11.5.2 Cyberattacks

Cyberattacks can be generally divided into three distinct groups, each of which targets one of the following aspects of the data: confidentiality, availability, and integrity (see Figure 11.10 for more details). There are various methods for each of these types that are explained briefly in the following. Note that coordinated cyberattacks can simultaneously target two or even all three aspects mentioned before. Cyberattacks can also be coordinated with physical attacks for more destructive effects on the CPPS (see Section 11.5.3 for more explanations). It must be noted that the principles of cyberattacks and their various methods are the cornerstone for understanding different aspects of CPPS security and resiliency.

11.5.2.1 Confidentiality

Confidentiality involves keeping away sensitive information from unauthorized persons or entities. This is typically achieved through different methods such as data encryption and access levels. As an important example, a man-in-the-middle attack seeks unauthorized access by secretly intercepting communications to maliciously capture and record valid sensitive data while transferring [30]. Common methods of such attacks include phishing, eavesdropping, traffic analysis, ransomware, and insider threats, all of which involve attempts to gain access to sensitive information or systems. Generally, preventive measures (see Section 11.6.1 for more explanations) such as personnel training, supply chain security, and access control are utilized so as to confront these attacks. Typically, confidentiality attacks are employed as an initial auxiliary but indispensable step before other types of cyber-physical attacks, allowing required information gathering and paving the way for their successful design and launch. In most of the attacks, the more complete the attacker's information is the more successful and impactful the intelligent attack would be. In addition, theft of personal and financial information based on confidentiality attacks can lead to privacy violations and direct economic losses. These types of effects are of more concern in the downstream sector (distribution networks) of the CPPS, including issues such as pricing attacks on demand response programs.

11.5.2.2 Availability

Availability refers to the accessibility of information or systems for authorized users whenever needed. By targeting the availability, different cyber-attacks primarily disrupt the information flow by blocking and delaying communication systems. The main objective of this attack is to affect the observability and controllability. From the observability perspective, hindering the information from the operators causes degraded monitoring and non-optimal decision-making. On the other hand, not having full controllability means that the calculated essential control commands cannot be properly executed, causing compromised CPPS operation. It must be noted that availability cyberattacks are not and cannot be usually stealthy.

As a key example of an availability attack, denial of service (DoS) [31] attempts to flood the cyber network and occupy its bandwidth by exploiting flaws in network protocols, thereby temporarily disrupting connections from the field to the control center and vice versa. In most of the papers in the CPPS security and resiliency literature, DoS availability cyber-attacks are mainly modeled as a simple failure of an element (link or node) in the cyber networks, blocking measured data and control command exchanging. Consequently, the control center will be unable to take appropriate countermeasures due to the lack of full observability and controllability, making the damage alleviation a struggle leading to a larger impact, especially in coordination with other types of cyber-physical threats. Although this type of attack is very common in reality and can have catastrophic effects at lower costs (relative to more intelligent and stealthy attacks, such as integrity), it has received limited attention in the CPPS security and resiliency field.

11.5.2.3 Integrity

Integrity refers to the trustworthiness of data, ensuring that it remains free from any manipulation or alteration during collecting, transferring, pre-processing, or storage. In the CPPS research field, intelligent, stealthy cyber-attacks targeting data integrity have obtained significant attention. These involve the manipulation of either measured data or calculated commands, respectively entitled false data/command injection attacks (FDIAs/FCIAs). In FDIAs, the manipulated measurements are injected into the raw data in order to mislead the SE module of the control center while being stealthy from its bad data detection module. Since almost all of the control center's calculations are developed based on the SE module output, the manipulated SE results will cause the nonoptimal decision-making of the operator, often leading to degraded CPPS security and resiliency. In other words, FDIAs aim to indirectly affect the control commands calculated by the operator (or various modules such as the OPF) by manipulating their inputs (modified measurements by the SE). On the other hand, in FCIAs, the attacker maliciously injects false control commands directly into circuit breakers, switch relays, transformer taps, and phase shifters without concerning themselves about being stealthy. FDIAs can be investigated from different perspectives, a comprehensive review of which can be found in [32]. Compared to FDIA, FCIAs have received less attention in the literature.

Two specific common forms of FDIA are load redistribution (LR) and replay attacks. The main goal of the LR attack is to alter the distribution of load measurements while keeping the total load unchanged so as to deceive the control center and impact the functionality and economics of the power grid. LR is a simplified form of FDIA that needs less data manipulation and can be concealed from the BDD more easily. In replay attacks, the attacker retransmits the previously recorded measurements to the control center either unchanged or with minimal modifications. This attack requires widespread accessibility for the data manipulation, but in case of successful implementation, it can fully conceal the real-time true state of the system, leading to incorrect decision-making. One advanced aspect of the FDIAs is the assumption regarding the attacker's knowledge about grid topology and parameters in designing the stealthy intelligent attack vector. This is related to confidentiality attacks as the prerequisite for integrity attack implementation, as previously mentioned. One method to deal with this issue is blind FDIAs, assuming no attacker's prior knowledge and employing data-driven methods to analyze the correlations among measured data and/or topology and parameters [33].

11.5.3 Coordinated Attacks

Coordinated attacks can involve the collaboration of multiple different or similar cyber or physical attacks for different purposes. In the CPPS security and resiliency field, most of the studies introduce a multi-level optimization model for the coordinated attack design where different methods such as machine learning and the KKT approach are mainly employed to solve the model. Various cyber-physical attacks are typically combined in two distinct ways in the literature (see Figure 11.11), as explained in the following.

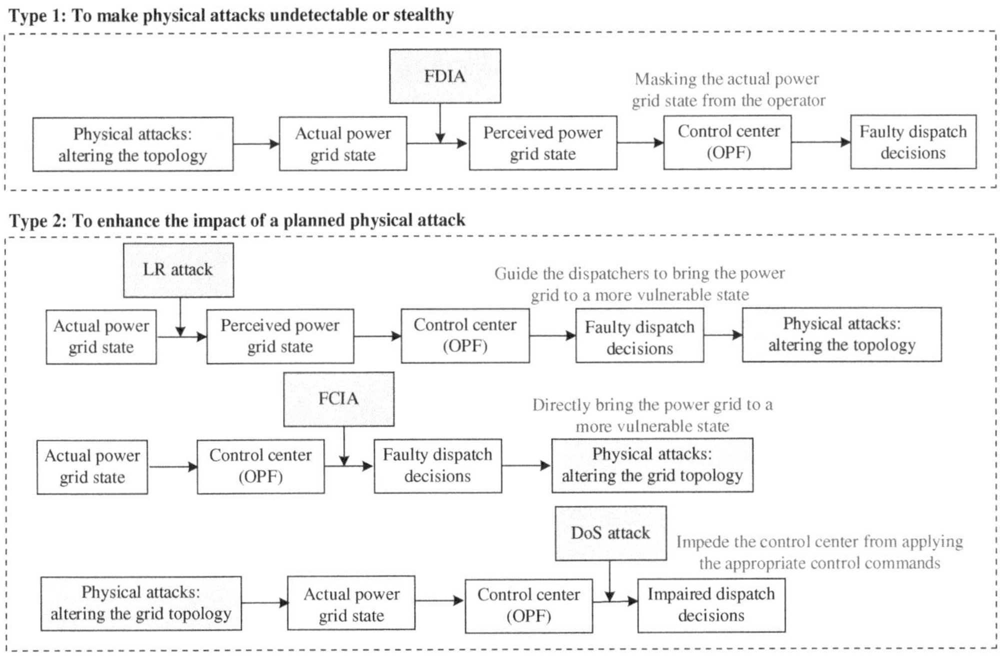

Figure 11.11 Overview and classification of different coordinated cyber-physical attacks.

The first type of coordinated attack is based on utilizing integrity attacks to make physical attacks undetectable or stealthy. The primary objective of individual physical attacks is generally to disrupt power infrastructure drastically. To do so, physical attacks are often accompanied by changes in the grid topology, which can be easily detected by the control center, as the topology is of great importance for the operators. To address this issue, the attacker can launch FDIAs following the implementation of a physical attack, masking the new state of the grid from the operator. In the common bi-level model of such coordinated attacks [34], a set of critical lines (most vulnerable points) is usually determined as the targets of physical attack at the upper level. Aiming to minimize the total resources that are used by the attacker for the FDIA, the optimal set of measurements that should be manipulated to make the physical attack undetectable is determined at the lower level. Considering incomplete attacker information and accessibility, the developed bi-level optimization can also be extended considering the constraints related to both visible and invisible areas from the attacker's perspective. Moreover, various constraints are also commonly included to prevent network islanding after physical attacks to practically remain stealthy from the control center viewpoint.

The second type of coordinated attack investigated in the literature is founded on the concept of utilizing different types of cyber-attacks in order to enhance the impact of a planned physical attack [35]. For example, the LR attack can be first utilized before the physical attack implementation to deceive the control center, resulting in wrong state determination, nonoptimal decision-making, and faulty dispatch decisions. Note that the attacker does not aim to keep the physical attack stealthy, similar to the first type of coordinated attack. However, it mainly aims to intelligently and indirectly (in case of an LR cyberattack) guide the dispatchers to bring the power grid to a more vulnerable state before the execution of the physical attack. Instead of an LR attack, FCIA can be modeled at the upper level to directly manipulate the control commands prior to the physical attack and prepare the condition for the physical attack. Using availability attacks can also enhance the impact of physical attacks. In this context, the availability attacks mainly impede the control center from applying the appropriate control commands for properly mitigating the impacts of physical attacks. This type of coordinated attack is also usually extended based on bi-level optimization models. However, a new separate FDIA can also be implemented after the physical attack to keep it stealthy, forming a tri-level optimization model.

11.6 Defense Framework

As previously mentioned, defense methods are essential to be effectively developed for preventing, detecting, and quickly responding to attacks, especially in today's world with a high reliance of society on digital technologies, which is empowered by the CPPS. In Section 11.5, different types of threats were introduced. As shown in Figure 11.12, in the present section, different defensive actions are discussed based on two main categories: preventive measures and mitigation actions.

11.6.1 Preventive Measures

Preventive measures are defensive actions that are mainly established with the goal of resolving the issues of identified vulnerabilities. This usually includes two main aspects: (1) minimizing the likelihood of a cyber-physical attack's successful occurrence and (2) reducing the potential impact of such threats in case of execution. Note that firewalls and encryption techniques are already widely used in the CPPSs regarding the first aspect, and additional measures like supply chain

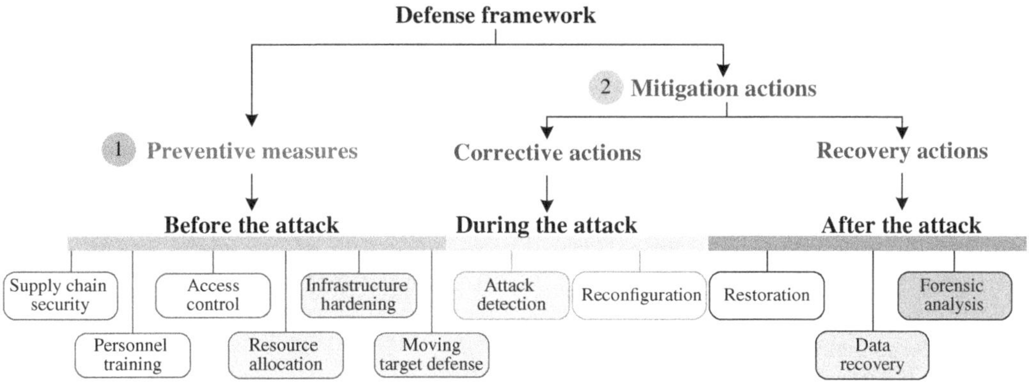

Figure 11.12 Overview and classification of different measures and actions in the defense framework of the CPPS.

security, access control, and personnel training are needed to be utilized for enhanced security and resiliency. Key measures regarding the second aspect include resource allocation, infrastructure hardening, and MTD. These preventive measures are briefly described in the following, emphasizing on the use of data-driven methods.

11.6.1.1 Supply Chain Security

Supply chain security ensures the safety, integrity, and reliability of all components within the CPPSs from the production sector to distribution networks. The supply chain involves many different entities or organizations in various sectors, i.e., manufacturing, distribution, delivery, and so on. A single product may involve numerous vendors providing various hardware, firmware, software, and services. Given the inherent complex cyber-physical coupling in CPPS, any vulnerability at any part of the supply chain can result in significant security and resiliency consequences. In this regard, the key point is to continuously assess the risks, set the procurement process standards, and provide effective personnel training. Despite all security measures throughout the supply chain, the attackers can still exploit a zero-day vulnerability. To deal with this issue, one effective solution is to use diversification methods, in which different software is independently provided by various vendors, each with its distinct architectures [36]. As a result, the likelihood of being exposed to the same zero-day vulnerability is significantly reduced, thereby making the occurrence of simultaneous attacks less likely. It must be noted that utilizing such solutions makes the interoperability of various components an issue that must be properly handled.

11.6.1.2 Access Control

Access control refers to a set of policies that are usually clarifying procedures for authentication and authorization. Authentication policies identify who can access the devices, systems, and information. For this purpose, various modern methods like one-time passwords, smart cards, and biometric techniques (e.g., fingerprints and facial recognition) can be utilized. Authorization policies clarify what activities the authenticated personnel are permitted to take and what their accessible resources and information are [37]. In CPPSs, this defensive measure needs to be individually considered for both cyber and physical sectors. Regarding the information layer, the users should be granted access only to the parts of the system that are relevant to their work. Regarding the physical grid, this defense measure plays a key role in safeguarding physical infrastructure from unauthorized access, especially when facing insider threats.

An insider threat is known as individuals (e.g., employees, contractors, and business partners) who, whether intentionally or unintentionally, exploit their authorized access to commit acts like data theft, sabotage, and physical manipulation, facilitating different cyber-physical threats and leading to various forms of destructive damages. The imposed issue of insider threat can be properly addressed by robust access controls, advanced monitoring of user activities, and detecting abnormal behaviors using data-driven methods. It must be emphasized that access control as a key preventive defensive resource can drastically decrease the occurrence probability of cyber confidentiality attacks, which is one of the main prerequisites for the successful design of other destructive cyber-physical threats (previously discussed in Section 11.5.2.1).

11.6.1.3 Personnel Training

Training individuals, particularly operators of the power grid, is crucial for the sake of CPPS security and resiliency. Such training programs must strongly emphasize enhancing their technical knowledge, especially in the field of cyber threats, due to its relatively new and unknown aspects. In this regard, operators should be thoroughly familiar with the different types of cyber-physical threats as well as the available modern measures for preventing and mitigating them (see Section 11.6.2 for more details). Note that the burden of real-time confrontation with various attacks is on these operators, i.e., actions toward intrusion detection, complete data recovery, and effective grid reconfiguration and restoration. To achieve training goals, an effective tool is to use the CPPS simulators and testbeds (as explained in Section 11.4.2). Such specifically designed testbeds allow the operators to experience a wide range of simulated events or cyberattacks and thereby enhance their skills for facing them. It must be noted that investing in personnel training programs is emphasized in various cybersecurity roadmaps of different organizations, such as [38]. This is due to the large impact of such programs on reducing human errors and decreasing the risk of unintentional insider threats, especially considering the low cost of such programs compared to other types of defensive measures.

11.6.1.4 Resource Allocation

Highlighting the attacker's knowledge and access level for designing intelligent attacks (whether stealthy or not), the studies have demonstrated that an attacker mainly avoids specific protected sensors due to the high intrusion costs. Thus, it is ideal to protect all field devices to prevent any attack or at least drastically enlarge the attack costs. However, it is not a practical solution in large-scale systems due to the techno-economic limitations of the defender. Inspired by this discussion, the defender should identify and prioritize critical devices (meters or actuators) and make decisions regarding the optimal allocation of the limited available defensive resources. Protecting specific field devices ensures that the attacker cannot find the most destructive attack vector due to its limited observability. Even in the case of blind FDIA with the attacker's limited knowledge assumption, the attacker must gain access to multiple unprotected meters to succeed. Generally speaking, optimal allocation of the protective resources can significantly decrease the possibility and increase the cost of successful attack design and execution.

As an example, for such protective resources, phasor measurement units (PMUs) can simultaneously and accurately provide the phasor measurements from geographically dispersed buses while synchronizing with global positioning system. This makes it significantly more difficult and costly for the attackers to manipulate the collected data. However, due to their high initial cost and the resulting limited number of installed PMUs, it is crucial for the defender to identify optimal locations for their placement to resolve the observability concern with minimum cost, using various methods like RL [39]. In the context of resource allocation, the majority of studies primarily utilized the multi-level optimization models, taking the attacker and defender interactions, especially a tri-level D-A-D scheme as previously discussed in Section 11.4.3. Multi-level optimization can also be used based on other configurations. For example, the interaction between each node and the attacker (lower level) can be modeled to pinpoint the vulnerable nodes considering the attacker's limited budget.

11.6.1.5 Infrastructure Hardening

A reliable power system is usually designed to be normal in case of $N-1$ (or $N-2$ in some cases) contingencies, having sufficient adequacy to handle commonly happening single failures. In order to be resilient against multiple simultaneous failures, which are caused by extreme natural events (like earthquakes) or cyber-physical deliberate attacks, there is a critical need to harden the infrastructure. The hardening solutions in the context of CPPSs include strengthening key elements of both the cyber network and physical grid in advance, specifically for facing cascading failures. This can be helpful for boosting survivability (decreasing the immediate impact and vulnerability) as well as facilitating quicker and more effective recovery, thus improving CPPS resiliency. To do so, the main options usually include increasing redundancy by considering extra components and connections in both physical and cyber networks (e.g., new transmission lines or communication links) and upgrading the components to be more resilient (reinforcing overhead lines or utilizing underground cables instead of them) [40].

As a frequently investigated topic, randomly or intelligently (e.g., based on node degree uniformity) adding links into the cyber network can improve the ability of the power grid to face events causing cascading failures. This is based on the enhanced observability and controllability provided by the reinforced cyber network. Such methods are designed to protect against specific threats, and they may also introduce additional vulnerabilities in different cases. In this field, a broad scope of literature has focused on different optimization models, game-theoretic methods, and machine learning techniques to find the best hardening mechanisms, especially focusing on cyber networks. It is also proved that the robustness of CPPS can be highly enhanced by deploying independent nodes in cyber networks (e.g., using a battery for power supply) with no other supporting node in other interdependent infrastructure. Also, adding the connective links between two interdependent networks (known as intra-links) can be useful for reducing vulnerability. Infrastructure hardening may also lead to compatibility issues, where legacy systems cannot effectively work with the modern allocated components.

11.6.1.6 Moving Target Defense

Consider the shell game in which someone hides a ball under one of multiple cups or shells, moves the shells around quickly, and asks an observer to guess which one the ball is under. Inspired by this idea, the defender in the MTD utilizes the confusion for the potential attackers (mainly by adding uncertainty) rather than eliminating all the vulnerabilities. The origins of different MTD types can be found in various fields. In reconfiguration-based MTD methods, the software and network configuration are usually rearranged with a fixed or variable time interval. Combining with other defensive layers is also possible for such schemes. For example, intrusion detection systems (IDSs) can be primarily utilized to detect the events before the MTD. In the diversity-based MTD, the defender usually migrates between the multiple diverse paths for routing as well as introducing different components. The key principle for deception-based MTD techniques lies in intentionally disseminating false information to mislead attackers.

In the context of CPPS, the main goal of MTD is to invalidate attackers' knowledge by dynamically changing the attack target and actively perturbing the system parameters (like the transmission lines reactance and software configuration). The resulting attacker's uncertainty makes it more difficult to effectively target the components and mask the attacks from the control center. The MTD method can be used against FDIAs to effectively invalidate attackers' knowledge. This is possible when the number of branches is at least twice the system states and the susceptance of more than N (number of buses) branches is perturbed [41]. Facing coordinated cyber-physical attacks, the MTD method can also be applied for attack localization and specifically identifying the physically disconnected lines [42]. Despite the key role of MTD in preventive defense, frequent altering of the parameters may introduce latency and misconfigurations, leading to a high degradation in CPPS performance. Integrating MTD with legacy systems may also lead to interoperability issues. This defense layer can occasionally be operated during an attack to mitigate its effects; however, because it is primarily preventive in nature, it is included in this section.

11.6.2 Mitigation Actions

Despite the important role of preventive measures, they have limitations in completely and perfectly preventing the occurrence of all cyber-physical threats in advance. On the other hand, emphasizing the preventive measures to block almost all threats is usually very costly. Therefore, mitigation actions must be designed, including corrective (during the degradation phase) and recovery (during the restoration phase) actions. In this context, the key actions include intrusion detection, data recovery, reconfiguration and restoration, and forensic analysis, all of which are explained in the following.

11.6.2.1 Attack Detection

Due to the inevitability of various attacks occurring, the defender must be ready to quickly and accurately detect them. In the chain of mitigation actions, the attack detection methods have gained the most attention in both academia and industry sectors due to their importance. This is because they could provide the opportunity for implementing corrective actions so as to alleviate the imposed impact of not-prevented attacks. This defense layer is mainly classified into power-side and information-side approaches.

Some of the detection methods (power-side) focus on inferring which measurements and their resulting estimated states do not completely comply with the known physical relationship among variables (e.g., power flow equations). In this context, both the BDD module and other additional residual-based detection methods can be primarily used while relying on the spatial-temporal correlation of the data. This means analyzing multiple different measurements simultaneously or evaluating a single sensor's values across different times. Methods utilizing this philosophy can be divided into two groups: model-based and data-driven. In model-based methods (e.g., extended Kalman filter), CPPS and its physical components are mathematically modeled to identify anomalies [43]. In data-driven methods, in contrast, a large dataset from historical data is utilized and analyzed using statistical methods or machine learning algorithms for pattern recognition without physical models [44].

Some other approaches (information-side), usually known as IDSs or security information and event management systems, mainly emphasize monitoring the network logs, analyzing the network traffic, and evaluating user behaviors for any known unauthorized intrusion and/or suspicious activity. In recent years, multiple new ML-based intrusion detection methods have been developed, commonly founded on either predefined patterns (commonly called signature-based) or deviations from established normal behavior (usually known as anomaly-based) [45]. Despite advancements in IDSs, they still lead to unnecessary alerts (false positives) and allow the attackers to bypass the detection mechanism (false negatives). Nevertheless, it is possible to effectively face such challenges, especially by incorporating emerging data-driven techniques.

11.6.2.2 Data Recovery

Data recovery refers to the process of quickly retrieving and restoring inaccessible, corrupted, and manipulated data. After cyber-physical threats, the timely and perfect data recovery following detecting their occurrence (as described in Section 11.6.2.1) is of great importance. The basic idea in data recovery methods is to minimize the data loss while ensuring the physical relationships among variables. In this context, the false and inaccessible detected data must be restored into their expected range without changing other intact (healthy) measurements that are not affected by the attack. Therefore, it is important to first identify the attack area and distinguish between false and true measurements. The most common and simple method of data recovery is replacing the attacked measurements with either their predicted or latest valid values. This method has the drawbacks of depending on the prediction error or not being updated and suffering from data incontinence among attacked and healthy measurements based on the physical equations. It is worth noting that the temporal variations of grid variables, which are due to the stochastic natural variations in the grid, mainly reflect a regular and traceable pattern within a short time period. Leveraging the high correlations of data with previous values, data-driven methods can be applied for data recovery, solving the aforementioned issues [46]. Nevertheless, the lack of a large required dataset for model training limits their effectiveness, especially when facing new intelligent attacks.

11.6.2.3 Reconfiguration and Restoration

These mitigation actions are designed to use proper corrective operation action as much as possible during the attack and restore the grid to normal condition as quickly and effectively as possible after the cyber-physical attack occurrence. This is an important part of resilient strategies in CPPS, which must be executed following intrusion detection and data recovery. Altering the network topology (usually known as reconfiguration) is initially applied to reroute the power supply or isolate faulted areas, followed by restoration procedures. Restoration is involved with optimal sequencing of repair tasks and the process of reconnecting various components and islands. In this context, the main goal is to minimize the energy loss considering the curtailed load and restoration time. It should be noted that the reconfiguration mostly refers to the corrective actions taken in Phase 2 during the attack. In contrast, the restoration mostly pertains to the recovery actions that are executed in Phase 3 after the attack (see Figure 11.2 in Section 11.2.2).

During the cascading failure phase after the attack, the reconfiguration and restoration decisions must be properly made so as to ensure further failures are prevented as much as possible and the attack impacts are restricted. During the recovery phase, several different constraints must be considered for improving the procedures and decreasing the restoration time. These constraints include different resource limitations, designed priority lists, previously set activity scheduling, and so on. From the CPSPS viewpoint, the allocation of repair crews and informing through social media can play key roles in improving the recovery process. The cyber nodes have top priority for restoring in order to maximize the grid observability for enhanced online decision-making and better controllability for automated fast execution of decisions. Some of the required policies in this regard include islanding detection and formation, black-start capability issues, switching to alternate routing, and so on. For more information, interested readers can refer to [47]. Within a large-scale complex power system with a high intricacy of cyber-physical coupling, the ML-based techniques, particularly RL, can be widely applied to develop a practical and effective restoration policy. Such techniques can also be helpful in automatically detecting faults and identifying their locations, thereby enabling quicker and more efficient restoration strategies [48].

11.6.2.4 Forensic Analysis

After a malicious attack (commonly after the recovery phase), a forensic analysis must be conducted to identify the flaws leading to the attack's success so as to provide the most accurate insights for the vulnerability assessment purposes. This may also include recognizing the insider threats and understanding their motives as well as collecting evidence for criminal justice goals. It is also anticipated to provide useful information for designing effective defense mechanisms against similar attacks in the future in order to enhance security and resiliency. Different forensics methods rely mainly on the required high volume of required data either before or after the attack. In addition to the need for data analytic methods, this also raises ethical and legal concerns regarding privacy issues, especially when personal data is involved [49]. The attacker could quickly and remotely remove or alter data to mislead forensic investigations. Therefore, the key is to rapidly isolate evidence after the attack by implementing effective access control mechanisms so as to ensure the authenticity of the collected data. This field especially requires an adequate number of personnel with specialized cybersecurity knowledge and skills.

11.7 Conclusion

The main purpose of the present chapter was to provide a broad, general, and macro view of the CPPS security and resiliency. Therefore, all explanations are tried to be conceptual, and relevant references have been incorporated throughout the chapter for further investigation. In this chapter, the need to move toward the CPPS was firstly discussed by outlining the major challenges they face, particularly focusing on the security and resiliency concerns. The chapter also provides preliminaries about data analytic methods, distinguishing between basic and advanced approaches.

The interdependency was then examined from failure propagation and impact evaluation viewpoints, and a comprehensive classification of modeling techniques was provided in this regard. The traditional direct methods often result in precise mathematical models that are yet hard to solve. These complicated models are usually handled using testbeds, allowing for accurate modeling without risking real-world systems. Emphasizing both evaluation and improving the CPPS security and resiliency, a variety of ML techniques can be involved for a broad range of tasks. Focusing on either preventive or mitigation strategies, game-theoretic methods and multi-level optimizations are also widely used for modeling interactions between defenders and attackers. CPPSs also deal with the social factors, making their modeling a much harder task.

This chapter highlights the vulnerability of CPPS by a brief and general overview of different cyber-physical threats. Physical attacks (especially targeting the critical components such as substations and power plants) can cause severe damage to the CPPS, even leading to cascading failures. As an essential preliminary step, cyber confidentiality attacks need to be launched first for data gathering and access provision before various attacks. Cyber availability attacks primarily block or delay information flow, affecting observability and controllability. Focusing on being stealthy, cyber integrity attacks are used to mislead control centers, whether with complete or incomplete information and access level. Combining different types of cyber and physical threats, general types of coordinated attacks can be formed with different motivations, i.e., using the cyberattack to either mask the physical event or enhance its impact.

A summary of key insights on different preventive measures that were implemented before the attack occurred and mitigation actions, including corrective actions during the attack and recovery actions after that, are also presented in this chapter. Some preventive measures are focused on minimizing the likelihood of a successful cyber-physical attack, including supply chain security (e.g., using diversity throughout the CPPS), access control (authentication and authorization aspects), and personnel training (e.g., utilizing CPPS testbeds). Other types of such measures aim to reduce the potential impacts of cyber-physical threats in case of execution, including resource allocation (e.g., PMU placement), infrastructure hardening (e.g., cyber link addition), and MTD (e.g., invalidating attackers' knowledge).

From the mitigation viewpoint, different helpful strategies such as intrusion detection and data recovery are highlighted during the attacks (as corrective actions), followed by reconfigurations and restoration (as recovery actions), and forensic analysis after the event. Emphasizing the stealthy integrity and confidentiality attacks, the applied detection methods were primarily categorized as power-side (model-based or data-driven) and information-side (based on signature or anomaly). Besides simple and inconsistent predicted or historical data replacement, modern data recovery techniques usually employ the ML methods to restore attacked data while minimizing data loss and preserving the integrity of healthy data. After the attack detection and data recovery, optimal reconfiguration and restoration decisions must be made so that the attack impacts are restricted and the CPPS is brought to its normal condition as soon as possible. Forensic analysis is the final crucial measure for identifying vulnerabilities, detecting the insider threats, and gathering evidence to support criminal justice goals.

Various challenges related to different aspects of CPPSs have been mentioned throughout the chapter. These challenges call for innovative development in the future, especially emphasizing the integration of data-driven methods. Regarding the complex interdependent nature of a CPPS, a collaborative effort is needed to establish widely accepted definitions and standards to facilitate adaptability and interoperability. Moreover, proper cyber-physical security and resilience metrics must be extended for assessment and enhancement purposes. Also, there are multiple promising future directions for addressing the challenges of social aspects in CPSPSs, especially for the modeling purpose through the integration of artificial intelligence. Moreover, in the presence of renewable energy sources with considerable penetration, various methods in this field must be robust and adaptable against different uncertainties. Despite the applications of different basic ML methods in the realm of CPPSs (e.g., attack detection), their practicality cannot be proven for real large-scale systems facing unknown attacks. On the other hand, utilizing various advanced data-driven methods (like RL) in the field of CPPSs is not mature, underscoring the critical need for further exploration in this research area. This is pertinent across all CPPS security and resiliency domains, including threat prevention, detection, mitigation, and recovery. In this regard, new innovative technologies like blockchain and edge computing hold promise, which require more in-depth investigations in the future.

References

1 Alvarez-Alvarado, M.S., Apolo-Tinoco, C., Ramirez-Prado, M.J. et al. (2024). Cyber-physical power systems: a comprehensive review about technologies drivers, standards, and future perspectives. *Computers and Electrical Engineering* 116: 109149.

2 Zografopoulos, I., Ospina, J., Liu, X., and Konstantinou, C. (2021). Cyber-physical energy systems security: threat modeling, risk assessment, resources, metrics, and case studies. *IEEE Access* 9: 29775–29818.

3 Jimada-Ojuolape, B. and Teh, J. (2020). Surveys on the reliability impacts of power system cyber–physical layers. *Sustainable Cities and Society* 62: 102384.

4 Xu, L., Guo, Q., Sheng, Y. et al. (2021). On the resilience of modern power systems: a comprehensive review from the cyber-physical perspective. *Renewable and Sustainable Energy Reviews* 152: 111642.

5 Yilma, B.A., Panetto, H., and Naudet, Y. (2021). Systemic formalization of Cyber-Physical-Social System (CPSS): a systematic literature review. *Computers in Industry* 129: 103458.

6 Alzaabi, F.R. and Mehmood, A. (2024). A review of recent advances, challenges, and opportunities in malicious insider threat detection using machine learning methods. *IEEE Access* 12: 30907–30927.

7 Azimi, M., Delkhosh, H., and Sheikh-El-Eslami, M.K. (2024). An extended vulnerability assessment method for interdependent cyber-physical power system: fast and precise solution. *IEEE Transactions on Smart Grid* 16: 445–448.

8 Alkhaleel, B.A. (2023). Machine learning applications in the resilience of interdependent critical infrastructure systems – a systematic literature review. *International Journal of Critical Infrastructure Protection* 44: 100646.

9 Ibrahim, M.S., Dong, W., and Yang, Q. (2020). Machine learning driven smart electric power systems: current trends and new perspectives. *Applied Energy* 272: 115237.

10 Dairi, A., Harrou, F., Bouyeddou, B. et al. (2023). Semi-supervised deep learning-driven anomaly detection schemes for cyber-attack detection in smart grids. In: *Power Systems Cybersecurity: Methods, Concepts, and Best Practices* (ed. H.H. Alhelou, Z.Y. Dong, and N. Chatziargyriou), 265–295. Cham: Springer International Publishing.

11 Shi, H., Xie, L., and Peng, L. (2021). Detection of false data injection attacks in smart grid based on a new dimensionality-reduction method. *Computers and Electrical Engineering* 91: 107058.

12 Kaur, D., Anwar, A., Kamwa, I. et al. (2023). A Bayesian deep learning approach with convolutional feature engineering to discriminate cyber-physical intrusions in smart grid systems. *IEEE Access* 11: 18910–18920.

13 Adawadkar, A.M.K. and Kulkarni, N. (2022). Cyber-security and reinforcement learning – a brief survey. *Engineering Applications of Artificial Intelligence* 114: 105116.

14 Habib, A.A., Hasan, M.K., Hassan, R. et al. (2023). Distributed denial-of-service attack detection for smart grid wide area measurement system: a hybrid machine learning technique. *Energy Reports* 9: 638–646.

15 Magoua, J.J. and Li, N. (2023). The human factor in the disaster resilience modeling of critical infrastructure systems. *Reliability Engineering & System Safety* 232: 109073.

16 He, S., Zhou, Y., Yang, Y. et al. (2024). Cascading failure in cyber–physical systems: a review on failure modeling and vulnerability analysis. *IEEE Transactions on Cybernetics* 54: 7936–7954.

17 Abdelmalak, M., Venkataramanan, V., and Macwan, R. (2022). A survey of cyber-physical power system modeling methods for future energy systems. *IEEE Access* 10: 99875–99896.

18 Yohanandhan, R.V., Elavarasan, R.M., Pugazhendhi, R. et al. (2022). A holistic review on Cyber-Physical Power System (CPPS) testbeds for secure and sustainable electric power grid–Part–I: background on CPPS and necessity of CPPS testbeds. *International Journal of Electrical Power & Energy Systems* 136: 107718.

19 Hunt, K. and Zhuang, J. (2024). A review of attacker-defender games: current state and paths forward. *European Journal of Operational Research* 313 (2): 401–417.

20 Marashi, K., Sarvestani, S.S., and Hurson, A.R. (2021). Identification of interdependencies and prediction of fault propagation for cyber–physical systems. *Reliability Engineering & System Safety* 215: 107787.

21 Hasan, M.K., Abdulkadir, R.A., Islam, S. et al. (2024). A review on machine learning techniques for secured cyber-physical systems in smart grid networks. *Energy Reports* 11: 1268–1290.

22 Li, X., Wang, P., Jin, X. et al. (2023). Reinforcement learning architecture for cyber–physical–social AI: state-of-the-art and perspectives. *Artificial Intelligence Review* 56 (11): 12655–12688.
23 Aghamohammadi, M.R. (2024). Risk and vulnerability assessment in power systems: a comprehensive review of challenges and outlook. In: *Future Modern Distribution Networks Resilience* (ed. M.T. Ameli, K. Jalilpoor, S. Azad, et al.), 23–66. Elsevier.
24 Ghiasi, M., Niknam, T., Wang, Z. et al. (2023). A comprehensive review of cyber-attacks and defense mechanisms for improving security in smart grid energy systems: past, present and future. *Electric Power Systems Research* 215: 108975.
25 Sami, N.M. and Naeini, M. (2024). Machine learning applications in cascading failure analysis in power systems: a review. *Electric Power Systems Research* 232: 110415.
26 Farahani, A., Delkhosh, H., Seifi, H., and Azimi, M. (2024). A new bi-level model for the false data injection attack on real-time electricity market considering uncertainties. *Computers and Electrical Engineering* 118: 109468.
27 Tatipatri, N. and Arun, S.L. (2024). A comprehensive review on cyber-attacks in power systems: Impact analysis, detection and cyber security. *IEEE Access* 12: 18147–18167.
28 Gao, X., Peng, M., and Chi, K.T. (2021). Cascading failure analysis of cyber–physical power systems considering routing strategy. *IEEE Transactions on Circuits and Systems II: Express Briefs* 70 (1): 136–140.
29 Abedi, A., Hesamzadeh, M.R., and Romerio, F. (2022). Adaptive robust vulnerability analysis of power systems under uncertainty: a multilevel OPF-based optimization approach. *International Journal of Electrical Power & Energy Systems* 134: 107432.
30 Elrawy, M.F., Hadjidemetriou, L., Laoudias, C., and Michael, M.K. (2023). Detecting and classifying man-in-the-middle attacks in the private area network of smart grids. *Sustainable Energy, Grids and Networks* 36: 101167.
31 Ortega-Fernandez, I. and Liberati, F. (2023). A review of denial-of-service attack and mitigation in the smart grid using reinforcement learning. *Energies* 16 (2): 635.
32 Reda, H.T., Anwar, A., and Mahmood, A. (2022). Comprehensive survey and taxonomies of false data injection attacks in smart grids: attack models, targets, and impacts. *Renewable and Sustainable Energy Reviews* 163: 112423.
33 Rahman, M., Yan, J., and Fapi, E.T. (2024). Adversarial artificial intelligence in blind false data injection in smart grid AC state estimation. *IEEE Transactions on Industrial Informatics* 20: 8873–8883.
34 Deng, R., Zhuang, P., and Liang, H. (2017). CCPA: coordinated cyber-physical attacks and countermeasures in smart grid. *IEEE Transactions on Smart Grid* 8 (5): 2420–2430.
35 Wang, X., Xue, F., Lu, S. et al. (2024). Coordinated cyber-physical attack on power grids based on malicious power dispatch. *International Journal of Electrical Power & Energy Systems* 155: 109678.
36 Touhiduzzaman, M., Hahn, A., and Srivastava, A.K. (2018). A diversity-based substation cyber defense strategy utilizing coloring games. *IEEE Transactions on Smart Grid* 10 (5): 5405–5415.
37 Nasr, P.M. and Yazdian-Varjani, A. (2017). Toward operator access management in SCADA system: deontological threat mitigation. *IEEE Transactions on Industrial Informatics* 14 (8): 3314–3324.
38 Martin, M., Reynolds, T., Sanghvi, A. et al. (2021). Power sector cybersecurity building blocks. NREL/TP-5R00-79396. National Renewable Energy Lab. (NREL), Golden, CO.
39 Zhang, M., Wu, Z., Yan, J. et al. (2022). Attack-resilient optimal PMU placement via reinforcement learning guided tree search in smart grids. *IEEE Transactions on Information Forensics and Security* 17: 1919–1929.

40 Tian, M., Dong, Z., Gong, L., and Wang, X. (2024). Line hardening strategies for resilient power systems considering cyber-topology interdependence. *Reliability Engineering & System Safety* 241: 109644.

41 Zhang, Z., Deng, R., Yau, D.K. et al. (2019). Analysis of moving target defense against false data injection attacks on power grid. *IEEE Transactions on Information Forensics and Security* 15: 2320–2335.

42 Yu, J., Li, Q., and Li, L. (2024). Localization of coordinated cyber-physical attacks in power grids using moving target defense and machine learning. *Electronics* 13 (12): 2256.

43 Nasiri, S., Seifi, H., and Delkhosh, H. (2023). A secure power system distributed state estimation via a consensus-based mechanism and a cooperative trust management strategy. *IEEE Transactions on Industrial Informatics* 20: 3002–3014.

44 Sayghe, A., Hu, Y., Zografopoulos, I. et al. (2020). Survey of machine learning methods for detecting false data injection attacks in power systems. *IET Smart Grid* 3 (5): 581–595.

45 Lansky, J., Ali, S., Mohammadi, M. et al. (2021). Deep learning-based intrusion detection systems: a systematic review. *IEEE Access* 9: 101574–101599.

46 Jorjani, M., Seifi, H., Varjani, A.Y., and Delkhosh, H. (2021). An optimization-based approach to recover the detected attacked grid variables after false data injection attack. *IEEE Transactions on Smart Grid* 12 (6): 5322–5334.

47 Aslani, B., Mohebbi, S., and Oughton, E. (2024). A systematic review of optimization methods for recovery planning in cyber-physical infrastructure networks: current state and future trends. *Computers & Industrial Engineering* 192: 110224.

48 Wei, F., Wan, Z., and He, H. (2019). Cyber-attack recovery strategy for smart grid based on deep reinforcement learning. *IEEE Transactions on Smart Grid* 11 (3): 2476–2486.

49 Kim, K., Alshenaifi, I.M., Ramachandran, S. et al. (2023). Cybersecurity and cyber forensics for smart cities: a comprehensive literature review and survey. *Sensors* 23 (7): 3681.

12

Application of Artificial Intelligence in Undervoltage Load Shedding in Digitalized Power Systems

An In-Depth Review

Nazanin Pourmoradi[1,2], *Sasan Azad*[1,2], *Mohammad Mehdi Amiri*[1,2], *and Miadreza Shafie-khah*[3]

[1] *Department of Electrical Engineering, Shahid Beheshti University, Tehran, Iran*
[2] *Electrical Networks Research Institute, Shahid Beheshti University, Tehran, Iran*
[3] *Research and Innovation Division, Nowocert, Dublin, Ireland*

12.1 Introduction

Power system stability during large faults is a critical research area. Protection and control techniques are continually evolving to address the increasing complexity of power systems [1]. Power blackouts remain a persistent global challenge, significantly impacting productivity and socioeconomic factors [2]. A notable example is the 2003 US–Canada blackout, which lasted 96 hours, affecting over 50 million people. Human error and overgrown vegetation on a transmission line contributed to this outage. The blackout report attributed the cascading grid collapse to the control operators' failure to assess the damage's extent [1]. Since then, researchers and utilities have focused on preventing voltage collapse caused by natural and technical faults [3].

Power blackouts occur when transmission networks degrade due to imbalances between demand and supply, particularly reactive power. Stressed power systems, characterized by long transmission distances, limited grid expansion, and increasing load demand, are most vulnerable to outages [4]. Traditionally, load shedding (LS) has been used to maintain power balance and prevent voltage collapse. LS can effectively mitigate potential instability and protect the system from collapse [5].

Event-driven LS uses a decision table with inputs (current operating conditions and contingencies) and outputs (LS locations and amounts). Based on input information, the corresponding LS action is immediately triggered, minimizing time delay. Response-driven LS (RLS) triggers LS actions based on critical electrical quantity responses (usually voltage and frequency) after a disturbance. RLS can be further categorized into undervoltage LS (UVLS) and underfrequency LS (UFLS). These strategies aim to counteract voltage or frequency decay when it falls below a threshold for an extended period. RLS uses a closed-loop control logic, requiring continuous real-time voltage or frequency measurements to guide its actions [5].

UFLS emphasizes active power balance, while UVLS emphasizes reactive power balance. It is particularly useful in dynamic generation systems, such as those equipped with automatic voltage regulators, tap changers, and synchronous condensers [4]. The cause of cascading grid failures is often attributed to voltage instability in recent blackout reports. Currently, UVLS schemes evaluate optimal LS using optimal power flow equations instead of conventional and adaptive LS methods. Due to the relatively slow dynamics of voltage stability, a static approach provides a more accurate estimation [4] of the system's voltage stability.

Data-Driven Energy Management and Tariff Optimization in Power Systems: Shaping the Future of Electricity Distribution through Analytics, First Edition.
Edited by Hamidreza Arasteh, Pierluigi Siano, Niki Moslemi, and Josep M. Guerrero.

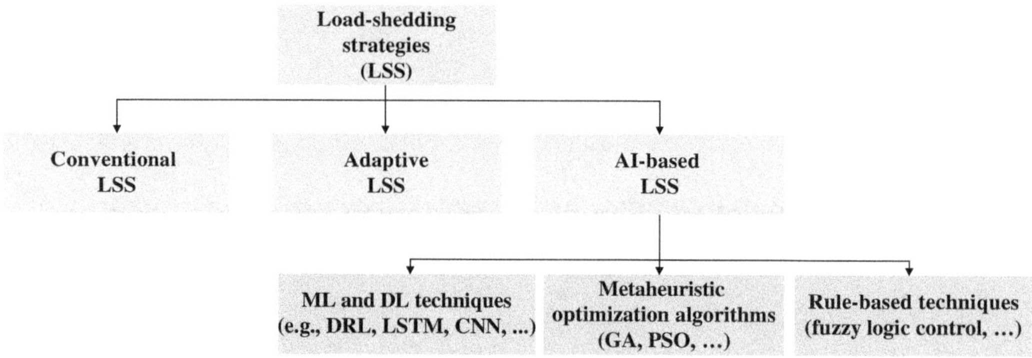

Figure 12.1 Load-shedding strategies.

12.2 Load-Shedding Strategies

Different LS methods have been used to prevent power system blackouts. After 50 years of searching for the best or most optimal solution, a large and diverse set of UFLS and UVLS algorithms has been developed [6]. This chapter classifies LS as conventional, adaptive, and computational intelligent LS schemes, as shown in Figure 12.1, and lists three fundamentals for the LS process:

- Determining when to shed load.
- Determining where to shed load.
- Determining how much load to shed.

Key considerations for effective LS implementation include:

- **Speed**: LS actions must be swift to prevent frequency decline from reaching critical levels.
- **Avoidance of Unwarranted Actions**: Unnecessary LS should be minimized.
- **Reliability and Redundancy**: The protection scheme must be reliable and redundant to avoid system-wide failure.
- **Optimal LS Amount**: The amount of load shed should be as minimal as possible while still ensuring grid protection and preventing the minimum allowable frequency from being exceeded.

12.2.1 Conventional LS

Conventional LS is a traditional method used in power systems to maintain stability during periods of high demand or system disturbances. It involves automatically disconnecting specific loads to reduce overall demand and prevent system collapse. Conventional LS operates by shedding a predetermined amount of load when the system frequency falls below a specified threshold. In many systems, the total amount of load shed is approximately 30% of the system's rated capacity. If the initial load shed is insufficient to stabilize the system, additional shedding stages are triggered as the frequency continues to decline. This process continues until the system overload is mitigated or all frequency-sensitive relays have activated [7].

12.2.2 Adaptive LS

Adaptive LS is a more advanced approach to LS that uses real-time system information to dynamically adjust the amount and location of load shed. Unlike conventional LS, which relies on predetermined shedding sequences and fixed amounts, adaptive LS can tailor its response to the specific conditions of the system at the time of the disturbance. This allows for more efficient and targeted LS, minimizing the impact on customers while maintaining system stability. By continuously monitoring system conditions and adjusting LS accordingly, adaptive LS can help to improve the overall resilience and reliability of power grids [7].

Adaptive LS employs various methods, including UFLS, data streams, genetic algorithms (GA), particle swarm optimization (PSO), and artificial neural networks (ANN). These approaches often utilize a power swing equation to determine the necessary amount of load to shed. The power swing equation is typically expressed as follows:

$$\Delta p = \frac{2H}{f} * \frac{\partial f}{\partial t} \tag{12.1}$$

where

ΔP represents the power imbalance between generation and load.
H is the inertia constant of the generator, a measure of its ability to store rotational energy.
f is the nominal frequency (Hz).
df/dt is the rate of change of frequency (Hz/s).

While adaptive LS improves upon traditional methods, variations in frequency behavior can still lead to suboptimal LS decisions.

12.2.3 AI-Based LS

The use of artificial intelligence (AI) to improve the efficiency and reliability of power system operations, including LS, has become increasingly popular. By analyzing vast amounts of data, identifying patterns, and making intelligent decisions, AI-based LS techniques can improve power system resilience and efficiency significantly. By leveraging the power of AI, utilities can enhance their ability to manage system disturbances, minimize disruptions to customers, and ensure a reliable and sustainable power supply. Key benefits of AI-based LS are as follows:

- **Real-Time Optimization**: AI can rapidly analyze system conditions and determine the optimal amount and location of load to shed to maintain stability while minimizing disruptions.
- **Predictive Analytics**: AI can forecast potential system disturbances and proactively implement LS measures to prevent cascading failures.
- **Adaptive Control**: AI-based systems can continuously learn and adapt to changing system conditions, ensuring that LS strategies remain effective over time.

Even with advanced techniques like adaptive LS, the tripping signal initiating the LS process remains primarily frequency-based or its derivative. This approach may not be ideal for industrial plants with varying generation capacity, flexible distribution networks, high-frequency decay rates, and a limited number of loads [7].

By shedding loads in real time based on operating conditions, a more effective scheme can be implemented, determining the exact amount of load to shed. During and after a disturbance, this scheme takes into account operating conditions, disturbance characteristics, durations, and transient responses of the system. As a result of highly nonlinear problems, techniques like neural networks (NN), GA, simulated annealing, and fuzzy logic (FL) can be used to solve, represent, plan, and act. The techniques can be customized by the user and can be optimized online for minimal LS. Table 12.1 provides a general comparison between different LS methods.

By combining the strengths of these various approaches, it's possible to develop more robust and efficient LS strategies tailored to specific power system requirements.

Table 12.1 Comparison between load-shedding methods.

Method	Pros	Cons
Conventional	Straightforward	Incapable of shedding the optimal amount of load. System complexity, modernity, and size are not considered
Adaptive	Estimates the amount of load shedding accurately. Robust	When severe contingencies arise, it is impossible to work appropriately It is not scalable to large, complex, and modern systems
AI	To shed the most amount of load, it can determine how much should be shed Suitable for large, complex, and modern systems	There is uncertainty about the convergence time

12.3 Principles of UVLS

According to several investigators, there are three major UVLS concerns that need to be addressed, which include the quantity of load sheds, the timing of the load shed construction, and the location of the load shed.

12.3.1 Amount of Load Shed

Optimizing the amount of load shed during a LS event is crucial. Excessive LS can lead to over-frequency and unnecessary customer interruptions, while insufficient shedding may fail to address voltage instability and potentially result in voltage collapse. Load characteristics play a significant role in assessing the stability of a power system. Typical loads, such as induction motors, thermostatically controlled loads (air conditioners or heaters), discharge-type lamps, and underload tap changers, are dynamic and time varying. Highly reactive loads are prioritized for shedding to avoid reactive power deficiencies that could lead to voltage collapse. Induction motors often constitute 20% of the load, as recommended by the Western Electricity Coordinating Council, making the impact of voltage instability dependent on the proportion of motor load and nonmotor load [8].

The amount of load shed must be optimized and can vary throughout a voltage collapse. The goal of UVLS techniques is to stabilize the system while minimizing LS. A previous study [9] presents a concrete approach to determine the optimal amount and location of load to shed. This method employs a multistage approach to solve the nonlinear optimization problem and minimize LS at each stage. Similarly, a GA-based technique was implemented in the Hydro-Québec system [10] to estimate the amount of load shed. However, this approach may be limited in its ability to handle a broader range of load behavior and scenarios, particularly for short-term voltage instability problems. A previous study [10] proposes a model that minimizes LS by incrementally checking system behavior through time-domain simulations. To address long-term voltage stability concerns, this model utilizes the quasi-steady-state simulation technique to identify the weakest buses, significantly reducing computational time. Metaheuristic algorithms, such as the chaotic slime mold algorithm [11], grasshopper optimization algorithm [12], and improved GA [13], have been employed in previous studies to determine the optimal amount and location of LS. These algorithms offer efficient and effective solutions for the complex optimization problems associated with LS.

12.3.2 Location for LS

Buses with a significant impact on voltage stability margins are known as weak buses. These buses are often located distant from generation sources and lack connected reactive power sources. Identifying weak buses is essential for long-term planning and operational studies in power systems, as remedial actions should be implemented at such locations. The most suitable bus for LS is typically the one with the highest $\frac{dV}{dQ}$value (rapid voltage decay), indicating increased vulnerability to voltage collapse [8]. A previous study [9] proposes a new multiport network approach using a multistage nonlinear programming methodology to determine the optimal location and amount of load to shed. The results obtained from this method were validated through modal analysis. However, modal analysis may have limitations in determining the maximum loading point and identifying critical areas, relying on the base case for analysis. The following section discusses various popular indices used in the literature, along with their mathematical expressions. These indices are valuable tools for timely LS location decisions.

12.3.3 Application of VSI for UVLS

VSIs provide valuable information about a system's state under various conditions [14]. As a result of VSIs, power systems can be assessed for their proximity to voltage collapse, loadability and security limits, and overall performance [15]. A variety of power system problems have been addressed using VSIs. The authors of Ref. [16] proposed the voltage collapse proximity indicator (VCPI), for example. Line outages and voltage collapse are assessed using this index. A power system operator can use VCPI to identify the weakest part of the system during emergencies, guiding them to take appropriate corrective action. Moreover, the authors of Ref. [17] proposed the fast voltage stability index (FVSI) as a method for determining a system's maximum loadability. The FVSI value approaches unity when the connected load reaches its maximum, which is known as the maximum loadability point of bifurcation. The modern VSI was introduced in Ref. [18] as a way of identifying weak buses and critical lines.

A number of recent studies have emphasized the importance of VSIs in preventing voltage collapse and blackouts. In Ref. [19], the VCPI was used to prevent voltage collapse in two steps for UVLS. A global index encompassing VCPI and critical bus voltage magnitudes was used in this method. Additionally, the authors of Ref. [1] used FVSI to assess system stability and determine optimal LS locations for UVLS, reducing assessment time. A new normalized voltage stability indicator, the P-index, was introduced in Ref. [20], which assesses the proximity of voltage collapse and determines the appropriate amount of LS. The studies did not, however, examine the dynamic aspects of voltage collapse. A UVLS method that is optimal for predicting LS amount and location while prioritizing the preservation of critical loads was proposed in Ref. [21]. However, it may suffer from suboptimal LS. To mitigate voltage instability and prevent voltage collapse following overloads in distribution systems, the study used the stability index (SI) and feedforward backpropagation NN (FFBPNN). As LS indicators, the authors of Ref. [22] proposed a novel UVLS technique using four VSIs: FVSI, VCPI, SI, and line stability index (L_{mn}). According to this method, VSI rankings are used to prioritize loads that are most likely to collapse. This method, however, was designed for a specific system. Table 12.2 summarizes the various VSIs used in LS.

12.4 AI-Based Methods

The scope of this chapter is predominantly centered on AI-based techniques, thus excluding conventional and adaptive LS methods from the primary focus of our research. AI-based techniques encompass a range of sophisticated methods, such as ANN, GA, FL control, adaptive neuro-fuzzy inference systems, PSO, and numerous other cutting-edge approaches, all designed to leverage AI for various applications. Numerous scholars have employed ANN in the context of LS within power systems [23, 24]. Nevertheless, it is important to note that ANNs tend to yield favorable outcomes primarily for well-defined and trained scenarios, displaying limitations in accurately predicting results for unknown or variable cases. Numerous scholars have employed FL control in LS applications.

Using a modified Hopfield neural network model, [25] optimized load shed and restored system stability margins to acceptable levels. However, this method was designed specifically for one particular system. We used FL control in Ref. [26] to calculate the minimum amount of LS needed during an emergency while ensuring a reliable loading margin was maintained.

Researchers have furthermore applied a range of metaheuristic optimization techniques for conducting LS. In contrast to traditional approaches, these algorithms are esteemed for their precision, adaptability, and robustness in managing intricate nonlinear systems [27]; existing methods such as PSO and GA have shown different strengths and limitations. The authors of Ref. [28] propose a hybrid approach combining ANN with back propagation and GA to develop LSS in power systems, aiming to minimize LS while maintaining system stability. The authors of Ref. [29] also employ GA by using a multiple-deme parallel GA to design LSS to prevent voltage instability. The primary limitation of GAs, which hinders their use in real-time applications, pertains to their slow responsiveness. Studies have revealed significant computation time associated with GA for LS calculations [30]. PSO has demonstrated its effectiveness as a robust and fast method for addressing nonlinear, multiobjective challenges. This technique has gained widespread acceptance in the field of power engineering. The authors of Ref. [31] address the necessity of maintaining a power generation–consumption balance in complex industrial environments during grid contingencies. It introduces an inventive LS algorithm that combines PSO with forecasted load information, offering an economical substitute to established LS procedures, which is validated through practicality assessments employing operational data from an industrial facility in Romania. The authors of Ref. [32] introduce another LSS utilizing a metaheuristic technique called modified discrete evolutionary programming (MDEP) to determine the optimal load to be shed to achieve power balance in controlled islanding scenarios. It is compared to conventional evolutionary programming and exhaustive search techniques, showing that the MDEP technique provides optimal LS with the shortest computational time. Other metaheuristic techniques proposed for LSS in the power system application include the firefly algorithm [33], ant colony optimization [34], and the ant lion optimizer [35]. These LSSs determine the optimal load and the specific location to be shed in order to uphold the secure operating state of the power system.

Lately, deep reinforcement learning (DRL) methods, which merge the state-perception capabilities of DL with the decision-making ability of RL, have gained recognition as a promising approach

Table 12.2 VSI applied for load shedding.

References	Index	Abbreviation	Equation	Stability threshold	Description
[1, 17]	Voltage collapse proximity indicator	VCPI	$\mathrm{VCPI} = \left\lvert 1\text{-} \frac{\sum_{m=1}^{N} V_m^*}{V_k} \right\rvert$ Where $V_m^* = \frac{Y_{km}}{\sum_{J=1}^{N} Y_{kJ}}$	$\mathrm{VCPI}_k \leq 1$	Can identify the most vulnerable sections in a system
[17]	Fast voltage stability index	FVSI	$\mathrm{FVSI} = \frac{4Z^2 Q_J}{V_i^2 X}$	$\mathrm{FVSI} \leq 1$	Can estimate the maximum load ability in a system
[20]	—	P-index	$\text{P-index}_j = \frac{-2\frac{P_{Lj}}{V_j}\frac{dV_j}{dP_{Lj}}}{1-2\frac{P_{Lj}}{V_j}\frac{dV_j}{dP_{Lj}}}$	$\text{P - index} \leq 1$	It can determine the maximum loading of the system and also determine the weakest bus of the system.
[22]	Stability index	SI	$\mathrm{SI} = \lvert V_S\rvert^4\text{-} \left[4(P_r X_{ij} - Q_S R_{IJ})^2 - 4(P_r R_{IJ} - Q_S X_{ij})^2\right] \lvert V_S\rvert^2$	$\mathrm{SI} \geq 0$	Able to determine the bus at which the voltage stability is minimum and most sensitive to voltage collapse
[22]	Line stability index	L_{mn}	$L_{mn} = \frac{4X_{ij} Q_r}{[V_s \sin(\theta - \delta)]^2}$	$L_{mn} \leq 1$	Can identify the exact location of voltage collapse in a system

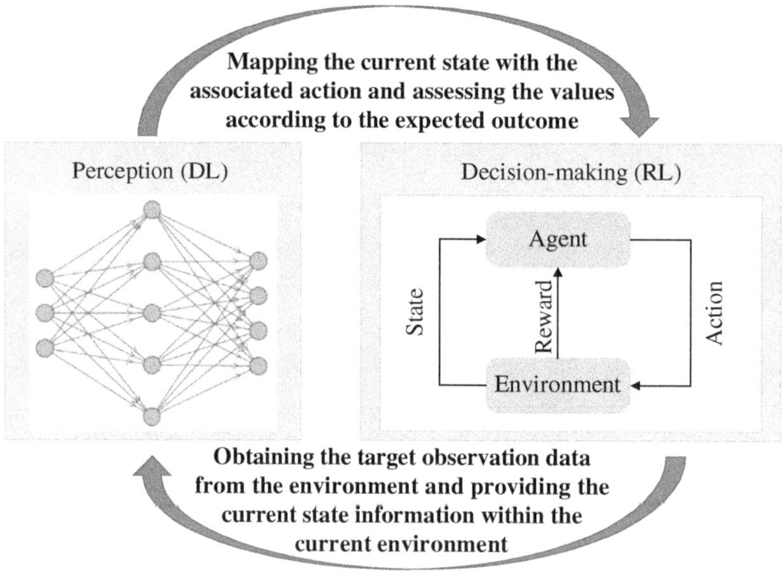

Figure 12.2 Foundational structure of DRL.

to address LS control challenges. DRL represents an AI technique more closely aligned with human cognitive processes and is often deemed genuine AI. Figure 12.2 [36] illustrates the foundational structure of DRL. DL acquires target observation data from the environment and offers the current environmental state information. Subsequently, RL links the current state to corresponding actions and assesses their values according to the expected return [37]. This iterative interaction process transforms decision-making into a step-by-step procedure.

An RL framework comprises an agent, which serves as the decision-maker, operating within an environment. The decision-making process in RL can be delineated as follows. During each time step, the agent acquires an observation from the environment, subsequently rendering decisions by the observation following the current policy. The environment, in turn, responds to the agent's action, transitioning to a new state and providing a reward signal for the evaluation of the action taken. The agent's objective revolves around maximizing rewards derived from interactions with the environment, which necessitates learning an optimal control strategy. RL can acquire optimal control behavior through persistent interaction with the environment and the gradient determined by the feedback reward signal [38]. Ref. [39] utilizes Deep Q-Network (DQN) for LS to tackle dynamic LS challenges in extensive power grids. This approach employs a convolutional long short-term memory (ConvLSTM) network to account for both spatial and temporal factors within the power system, ultimately enhancing voltage recovery performance. The DQN is a renowned DRL algorithm that excels in learning discrete action strategies but struggles with continuous action strategies. Hence, Ref. [40] employs power system operation data to design a novel reward function based on voltage. It utilizes DRL with the deep deterministic policy gradient algorithm, a continuous training method, to formulate an LSS for controlling grid voltage during power system emergencies. Ref. [41] introduces a novel derivative-free DRL algorithm, PARS, designed to address computational inefficiency and scalability challenges when applying DRL to power system voltage stability control with LS. Refs. [42, 43] also propose a knowledge-enhanced DRL method for intelligent LS. Table 12.3 compiles a comprehensive list of references encompassing AI-based LS approaches, providing a thorough repository of sources on this subject.

Table 12.3 Load-shedding literature table.

Reference	Input data	Method	Brief summary of the works' methodology	Year
[45]	Generator's active and reactive power, voltage parameters, the active and reactive loads, transmission line current	Risk-averse DL method	This article presents a risk-averse DL approach for real-time emergency LS, which involves training a deep neural network to minimize the occurrence of load undercutting events, which can lead to post-control instability and higher costs.	2022
[46]	Load data (active and reactive power flow of buses, voltage at buses, the line reactance value, etc.)	Hybrid algorithm based on GA and PSO	The paper uses a hybrid algorithm to optimize LS in response to contingency states, utilizing the FVSI to identify weak buses and significantly improving voltage profiles with reduced LS and faster convergence compared to GA alone.	2018
[39]	Both competitive Q-network and Q-network input = current state	DQN	The work's methodology involves the development of a DQN for load shedding that considers spatial and temporal information of a dynamic power grid using ConvLSTM networks and a novel reward function design, aiming to provide real-time LS decisions.	2022
[29]	The load buses' active and reactive powers, AVR voltage output, transient direct axis and quadrature axis EMF of generators, and PQ bus voltages	GA and neural networks	The work methodology involves using MDPGA to design LSS, triggered by a low voltage stability margin index, with computational speed improvements from a modular neural network and precise training through a simultaneous equilibrium tracing technique.	2021
[47]	Bus voltages, active and reactive powers acquired by PMUs from the buses	Deep feedback learning machine (DFLM)	The methodology employs a data-driven UVLS scheme, integrating a DFLM for precise prediction of voltage violations post-UVLS, with adaptive decision-making aided by stepwise constraint relaxation and incremental DFLM adaptation strategies.	2021
[48]	Line reactance and impedance, voltage and reactive power at the sending end, line loss	EP	The methodology involves the development of a computational intelligence-based technique, specifically EP, to ensure power transmission systems operate within acceptable voltage limits.	2019

Deep learning (DL) models often face limitations due to their inability to retain previously learned knowledge. This necessitates retraining for new tasks, which can be time-consuming. Additionally, prediction accuracy can be compromised when encountering new scenarios without sufficient sample data. Transfer learning (TL) addresses these challenges by leveraging previously learned features to enhance the performance of pretrained DL models on new tasks, as explored in [44].

12.5 Case Study

This section presents an example of LS using AI. For this purpose, a LS model based on a convolutional neural network (CNN) is presented. To test the effectiveness of the model, the IEEE 39 bus system has been used. The diagram of the studied system includes 10 generators, 46 transmission lines, and 39 buses. The LS model consists of three parts: database generation, offline training stage, and online application stage. In the following, these steps are briefly explained.

12.5.1 Database Generation

The simulations in this study are conducted on the modified IEEE 39-bus system. To implement the LS model, a database containing various operating points is generated as follows. The operating conditions are obtained by random sampling the load within the range of its practical changes. Specifically, according to the base load level for each bus, random sampling of load changes is performed. The range of load changes is between 0.65 and 1.25 of the base load. Then, according to the set of contingency faults, stable or unstable labeling is done for each operating point with time domain simulation. Finally, for the operating points that have unstable tags, the optimal LS value is determined according to available LS resources. To generate the database, the location of the faults is bus 17 and bus 25, and the duration of the fault is 0.2 seconds. Based on the preliminary sensitivity analysis in [45, 49], buses 4, 8, 20, and 39 are selected as candidate buses for LS. Moreover, in practice, it is inappropriate in optimal emergency LS to shed all the loads on a bus. In this study, the maximum LS percentage at all candidate buses is set to 90% of the load. GA is used as the heuristic algorithm to find the optimal emergency LS strategy for each generated sample. For model training, 80% of the operating points are randomly selected, and the remaining 20% are used for model testing.

12.5.2 Offline Training

In this stage, the database created in the previous step is split into two parts: training and testing, with a ratio of 4:1. The training database will be utilized to train the LS model. The CNN model in this study comprises two convolutional layers, two max pooling layers, two fully connected layers for feature interpretation, and the output layer. The convolutional layers are built using kernels of size 3, with each activation block employing the rectified linear unit (ReLU) function. The pooling layer in the CNN structure utilizes the maximum value function, with a pooling size and strides of 2 and 1, respectively. Batch normalization is incorporated to enhance learning convergence and domain adaptation. Dropout is applied before the fully connected layer to mitigate overfitting. ReLU activation is used for the dense layers.

In the offline training stage, the CNN-based assessment model is trained to minimize the discrepancy between the model predictions and the actual states, and the learning parameters are obtained. To achieve this objective, a loss function and an optimization algorithm for the learning parameters are essential. In this case, the model predictions and actual states are compared through the loss function, and the optimization algorithm seeks to reduce the loss function by iteratively updating the learning parameters. In this study, Huber is used as a loss function, and Adam's algorithm is used for optimization. It is necessary to explain that the learning rate and the number of epochs in this study are 0.001 and 150, respectively. Also, the batch size is 100.

12.5.3 Online Application

During the online application stage, PMUs gather real-time measurements. Following the data collection, the measured values are input into the model, which has been optimized during the offline training phase. Consequently, the LS result for the power system is determined promptly.

To evaluate the model accuracy, the root mean square error (RMSE) and mean absolute percentage error (MAPE) criteria have been used [50], which are presented in Eqs. (12.2) and (12.3), respectively. L_{Actual} is the actual load, $L_{Forecasted}$ is the forecasted load, and N is the number of time steps.

$$\text{RMSE} = \sqrt{\frac{\sum_{i=1}^{N} (L_{Forecasted} - L_{Actual})^2}{N}} \tag{12.2}$$

$$\text{MAPE} = \sum_{i=1}^{N} \left| \frac{L_{Actual} - L_{Forecasted}}{L_{Actual}} \right| \times \frac{100\%}{N} \tag{12.3}$$

Finally, the result of model evaluation for LS is shown in Table 12.4. Also, the result of the CNN model has been compared with other algorithms. Also, the results of the CNN-based LS model were compared with other deep learning networks such as LSTM and GRU, using evaluation indices including MAPE and RMSE. The results show that CNN has performed better than GRU and LSTM.

12.6 Future Challenges and Transfer Learning

AI-based methods for LS have attracted more attention from authors in the past years due to advantages such as determining the optimal amount of LS. Meanwhile, methods based on ML and reinforcement learning are more interesting than other methods based on AI for power systems. However, these methods also face challenges such as unlearned topologies and unlearned faults for implementation in power systems. ML and reinforcement learning have been used in the past for optimal LS. Although an optimal LS model can be trained for a particular topology, it often does

Table 12.4 Comparison of prediction accuracy of different approaches.

Approach	MAPE (%)	RMSE (%)
CNN	4.21	3.92
GRU	5.63	4.86
LSTM	6.15	4.95

not work well for other topologies. Power system topology can change frequently during operation due to maintenance and control actions. LS models are typically trained for various potential faults, and each model is applied to corresponding fault scenarios. It is generally assumed that the training data (LS database) and the unknown data (online measurements) follow the same distribution, enabling satisfactory model application accuracy. However, this assumption may not always hold true, especially when the power system encounters significant unexpected changes or unlearned faults. For example, if the training database only covers a specific set of faults, a new, unlearned fault may arise during online application. In this case, LS models may suffer from a decrease in accuracy. According to the above, topology changes and unknown faults can cause serious challenges in the application of LS models in real power systems. To solve these challenges, one of the solutions is to use transfer learning and model retraining. When the topology changes and unknown faults occur, the model encounters new and unknown data, which reduces its accuracy. In this situation, one of the best solutions is to retrain the model with the least data. Retraining with low data is important in two ways:

- Generating a large database is difficult, time consuming, and expensive.
- LS is used as a last resort to prevent network instability, and LS models should be available as soon as possible if updates are needed. In this situation, using a small database allows updating in less time.

TL methods can generally be divided into three categories: parameter-based, feature-based, and instance-based methods [51]. Each of these categories will be discussed in detail in the following.

Instance-based TL: The instance-based TL method involves reweighting labeled data from the source domain to be applied in the target domain, presenting an approach that is straightforward, efficient, and theoretically comprehensible. The fundamental concept of these methods is to reweight the labeled training data to address the discrepancies between source and target distributions. This reweighting is implemented by multiplying the individual loss of each training instance by a positive weight during the training process [51]. The general framework of the instance-based TL is illustrated in Figure 12.3.

Parameter-based TL: Parameter-based TL involves transferring knowledge at the model or parameter level. This approach is often implemented by freezing specific neural layers in the source domain model and retraining the final one or two layers using data from the target domain system [52]. The frozen layers extract features from the input data, which are then mapped to the target state using weights and biases in the fully connected layers. Figure 12.4 illustrates the parameter-based TL technique proposed in this study.

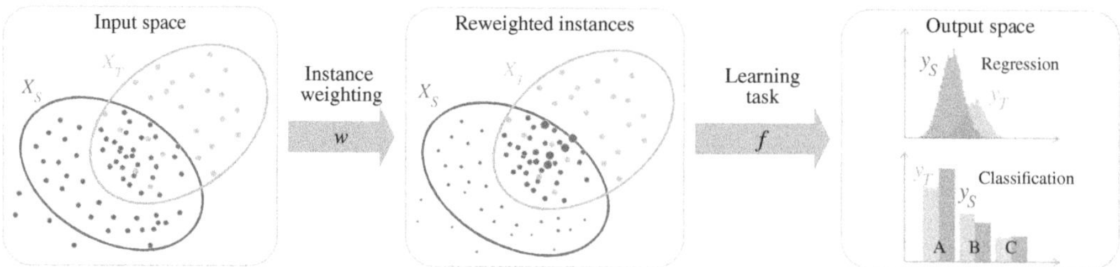

Figure 12.3 Instance-based TL framework.

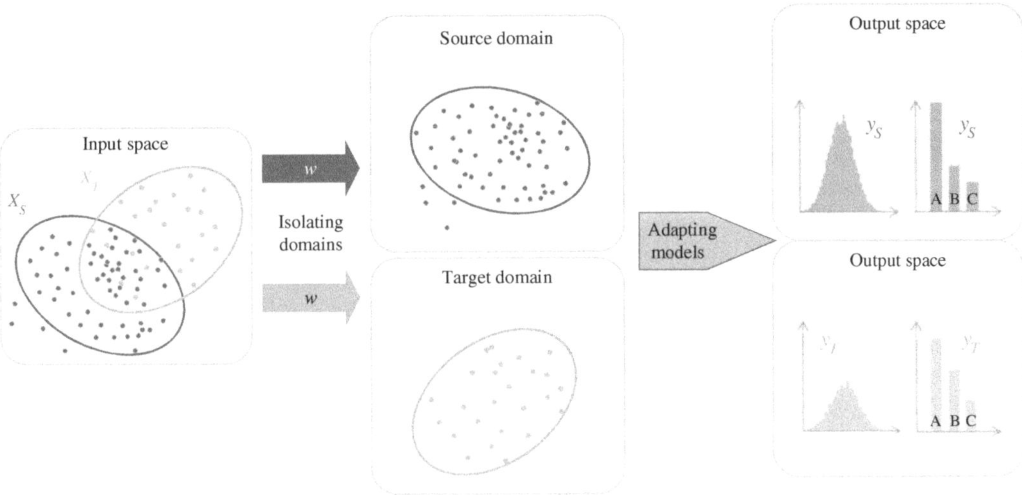

Figure 12.4 Parameter-based TL framework.

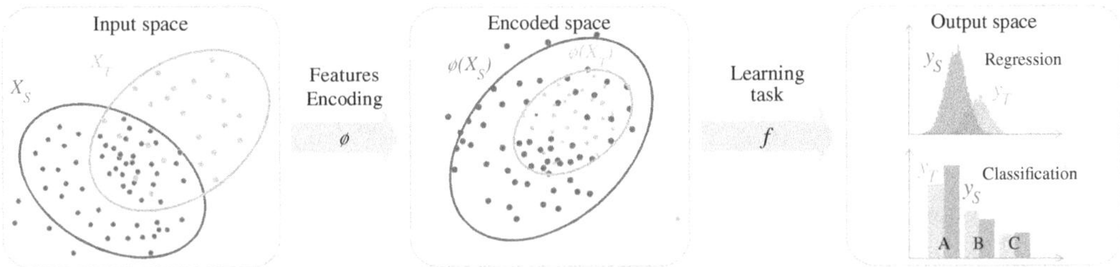

Figure 12.5 Feature-based TL framework.

Feature-based TL: Feature-based methods rely on identifying common features that exhibit similar behavior concerning the task across both the source and target domains. A new feature representation, commonly known as the encoded feature space, is created using a projecting application ϕ that aims to align the source and target distributions. The learning task is then performed within this encoded feature space [53]. The feature-based TL framework is depicted in Figure 12.5.

12.7 Conclusion

This chapter has provided a comprehensive review of various UVLS methods found in the literature. It has critically examined the three primary categories of UVLS methods—conventional, adaptive, and AI—highlighting their respective advantages and disadvantages. UVLS is considered the last resort for mitigating power imbalances during disturbances. Consequently, proposed UVLS techniques must be robust and swift to effectively address critical system emergencies. In recent years, methods based on ML and reinforcement learning have received more attention due to their high accuracy and high speed. But it should be noted that topology changes and unknown errors are a major challenge for implementing these methods in real-world power systems. In this situation, one of the solutions is to combine model retraining with transfer learning, which can provide conditions for retraining with little data. Further, this chapter has reviewed the VSIs application for UVLS methods. Finally, this chapter presents a case study for intelligent LS using CNN.

References

1 Kisengeu, S.M., Muriithi, C.M., and Nyakoe, G.N. (2021). Under voltage load shedding using hybrid ABC-PSO algorithm for voltage stability enhancement. *Heliyon* 7 (10): e08138.

2 Naidoo, P. and Vollgraaff, R. (2019). South Africa's record rolling blackouts raise recession risk.

3 Alhelou, H.H., Hamedani-Golshan, M.E., Njenda, T.C., and Siano, P. (2019). A survey on power system blackout and cascading events: research motivations and challenges. *Energies* 12 (4): 682.

4 Rai, S., Kumar, Y., and Agnihotri, G. (2014). Under voltage load shedding for contingency analysis to optimize power loss and voltage stability margin. *International Journal of Electronic and Electrical Engineering* 3: 57–64.

5 Liu, J. (2022). Real-time event-driven load shedding for power system transient stability control using deep learning techniques. Doctoral dissertation. UNSW Sydney.

6 Yang, H., Li, N., Sun, Z. et al. (2021). Real-time adaptive UVLS by optimized fuzzy controllers for short-term voltage stability control. *IEEE Transactions on Power Systems* 37 (2): 1449–1460.

7 Sundarajoo, S. and Soomro, D.M. (2023). Under voltage load shedding and penetration of renewable energy sources in distribution systems: a review. *International Journal of Modelling and Simulation* 43 (6): 1002–1020.

8 Larik, R.M., Mustafa, M.W., and Aman, M.N. (2019). A critical review of the state-of-art schemes for under voltage load shedding. *International Transactions on Electrical Energy Systems* 29 (5): e2828.

9 Wang, Y., Pordanjani, I.R., Li, W. et al. (2011). Strategy to minimise the load shedding amount for voltage collapse prevention. *IET Generation, Transmission & Distribution* 5 (3): 307–313.

10 Van Cutsem, T., Moors, C., and Lefebvre, D. (2002). Design of load shedding schemes against voltage instability using combinatorial optimization. *2002 IEEE Power Engineering Society Winter Meeting. Conference Proceedings (Cat. No. 02CH37309)*, New York (27–31 January 2002), Vol. 2, pp. 848–853. IEEE.

11 Abid, M.S., Apon, H.J., Ahmed, A., and Morshed, K.A. (2022). Chaotic slime mould optimization algorithm for optimal load-shedding in distribution system. *Ain Shams Engineering Journal* 13 (4): 101659.

12 Ahmadipour, M., Othman, M.M., Salam, Z. et al. (2023). Optimal load shedding scheme using grasshopper optimization algorithm for islanded power system with distributed energy resources. *Ain Shams Engineering Journal* 14 (1): 101835.

13 Wang, K., Kang, L., and Yang, S. (2022). A coordination optimization method for load shedding considering distribution network reconfiguration. *Energies* 15 (21): 8178.

14 Danish, M.S.S., Senjyu, T., Danish, S.M.S. et al. (2019). A recap of voltage stability indices in the past three decades. *Energies* 12 (8): 1544.

15 Zaheb, H., Danish, M.S.S., Senjyu, T. et al. (2020). A contemporary novel classification of voltage stability indices. *Applied Sciences* 10 (5): 1639.

16 Moghavvemi, M. and Faruque, M.O. (1999). Power system security and voltage collapse: a line outage based indicator for prediction. *International Journal of Electrical Power & Energy Systems* 21 (6): 455–461.

17 Musirin, I. and Rahman, T.A. (2002). Estimating maximum loadability for weak bus identification using FVSI. *IEEE Power Engineering Review* 22 (11): 50–52.

18 Mokred, S., Wang, Y., and Chen, T. (2023). Modern voltage stability index for prediction of voltage collapse and estimation of maximum load-ability for weak buses and critical lines identification. *International Journal of Electrical Power & Energy Systems* 145: 108596.

19 Klaric, M., Kuzle, I., and Tesnjak, S. (2004). Undervoltage load shedding using global voltage collapse index. *IEEE PES Power Systems Conference and Exposition, 2004*, New York (10–13 October 2004), pp. 453–459. IEEE.

20 Kamel, M., Karrar, A.A., and Eltom, A.H. (2017). Development and application of a new voltage stability index for on-line monitoring and shedding. *IEEE Transactions on Power Systems* 33 (2): 1231–1241.

21 Sundarajoo, S. and Soomro, D.M. (2023). Optimal load shedding for voltage collapse prevention following overloads in distribution system. *International Journal of Engineering* 36 (7): 1230–1238.

22 Yusof, N.A., Mohd Rosli, H., Mokhlis, H. et al. (2017). A new under-voltage load shedding scheme for islanded distribution system based on voltage stability indices. *IEEJ Transactions on Electrical and Electronic Engineering* 12 (5): 665–675.

23 Hsu, C.T., Chuang, H.J., and Chen, C.S. (2011). Adaptive load shedding for an industrial petroleum cogeneration system. *Expert Systems with Applications* 38 (11): 13967–13974.

24 Hsu, C.T., Kang, M.S., and Chen, C.S. (2005). Design of adaptive load shedding by artificial neural networks. *IEE Proceedings-Generation, Transmission and Distribution* 152 (3): 415–421.

25 Arya, L.D., Pande, V.S., and Kothari, D.P. (2005). A technique for load-shedding based on voltage stability consideration. *International Journal of Electrical Power & Energy Systems* 27 (7): 506–517.

26 Poshtan, M. and Farinwata, S. (2006). Intelligent undervoltage load curtailment. *2006 IEEE International Conference on Fuzzy Systems*, Vancouver, British Columbia, Canada (16–21 July 2006), pp. 1721–1728). IEEE.

27 Kisengeu, S.M., Nyakoe, G.N., and Muriithi, C.M. (2020). Under voltage load shedding using hybrid metaheuristic algorithms for voltage stability enhancement: a review. *2020 IEEE PES/IAS PowerAfrica*, Nairobi, Kenya (25-28 August 2020), pp. 1–5. IEEE.

28 Anh, Q.H., Tan, P.T., and An, N.T. (2020). A hybrid artificial neural network-genetic algorithm for load shedding. *International Journal of Electrical and Computer Engineering* 10 (3): 2250.

29 Gholami-Rahimabadi, A., Razmi, H., and Doagou-Mojarrad, H. (2021). Multiple-deme parallel genetic algorithm based on modular neural network for effective load shedding. *Soft Computing* 25: 13779–13794.

30 Al-Hasawi, W.M. and El Naggar, K.M. (2002). Optimum steady-state load-shedding scheme using genetic based algorithm. *11th IEEE Mediterranean Electrotechnical Conference (IEEE Cat. No. 02CH37379)*, Cairo, Egypt (7–9 May 2002), pp. 605-609. IEEE.

31 Baiceanu, F.C., Ivanov, O., Beniuga, R.C. et al. (2023). A continuous multistage load shedding algorithm for industrial processes based on metaheuristic optimization. *Mathematics* 11 (12): 2684.

32 Saharuddin, N.Z., Zainal Abidin, I., Mokhlis, H., and Shair, E.F. (2021). Load shedding scheme based metaheuristic technique for power system controlled islanding. *Indonesian Journal of Electrical Engineering and Computer Science* 23 (3): 1306–1314.

33 Yang, X.S. (2009). Firefly algorithms for multimodal optimization. *International Symposium on Stochastic Algorithms, Lecture Notes in Computer Sciences* 5792: 169–178.

34 Church, C., Morsi, W.G., El-Hawary, M.E. et al. (2011). Voltage collapse detection using ant colony optimization for smart grid applications. *Electric Power Systems Research* 81 (8): 1723–1730.

35 Yasin, Z.M., Samón, I.N., Mohamad, H. et al. (2017). Optimal undervoltage load shedding using ant lion optimizer. *International Journal of Simulation: Systems, Science and Technology* 17 (41): 47.1–47.6.

36 Zhang, Z., Zhang, D., and Qiu, R.C. (2019). Deep reinforcement learning for power system applications: an overview. *CSEE Journal of Power and Energy Systems* 6 (1): 213–225.

37 Li, S.E. (2023). Deep reinforcement learning. In: *Reinforcement Learning for Sequential Decision and Optimal Control* (ed. S.E. Li), 365–402. Springer Nature Singapore: Singapore.

38 Cao, D., Hu, W., Zhao, J. et al. (2020). Reinforcement learning and its applications in modern power and energy systems: a review. *Journal of Modern Power Systems and Clean Energy* 8 (6): 1029–1042.

39 Zhang, J., Luo, Y., Wang, B. et al. (2021). Deep reinforcement learning for load shedding against short-term voltage instability in large power systems. *IEEE Transactions on Neural Networks and Learning Systems* 34 (8): 4249–4260.

40 Li, J., Chen, S., Wang, X., and Pu, T. (2021). Load shedding control strategy in power grid emergency state based on deep reinforcement learning. *CSEE Journal of Power and Energy Systems* 8 (4): 1175–1182.

41 Huang, R., Chen, Y., Yin, T. et al. (2021). Accelerated derivative-free deep reinforcement learning for large-scale grid emergency voltage control. *IEEE Transactions on Power Systems* 37 (1): 14–25.

42 Chen, H., Zhuang, J., Zhou, G. et al. (2023). Emergency load shedding strategy for high renewable energy penetrated power systems based on deep reinforcement learning. *Energy Reports* 9: 434–443.

43 Hu, Z., Shi, Z., Zeng, L. et al. (2023). Knowledge-enhanced deep reinforcement learning for intelligent event-based load shedding. *International Journal of Electrical Power & Energy Systems* 148: 108978.

44 Xie, J. and Sun, W. (2021). A transfer and deep learning-based method for online frequency stability assessment and control. *IEEE Access* 9: 75712–75721.

45 Liu, J., Zhang, Y., Meng, K. et al. (2022). Real-time emergency load shedding for power system transient stability control: A risk-averse deep learning method. *Applied Energy* 307: 118221.

46 Larik, R.M., Mustafa, M.W., Aman, M.N. et al. (2018). An improved algorithm for optimal load shedding in power systems. *Energies* 11 (7): 1808.

47 Zhu, L. and Luo, Y. (2021). Deep feedback learning based predictive control for power system undervoltage load shedding. *IEEE Transactions on Power Systems* 36 (4): 3349–3361.

48 Abd Shukor, S.F., Shukor, A., Musirin, I. et al. (2020). Intelligent based technique for under voltage load shedding in power transmission systems. *Indonesian Journal of Electrical Engineering and Computer Science* 17 (1): 110–117.

49 Li, Q., Xu, Y., and Ren, C. (2020). A hierarchical data-driven method for event-based load shedding against fault-induced delayed voltage recovery in power systems. *IEEE Transactions on Industrial Informatics* 17 (1): 699–709.

50 Shobeiry, S.M., Azad, S., and Ameli, M.T. (2023). Smart short-term load forecasting through coordination of LSTM-based models and feature engineering methods during the COVID-19 pandemic. *Electric Power Components & Systems* 51 (2): 171–187.

51 Hijazi, M., Dehghanian, P., and Wang, S. (2023). Transfer learning for transient stability predictions in modern power systems under enduring topological changes. *IEEE Transactions on Automation Science and Engineering* 21 (3): 3274–3288.
52 Cui, H., Wang, Q., Ye, Y. et al. (2022). A combinational transfer learning framework for online transient stability prediction. *Sustainable Energy, Grids and Networks* 30: 100674.
53 Niu, S., Hu, Y., Wang, J. et al. (2020). Feature-based distant domain transfer learning. *2020 IEEE International Conference on Big Data (Big Data)* Atlanta, GA (10–13 December 2020), pp. 5164–5171. IEEE.

Index

Data-Driven Energy Management and Tariff Optimization in Power Systems: Shaping the Future of Electricity Distribution through Analytics, First Edition.
Edited by Hamidreza Arasteh, Pierluigi Siano, Niki Moslemi, and Josep M. Guerrero.

d